W9-CRN-088

Bernd S. W. Schröder

Ordered Sets
An Introduction

Birkhäuser
Boston • Basel • Berlin

Bernd S. W. Schröder
Program of Mathematics and Statistics
Lousiana Tech University
Ruston, LA 71272
U.S.A.

Library of Congress Cataloging-in-Publication Data

Schröder, Bernd S. W. (Bernd Siegfried Walter), 1966-
  Ordered sets : an introduction / Bernd S. W. Schröder.
    p. cm.
  Includes bibliographical references and index.
  ISBN 0-8176-4128-9 (acid-free paper) – ISBN 3-7643-4128-9 (acid-free paper)
    1. Ordered sets. I. Title.

QA171.48.S47 2002
511.3'2–dc21   2002018231
                                                            CIP

AMS Subject Classifications: 06-01, 06-02

Printed on acid-free paper.
©2003 Birkhäuser Boston

Birkhäuser

ISBN 0-8176-4128-9   SPIN 10721218
ISBN 3-7643-4128-9

Typeset by the author.
Printed in the United States of America.

9 8 7 6 5 4 3 2 1

Birkhäuser   Boston • Basel • Berlin
A member of BertelsmannSpringer Science+Business Media GmbH

Contents

Concept Map of the Contents

This text can be read linearly, cover-to-cover as is the case with any good mathematics text. This approach however would delay readers who are interested in the content of the later chapters. As it turns out, such a delay is not necessary.

The concept map in Figure 1 on page x indicates how content can be organized to satisfy a variety of interests. The only common requirement is satisfactory coverage of the (purposely lean) core.

The arrows denote possible straight paths from one segment to another. The connections only indicate some reasonable paths through the text. There may be others, and the indicated paths need not be completely straight (because of interrelations with other topics). Some omitted content "between" the subject blocks may need to be acquired by selective reading at the appropriate place in the later chapter. These gaps should not be a deterrent. For definitions etc. to fill such gaps, please consult the index. Aside from the references in the concept map, the author also highly recommends [19, 26, 218, 220, 224, 227] for further study.

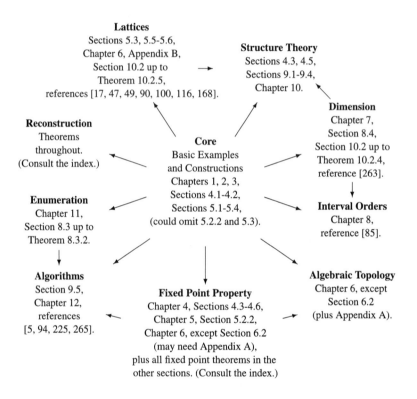

Figure 1: Concept map of the contents. After the common core, the reader can follow any of the arrows to topics of interest without encountering insurmountable gaps.

Preface

Order theory can be seen formally as a subject between lattice theory and graph theory. Indeed, one can say with good reason that lattices are special types of ordered sets, which are in turn special types of directed graphs. Yet this would be much too simplistic an approach. In each theory the distinct strengths and weaknesses of the given structure can be explored. This leads to general as well as discipline-specific questions and results. Of the three research areas mentioned, order theory undoubtedly is the youngest. The first specialist journal *Order* was launched in 1984 and much of the research that guided my own development started in the 1970s.

When I started teaching myself order theory (via a detour through category theory) I was only dimly aware of lattices and graphs. (I was working on a Ph.D. in harmonic analysis and probability theory at the time.) I was attracted to the structure, as it apparently fits the way I think. It was possible to learn the needed basics from research papers as well as surveys. From there it was immensely enjoyable to start working on unexplored problems. This is the beauty of a fresh field. Interesting results are almost asking to be discovered. I hope that the reader will find the same type of attraction to this area (and will ultimately make interesting contributions to the field).

Yet there also is a barrier to entering such a new field. In new fields, standard texts are not yet available. I felt it would be useful to have a text that would expose the reader to order theory as a discipline without too quickly focusing on one specific subarea. In this fashion a broader picture can be seen. This is my attempt at such a text. It contains all that I know about the theory of ordered sets. From here, articles on ordered sets as well as the standard references I had available starting out (which are primarily Rival's Banff conference volumes [218, 220, 225], but also Birkhoff's classic [17] on lattice theory, Fishburn's text [85] on interval orders, and later Davey and Priestley's text [49] on lattices and Trotter's monograph [263] on dimension theory) should be easily accessible. The idea was to describe what I consider the basics of ordered sets without the work becoming totally idiosyncratic. Some of the salient features of this text are bulleted below.

- **Theme-driven approach.** Most of the topics in this text are introduced by investigating how they relate to research problems. We will frequently revisit the open problems that are explained early in the text. Further problems are added as we progress. In this fashion, I believe, the reader will be able to form the necessary intuitions about a new structure more easily than if there were no common undercurrent to the presentation. I have deliberately tried to avoid the often typical beginning of a text in discrete mathematics. There is no deluge of pages upon pages of basic definitions. New notions are introduced when they first arise and they are connected with known ideas as soon as possible.

- **Connections between topics.** Paraphrasing W. Edwards Deming, one can say that *"If we do not understand our work as part of a process, we do not understand our work at all."* This statement applies to industry, education and also research. Many powerful results have been proved by connecting several seemingly different branches of mathematics. Consequently I tried to show some of the connections between the different areas of ordered sets, as well as connections of the theory of ordered sets to areas such as algebraic topology, analysis, and computer science, to name several. These types of cross-connections have always been fascinating to me. Consider, for example, the use of algebraic topology in Chapter 6. Its connections to ordered sets yield results of a caliber that appears impossible to achieve by staying within a single discipline.

   Along the same lines it must immediately be said that I cannot claim to know all connections between order and other fields. Thus this text is by no means a complete guide to interdisciplinary work that involves order.

- **Breadth and Flexibility.** One of my professors once said that mathematics is a field that one can study for 80 years without repeating a topic, but unfortunately also without contributing anything new. What we don't know will always exceed what we know, and order is no exception. On the other hand, there is much to be said for a broad education. The more one knows, the more connections one can make and the more potential one has to make good contributions.

   To allow for the benefits of broad training without the (very real) drawback of being overloaded with information, the reader may refer to the concept map on p. ix. The concept map shows how sections can be grouped around the core to satisfy a variety of purposes. The core was purposely kept lean.

This organization makes it easier to tackle the breadth of the text and also makes the text more flexible. Follow your strongest interests first, then obtain more information about other things. So, I hope the reader will come back to this text frequently to learn more and to use this knowledge as a springboard towards new work in ordered sets.

If the reader decides to stray from the linear presentation of the text, some results that use examples from earlier sections may need to be skipped (or better, acquired by selective reading).[1] Still, the reader should be able to acquire sufficient understanding of the chapter by relying on work that connects to topics already read. The reader can be guided by his/her primary interests and will still be exposed to many of the cross-connections mentioned above. (This remark should not dissuade the more intrepid readers from going cover-to-cover.)

- **Depth.** The greatest depth one can achieve in any research topic is to understand the open research problems and to be aware of most or all of the results pertaining to their solutions. For at least three problems, the fixed point property, the order reconstruction problem and the automorphism problem, the reader is exposed to essentially all that I know about these topics. I will never claim comprehensive knowledge of a topic. For these problems, however, I am virtually certain that I have gone as far as the theory does to date. More open problems are given in the text as well as in the sections behind the exercises. In this fashion the text allows the reader to attempt research even after reading a few chapters.

- **Open Problems.** The idea to work with a set of open problems certainly is inspired by the early issues of the journal *Order*. Open problems show the frontiers of a field and are thus in some way the life of a branch of mathematics. Therefore I tried to at least show the reader what were and are considered major open problems in ordered sets. The main open problems that are part of the text come from Order's problem list or are inspired by it. I can safely say that they are of interest to the community at large. Naturally the text is biased towards the problems that I know more about. For other problems in the text I have tried to include references to more literature.

The open problems after the exercise sections and open problem 10.3.7 reflect my own curiosity. These problems range from special cases of the major problems, to simpler proofs of known results, to some things that "I simply would like to know." The only way I have to gauge their difficulty is to say that I have not solved them (yet?). I suspect that some may turn out to be quite difficult (material for MS or Ph.D. theses). I sincerely hope that none of them turn out to be trivial or that their solutions were overlooked by me. Yet this possibility can never be excluded, especially since, to provide

---

[1] The index, which I tried to make as redundant as possible, should help in filling holes, clarifying unknown notation, etc.

a more complete picture of order, I sometimes stretched myself into areas that I am not as familiar with. Either way, I would be interested in solutions as they arise.

More open problems about order can be found in the collections [218, 220, 225].

- **Standard topics.** There are certain standard topics that mathematicians unfamiliar with order may automatically identify with order theory. The most worked-on parts of the theory of ordered sets appear to be lattice theory and dimension theory. Interval orders also have received a good bit of attention due to their applicability in modeling schedules. There are textbooks available in each of these areas (cf. [17, 49, 85, 263]) and any exposure of these areas as *part* of a text must necessarily be incomplete. Therefore in these areas only the basics and some relations to the main themes of the text are explored. A deeper study of these areas has to be relegated to further specialization. The same can be said for the treatment of constraint satisfaction problems in Chapter 12.

- **Ordering of topics and history.** Like many authors, I want the text to be readable, not a list of achievements in exact chronological order. Thus, whenever necessary, logic supersedes historical order. Historically later results are used freely to prove historically earlier results when this appeared appropriate. The overall presentation is intended to be linear.

  The notes at the end of each chapter (if they refer to historical developments) reflect a very limited historical view at best. I could do no better, as many of the topics this text touches upon (such as reconstruction in general, lattices, dimension theory, constraint satisfaction, to name several) are in fact the tips of rather large icebergs, which in many cases I cannot fully fathom myself.

  The reader who is interested in the history of a subject is advised to look at surveys I have referenced or to run searches of the *Mathematical Reviews* database with the appropriate key word. Electronic search tools are perfect for such tasks. The reader should however be aware that terminology is changing over time.[2]

- **Reading with a pencil** is mandatory, more so than in a regular mathematics text. The diagram of an ordered set is a very strong visual tool. Many proofs that appear difficult at first become clear when drawing diagrams of the described situations. Indeed much of the appeal of the theory of ordered sets derives from its strongly visual character.

---

[2]In fact, I have spent some time reinventing certain results because I was unaware of the particular vocabulary of an area. While this appears inefficient, the only alternative would be to always just stick with what I know well. This I consider an unacceptable proposition.

Mathematics is learned by doing, by confronting a topic and acquiring the tools to master it. It is time to do so. I hope reading this text will be an enjoyable and intellectually enriching contribution to the reader's mathematical life. Readers interested in code for Chapter 12 or who have solutions to open problems or suggestions for exercises are encouraged to contact the author. I am planning to post updates on my web site.

Ruston, LA, October 24, 2002

– Bernd Schröder

Acknowledgements

This book is dedicated to my family. It has been written largely on their time. Without the support of my wife Claire, my children Samantha, Nicole, Haven and Mlle and of my parents-in-law Merle and Jean, the writing would not have been possible. Moreover, without the help of my parents Gerda and Siegfried and the sacrifices they made I would have never reached the starting point in the first place.

In addition to the immeasurable contribution of my family, many individuals made contributions of large magnitude. I would like to thank all of them here and I hope I am not forgetting anyone.

Joseph Kung was a kind and resourceful editor. Ann Kostant and Tom Grasso, my contacts at Birkhäuser, gave me good insights into the publishing process, moral support and a much needed extension of a deadline. Elizabeth Loew was a great resource on typographical matters. Laura Ogden, Anita Dotson and their staff at Louisiana Tech's Interlibrary Loan patiently and efficiently filled a never-ending stream of orders. Jean-Xavier Rampon was a wonderful host for a research visit to Nantes in the summer of 2000: aside from our work, much progress was made on this text. Jonathan Farley started the whole project by making me aware of Birkhäuser's search for authors. Without that one forwarded e-mail the opportunity would have passed by. Along these lines I owe special thanks to my Academic Director, Dr. Gene Callens, for the (very reasonable) advice against writing a book before being tenured and also for respecting my decision to disregard his advice.

Many people have helped me with questions regarding parts of the book. Mike Roddy let me use his improvement of the proof of Theorem 10.2.11 and he inspired part of the cover. Khaled Al-Agha helped with some parts of Chapter 7 and a number of questions were answered by Tom Trotter. Ralph McKenzie helped with Exercise 22 and Remark 15 in Chapter 10. Moreover his help with the proof of Hashimoto's theorem (Lemma 10.4.8!) was invaluable. My graduate student Joshua Hughes read the early chapters and eliminated a good number of typographical errors.

Finally I would like to thank the readers and the technical staff who helped this book through the final stages.

1
The Basics

There are few prerequisites to this text. The reader should be familiar with real numbers, functions, sets and relations. Moreover the elusive property known as "mathematical maturity" should have been developed to the point that the reader can read and understand proofs and produce simple proofs. A text that develops these skills is for example [103]. A background in graph theory helps, but is not necessary.

1.1 Definition and Examples

This section introduces some of the basic definitions and notations to be used throughout the text. The author has tried to keep the length of this section to a minimum. Terminology is introduced "in context" whenever possible. When encountering unknown notation, terminology, symbols later in the text, the reader should consult the index to locate the definition.

The central concept in this book is the concept of an ordered set, which is a set equipped with a special type of binary relation. Recall that abstractly a binary relation on a set P is just a subset $R \subseteq P \times P = \{(p, q) : p, q \in P\}$. $(p, q) \in R$ simply means that "p is related to q under R". A binary relation R thus contains all the pairs of points that are related to each other under R. For any binary relation R in this text we will write pRq instead of $(p, q) \in R$ whenever p is related to q via R. The relations of most interest to us are order relations.

Definition 1.1.1 *An* **ordered set** *(or* **partially ordered set** *or* **poset***) is an ordered pair* $(P, \leq)$ *of a set P and a binary relation* $\leq$ *contained in* $P \times P$, *called the* **order** *(or the* **partial order***) on P, such that*

1. *The relation* $\leq$ *is reflexive. That is, each element is related to itself;*

$$\forall p \in P : p \leq p.$$

2. *The relation* $\leq$ *is antisymmetric. That is, if p is related to q and q is related to p, then p must equal q;*

$$\forall p, q \in P : [(p \leq q) \wedge (q \leq p)] \Rightarrow (p = q).$$

3. *The relation* $\leq$ *is transitive. That is, if p is related to q and q is related to r, then p is related to r;*

$$\forall p, q, r \in P : [(p \leq q) \wedge (q \leq r)] \Rightarrow (p \leq r).$$

The elements of P are called the **points** of the ordered set. Order relations introduce a hierarchy on the underlying set. The statement "$p \leq q$" is read "p is less than or equal to q" or "q is greater than or equal to p". The antisymmetry of the order relation ensures that there are no two-way ties ($p \leq q$ and $q \leq p$ for distinct p and q) in the hierarchy. The transitivity (in conjunction with antisymmetry) ensures that no cyclic ties ($p_1 \leq p_2 \leq \cdots \leq p_n \leq p_1$ for distinct $p_1, p_2, \ldots, p_n$) exist. We will also use the notation $q \geq p$ to indicate that $p \leq q$. In keeping with the idea of a hierarchy, we will say that $p, q \in P$ are **comparable** and write $p \sim q$ iff $p \leq q$ or $q \leq p$. We will write $p < q$ for $p \leq q$ and $p \neq q$. In this case we will say p is **(strictly) less than** q An ordered set will be called finite (infinite) iff the underlying set is finite (infinite).

In many disciplines one investigates sets that are equipped with a structure. In cases where no confusion is possible it is customary to not mention the structure explicitly. (Consider for example references like [56] on analysis or [275] on topology.) Since mostly there will be no confusion possible, we will do the same and often not mention the order explicitly. A phrase such as "Let P be an ordered set" normally will mean that P carries an order that is usually denoted by $\leq$. In case we have several orders under consideration, they will be distinguished using subscripts or different symbols. When we have to talk about orders, we will automatically assume that a property of an order is defined in the same way as a property of the ordered set and vice versa.

Example 1.1.2 As we will see in the following sections, ordered structures are present throughout mathematics. To start out, some examples of ordered sets are

1. The natural numbers $\mathbb{N}$, the integers $\mathbb{Z}$, the rational numbers $\mathbb{Q}$ and the real numbers $\mathbb{R}$ with their usual orders are ordered sets.

2. Any set of sets is ordered by set inclusion $\subseteq$. Similarly, geometric figures (circles in the plane, balls or simplices in higher-dimensional spaces) are ordered by inclusion. Note that if we consider each geometric object as a set of points, we are back to sets ordered by set containment.

3. The simplest example of a set system ordered by inclusion is the **power set** $\mathcal{P}(X)$ of a set X.

4. Every ordinal number in set theory is an ordered set.

5. The vector space $C([0, 1], \mathbb{R})$ of continuous functions from $[0, 1]$ to $\mathbb{R}$ can be ordered as follows: For $f, g \in C([0, 1], \mathbb{R})$ we say $f \leq g$ iff for all $x \in [0, 1]$ we have $f(x) \leq g(x)$ (where the latter $\leq$ is the order of the real numbers).[1]

6. The natural numbers can also be ordered as follows: For $p, q \in \mathbb{N}$ we say $p \sqsubseteq q$ iff p divides q.

7. If J is a set of intervals of the real line, we can order J by

$$[a, b] \leq_{\text{int}} [c, d] \qquad \text{iff} \qquad b \leq c \quad \text{or} \quad [a, b] = [c, d].$$

8. If $(P, \leq_P)$ is an ordered set, the **dual** P^d of P is the set P with the order $\leq_{pd} := \{(a, b) : b \leq_P a\}$.

9. The lexicographic order on $\mathbb{N}^n$ is defined by $(x_1, \ldots, x_n) \leq (y_1, \ldots, y_n)$ iff $(x_1, \ldots, x_n) = (y_1, \ldots, y_n)$ or there is a $k \in \{1, \ldots, n\}$ with $x_i = y_i$ for $i < k$ and $x_k < y_k$.

Proofs that the structures in Example 1.1.2 truly are ordered sets are not too hard. We will limit ourselves here to proving that the lexicographic order for $\mathbb{N}^n$ is indeed an order.

Proof that Example 1.1.2, part 9 is an ordered set. We will have to prove that $\leq$ is reflexive, antisymmetric and transitive. Let $(x_1, \ldots, x_n)$, $(y_1, \ldots, y_n)$, and $(z_1, \ldots, z_n)$ be elements of $\mathbb{N}^n$.
Reflexivity. By definition we have that $(x_1, \ldots, x_n) \leq (x_1, \ldots, x_n)$.
Antisymmetry. Let

$$(x_1, \ldots, x_n) \leq (y_1, \ldots, y_n) \quad \text{and} \quad (x_1, \ldots, x_n) \geq (y_1, \ldots, y_n)$$

and assume $(x_1, \ldots, x_n) \neq (y_1, \ldots, y_n)$. Then there are $k_1, k_2 \in \{1, \ldots, n\}$ such that

1. $x_i = y_i$ for $i < k_1$ and $x_{k_1} < y_{k_1}$ and

2. $y_j = x_j$ for $j < k_2$ and $y_{k_2} < x_{k_2}$.

---

[1] Here it should be clear from the context which order (for numbers or for functions) is meant.

Assume without loss of generality that $k_1 \leq k_2$. Then $x_{k_1} < y_{k_1}$ by 1 and $y_{k_1} \leq x_{k_1}$ by 2, which is a contradiction. Thus we must have

$$(x_1, \ldots, x_n) = (y_1, \ldots, y_n).$$

This proves antisymmetry.

Transitivity. Let

$$(x_1, \ldots, x_n) \leq (y_1, \ldots, y_n) \quad \text{and} \quad (y_1, \ldots, y_n) \leq (z_1, \ldots, z_n).$$

If any two of the three tuples are equal, then there is nothing to prove. Hence we can assume the three tuples are mutually distinct. In this case there are $k_1, k_2 \in \{1, \ldots, n\}$ such that

1. $x_i = y_i$ for $i < k_1$ and $x_{k_1} < y_{k_1}$, and

2. $y_j = z_j$ for $j < k_2$ and $y_{k_2} < z_{k_2}$.

Let $k := \min\{k_1, k_2\}$. Then for $i < k$ we have $x_i = y_i = z_i$ and for the index k we have $x_k \leq y_k \leq z_k$ with at least one of the inequalities being strict. Thus $x_k < z_k$ and $(x_1, \ldots, x_n) \leq (z_1, \ldots, z_n)$, which concludes our proof of transitivity. ∎

It is a good exercise for the reader to prove that every example in 1.1.2 is an ordered set. Not every relation that looks like it induces a hierarchy is an order relation however.

Proposition 1.1.3 (Cf. [270].) *If $\mathcal{J}$ is a set of subsets of an ordered set P, we can define $A \sqsubseteq B$ iff*

1. *For all $a \in A$ there is an element b in B such that $a \leq b$, and*

2. *For all $b \in B$ there is an element a in A such that $b \geq a$,*

Then $\sqsubseteq$ is reflexive and transitive, but not necessarily antisymmetric.

Proof. Let P be an ordered set, $\mathcal{J}$ a set of subsets of P and let $A, B, C \in \mathcal{J}$ be arbitrary. Since for all $a \in A$ we have $a \leq a$, we obtain $A \sqsubseteq A$ for all $A \in \mathcal{J}$. Thus $\sqsubseteq$ is reflexive.

If $A \sqsubseteq B \sqsubseteq C$, then for all $a \in A$ there is a $b \in B$ with $a \leq b$. For b there is a $c \in C$ such that $b \leq c$ and hence $a \leq c$. Similarly, for all $c \in C$ there is a $b \in B$ with $c \geq b$. For b there is an $a \in A$ such that $b \geq a$ and hence $c \geq a$. Thus $A \sqsubseteq C$ and $\sqsubseteq$ is transitive.

However $\sqsubseteq$ is in general not antisymmetric. Consider the sets $\{1, 2, 4\}$ and $\{1, 3, 4\}$ as subsets of $\mathbb{N}$ with the natural order. Then $\{1, 2, 4\} \sqsubseteq \{1, 3, 4\}$ and $\{1, 2, 4\} \sqsupseteq \{1, 3, 4\}$, but clearly the two sets are not equal. ∎

1.2   The Diagram

Having defined ordered sets as sets equipped with a certain type of relation, we are ready to investigate these entities. Yet it would be helpful to have a visual aid to work with ordered sets. A picture often says more than a thousand words. Such a visual aid is inspired by graph theory, so let us quickly review what a graph is.

Definition 1.2.1 *A* **graph** *G is a pair (V, E) of a set V (of* **vertices***) and a set $E \subseteq \{\{a, b\} : a, b \in V\}$ (of* **edges***).*

This is a perfectly fine definition. However when working with graphs most people think not of the set theoretical entities of Definition 1.2.1. Instead they visualize an entity such as shown in Figure 1.1 a). The connection is simple: For each vertex $v \in V$ we put a point in the plane (or in 3-space) to represent the vertex. For any two vertices $v, w \in V$ we join the corresponding points with a line (an edge) that does not touch any other points iff $\{v, w\} \in E$. In this fashion we have a good visual tool for the work with graphs and also a way to translate real life problems into mathematics (road networks for example can be modeled using graphs). We could now do the same thing for orders. Put points in the plane or 3-space and join related points with edges. Arrows could indicate the way in which points are related. This idea would have two shortcomings. First, the hierarchical structure of the order may be hard to detect and second, there will be many lines that can be considered superfluous because of transitivity. We shall tackle the second problem first.

Definition 1.2.2 *Let P be an ordered set. Then $p \in P$ is called a* **lower cover** *of $q \in P$ (and q is called an* **upper cover** *of p) iff $p < q$ and for all $z \in P$ we have that $p \le z \le q$ implies $z \in \{p, q\}$. In this case we write $p \prec q$. Points p and q that satisfy $p \prec q$ or $q \prec p$ will also be called* **adjacent.**

Example 1.2.3 To become familiar with the covering relation, consider.

1. In the power set $\mathcal{P}(\{1, \ldots, 6\})$ the set $\{1, 3\}$ is a lower cover of $\{1, 3, 5\}$, but it is not a lower cover of $\{1, 2, 3, 4\}$.

2. In $\mathbb{Z}$ each number k has exactly one upper cover $(k+1)$ and one lower cover $(k-1)$.

3. Whether two elements are covers of each other depends on the surrounding universe. 2 is not an upper cover of 0 in $\mathbb{Z}$, but it is an upper cover of zero in the set of even numbers. Similarly $\{1, 2, 3, 4\}$ is an upper cover of $\{1, 3\}$ in the set of subsets of $\{1, \ldots, 6\}$ that have an even number of elements.

4. In infinite ordered sets, elements may or may not have covers. Consider that in $\mathbb{R}$ and $\mathbb{Q}$ no two elements are covers of each other.

The covering relation carries no superfluous information. It is the smallest relation that carries all the information for a given finite order. To visually incorporate the hierarchy, we only have to impose an up-down direction. The resulting tool that is analogous to the sketch of a graph is the Hasse diagram. Its main use is, because of the difficulty indicated in Example 1.2.3, part 4, for finite ordered sets.

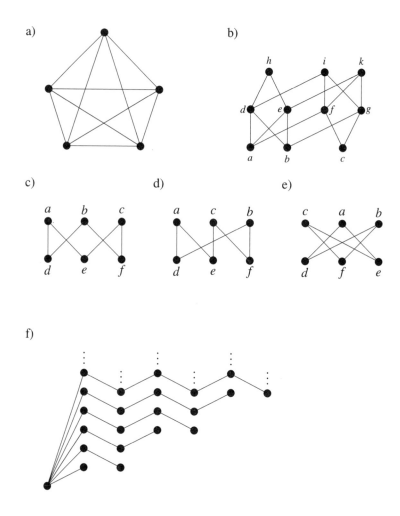

Figure 1.1: a) The complete graph with five vertices K_5, b) The diagram of an ordered set (the set $P2$ in [235]), c)–e) Three diagrams for the same ordered set (points to be identified by an isomorphism have the same letters), f) The "diagram" of an infinite ordered set (this "spider", as Rutkowski named it, is taken from [270], Remark 5.1).

Definition 1.2.4 *The* **(Hasse) diagram**[2] *of a finite ordered set P is the ordered pair* $(P, \prec)$, *where* $\prec$ *is the lower cover relation as defined in Definition 1.2.2.*

Again a perfectly reasonable mathematical definition which gives rise to the following possibility for visualization (cf. Figure 1.1, b-e).

1. Draw the points of P in the plane (or in 3-space, or on some surface) such that if $p \leq q$, then the point for q has a larger y-coordinate (z-coordinate) than the point for p,

2. If $p \prec q$, draw a line or curve (an edge) between the points corresponding to p and q such that

   - The slope of the edge (or $\frac{\partial}{\partial z}$ if we are in 3-space and just assume that our connecting curves are somewhat smooth) does not change its sign and

   - The edge does not touch any points of P except those corresponding to p and q.

Do not join any other pairs of points by edges.

For purpose of illustration consider Figure 1.1, part b). We can easily read off the comparabilities in the diagram shown. For example, $a \leq d$, since there is an edge from a to d and a is lower than d. We also have $a \leq k$, since there is an edge going up from a to e and another going up from e to k. This means $a \prec e \prec k$ and by transitivity $a \leq k$. The knowledge that order relations are transitive allows us to avoid drawing some edges that would only clutter the picture. Just imagine all the edges (a, h), (a, i), (a, k), (b, h), (b, i), (b, k), (c, i), and (c, k) added to the picture. It would be quite confusing. Also note that from Figure 1.1 part b) we see that $c \not\leq h$. Indeed, even though there is an edge from c to f, an edge from f to a, an edge from a to d and an edge from d to h, not all the edges are traversed in an upward direction in the trail just described. The edge from f to a is traversed downwards (from f to a), meaning that $c \prec f \succ a \prec d \prec h$. Transitivity cannot be applied to this sequence (and not to any other such sequence) and so $c \not\leq h$.

Drawing diagrams is not canonical. The same ordered set can be drawn in different ways according to the investigator's preferences or needs. As an example, consider Figure 1.1, parts c)–e). Each picture depicts the same ordered set, yet they do look distinctly different.

Diagrams can also be drawn for infinite ordered sets, but are then in need of explanation. The infinite ordered set depicted in Figure 1.1 f) consists of one "zig-zag" with n elements for each $n \in \mathbb{N}$ such that all "zig-zags" have the same left endpoint. How to draw a diagram of a certain set is often a matter of taste. For some discussion on the subject cf. [4]. For drawing diagrams on surfaces cf. [77]

---

[2]Named after the German algebraic number theorist Helmut Hasse who used diagrams to picture the ordered sets of subfields or field extensions.

and for a multitude of results regarding diagrams cf. [221]. We will use diagrams extensively as visual tools. Drawings of diagrams in this text are biased towards the author's taste.

From a relation-theoretic point-of-view the diagram is a certain subset of the order relation, formed according to the rule that only pairs (a, b) are selected for which there is no intermediate point i such that (a, i) and (i, b) are also in the relation. To formalize how to recover the original relation from the diagram we merely need to formulate the reading process indicated above as a mathematical construction.

Definition 1.2.5 *Let $\prec$ be a binary relation on the finite set P. Then the* **transitive closure** $\prec^t$ *of $\prec$ is defined by $a \prec^t b$ iff there is a sequence $a = a_1, a_2, \ldots, a_n = b$ such that $a_1 \prec a_2 \prec \cdots \prec a_n$.*

The name is justified by the following result and by Exercise 7.

Proposition 1.2.6 *The transitive closure of a binary relation $\prec$ on a finite set P is transitive.*

Proof. Let $a, b, c \in P$ with $a \prec^t b \prec^t c$. There are $a = a_1, a_2, \ldots, a_n = b$ with $a_1 \prec a_2 \prec \cdots \prec a_n$ and $b = a_n, a_{n+1}, \ldots, a_{n+m} = c$ with $a_n \prec a_{n+1} \prec \cdots \prec a_{n+m}$. Therefore $a = a_1 \prec a_2 \prec \cdots \prec a_{n+m} = c$ and hence $a \prec^t c$. ∎

For finite sets, the transitive closure allows us to translate back from the diagram to the order relation.

Proposition 1.2.7 *Let P be a finite ordered set with order $\leq$ and let $\prec$ be its lower cover relation. As usual, $=$ denotes the equality relation. Then $\leq$ is the transitive closure of the union of the relations $\prec$ and $=$.*

Proof. Let $\leq'$ denote the transitive closure of $\prec \cup =$. (Note that despite the suggestive notation, at this stage we do not even know if $\leq'$ is an order relation.) We will prove that $\leq = \leq'$. Since $\prec \cup =$ is contained in $\leq$ and since $\leq$ is transitive, we must have that $\leq' \subseteq \leq$ (cf. Exercise 7).

To prove the other inclusion, assume there are $a, b \in P$ with $a \leq b$ and $a \not\leq' b$. Then we must have $a < b$, since $a = b$ implies $a \leq' b$. Since P is finite we can find points $c, d \in P$ with $c \leq d$ and $c \not\leq' d$ such that for all $c < z < d$ we have $c \leq' z$ and $z \leq' d$. Since $c \prec d$ would mean $c \leq' d$, there must be a $z \in P$ with $c < z < d$. But then $c \leq' z \leq' d$ and since the transitive closure of a relation is transitive, we infer $c \leq' d$, a contradiction. Thus a, b as described above cannot exist and we conclude that $\leq = \leq'$. ∎

We will discuss some algorithmic ramifications of computing the diagram and transitive closures in Sections 12.1 and 12.2.

1.3 Order-Preserving Mappings/Isomorphism

Figure 1.1 c), d) and e) shows three different pictures of the same ordered set. Had we assigned different labels to the points, we could have depicted three ordered sets that appear "different and yet the same". This phenomenon is similar to the topological result (which seems to have become folklore), that "a donut is ho-motopic/isomorphic to a teacup". Indeed we can find a "continuous deformation" that has diagrams at each stage and, say, turns set c) into set d). The formal back-ground lies in the investigation of structure-preserving maps (or "morphisms") which is strongly represented in algebra and topology (cf. the strong role of struc-ture homomorphisms in algebra and of continuous functions in topology). Since the underlying structure we are interested in is the order, the following definition is only natural.

Definition 1.3.1 *Let $(P, \leq_P)$ and $(Q, \leq_Q)$ be ordered sets and let $f : P \to Q$ be a map. Then f is called an* **order-preserving function** *iff for all $p_1, p_2 \in P$ we have*

$$p_1 \leq_P p_2 \Rightarrow f(p_1) \leq_Q f(p_2).$$

(We will usually not index the orders in such a situation. Also we will use the words "function" and "map" interchangeably. Finally let it be noted that order-preserving maps are sometimes also referred to as **isotone maps**.*)*

Example 1.3.2 We continue with some examples of order-preserving and non-order-preserving maps. Note that the same map can be order-preserving or not, depending on the orders of domain and range.

1. The function $f : \mathbb{N} \to \mathbb{N}$, defined by $f(x) = 5x$ is an order-preserving map if both domain and range carry the natural order. It is also order-preserving if both domain and range carry the order $\sqsubseteq$ of part 6 in Example 1.1.2.

2. The function $f : \mathbb{N} \to \mathbb{N}$, defined by $f(x) = x + 1$, is order-preserving if both domain and range carry the natural order. However, it is not order-preserving if both domain and range carry the order $\sqsubseteq$ of part 6 in Example 1.1.2.

3. Let S be a set and let $\mathcal{P}(S)$ be its power set ordered by set inclusion. The map $\mathcal{I} : \mathcal{P}(S) \to \mathcal{P}(S)^d$ defined by $\mathcal{I}(X) := S \setminus X$ is an order-preserving map from $\mathcal{P}(S)$ to its dual.

4. Let B be the ordered set in part b) of Figure 1.1 and let C be the ordered set in part c) of Figure 1.1. Then the function $F : B \to C$ defined by $F(h) = a$, $F(a) = F(b) = F(d) = F(e) = d$, $F(f) = F(g) = F(i) = F(k) = b$, and $F(c) = f$ is order-preserving. We use reflexivity of order relations here. This property (though essentially taken for granted and not noted on the diagram) allows us to collapse related points into one.

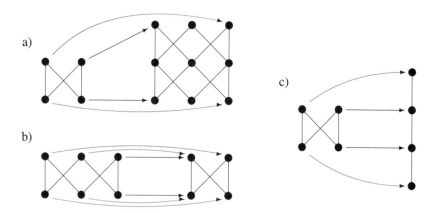

Figure 1.2: Graphical representation of some order-preserving maps.

5. Consider Figure 1.2. The sets where the arrows start are the domains of the maps. The arrows indicate where each individual point is mapped. All the maps thus given in Figure 1.2 are order-preserving.

If Definition 1.3.1 is a reasonable definition for structure-preserving maps on ordered sets, then the composition of two order-preserving maps should again be order-preserving.

Proposition 1.3.3 *Let P, Q, R be ordered sets and let $f : P \to Q$ and $g : Q \to R$ be order-preserving maps. Then the map $g \circ f : P \to R$ is also order-preserving,*

Proof. Let $p_1, p_2 \in P$ with $p_1 \leq p_2$. Then since f is order-preserving we have $f(p_1) \leq f(p_2)$ and since g is order-preserving $g(f(p_1)) \leq g(f(p_2))$. Thus $g \circ f$ is order-preserving.   ∎

The above examples of order-preserving maps show that these maps preserve the order "one way". However, even the existence of a bijective order-preserving map between two sets, such as in part c) of Figure 1.2, is not a guarantee that both sets are "essentially the same". (This is similar to the situation in algebra or topology.) What is missing in part c) of Figure 1.2 is that the inverse function is not order-preserving.

Suppose now the two ordered sets P and Q have a bijection Φ between them such that Φ as well as Φ^{-1} preserve the order. When investigating an ordered structure, the underlying set normally only gives us the substance that we mold into structures. Which individual point is placed where in the structure is thus quite unimportant. Hence for many purposes in order theory, P and Q as above are indistinguishable, as they have the same order-theoretical structure. This is the concept of (order-)isomorphism.

Definition 1.3.4 *Let P, Q be ordered sets and let $\Phi : P \to Q$. Then Φ is called an (**order-)isomorphism** iff*

1. *Φ is order-preserving,*

2. *Φ has an inverse Φ^{-1},*

3. *Φ^{-1} is order-preserving.*

*The ordered sets P and Q are called (**order-)isomorphic** iff there is an isomorphism $\Phi : P \to Q$.*

The following characterization of isomorphisms reinforces the notion that isomorphic ordered sets can be regarded as "the same".

Proposition 1.3.5 *Let P, Q be ordered sets. Then $f : P \to Q$ is an order-isomorphism iff*

1. *f is bijective.*

2. *For all p_1, $p_2 \in P$ we have*

$$p_1 \leq p_2 \Leftrightarrow f(p_1) \leq f(p_2).$$

Proof. If $f : P \to Q$ is an order-isomorphism, then 1 and 2 are trivial. To prove that 1 and 2 imply f is an isomorphism, we only need to prove that $f^{-1} : Q \to P$ is order-preserving. Let $q_1, q_2 \in Q$ be such that $q_1 \leq q_2$. Then there are $p_1, p_2 \in P$ such that $f(p_i) = q_i$ for $i = 1, 2$. Now $f(p_1) = q_1 \leq q_2 = f(p_2)$ implies $p_1 \leq p_2$, that is, $f^{-1}(q_1) \leq f^{-1}(q_2)$. Thus f^{-1} is order-preserving. ∎

Proposition 1.3.6 *Let P, Q, R be ordered sets and let $\Phi : P \to Q$ and $\Psi : Q \to R$ be order-isomorphisms. Then $\Psi \circ \Phi$ is an order-isomorphism.*

Proof. Left as Exercise 14. ∎

The strength of using isomorphisms is that structures that at first appear different can turn out to be equal for all intents and purposes. Thus even structures that appear different can have the same properties. For example, the ordered sets in Figure 1.3 appear quite different at first, yet they are isomorphic via the map (from the left set to the right set) $1 \mapsto 1, 2 \mapsto 2, 3 \mapsto 4, 4 \mapsto 3, 5 \mapsto 5, 6 \mapsto 8$, $7 \mapsto 6, 8 \mapsto 7, 9 \mapsto 10, 10 \mapsto 9, 11 \mapsto 11$. Verification that the indicated map is an isomorphism is a good exercise for the reader. While the experienced reader can (and should) classify this task as "tedious, but routine", it gives an indication how hard it is to find an isomorphism. How many calculations would have been necessary to find the indicated map, had it not been given above? Even worse, what if we start out with sets of which we do not know if they are isomorphic

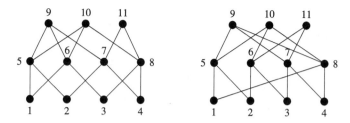

Figure 1.3: Two isomorphic ordered sets.

or not? How do we know how many and what types of checks to perform until we can be sure no isomorphism exists? These questions show that it can be quite difficult to decide if two ordered sets are isomorphic or not. In fact it is still an open problem *how* difficult this decision is. We will discuss this issue in Remark 8 in Chapter 12.

1.4   Fixed Points

A property that has attracted a good bit of attention in order theory is the fixed point property, defined as follows.

Definition 1.4.1 *We will call a function* $f : D \to D$ *whose domain and range are equal a* **self map** *on D.* $f : D \to D$ *is an* **order-preserving self map** *on D iff D carries an order* $\leq$ *and* $a \leq b$ *implies* $f(a) \leq f(b)$.

Definition 1.4.2 *Let P be an ordered set and let* $f : P \to P$ *be an order-preserving self map. Then* $p \in P$ *is called a* **fixed point** *of f iff* $f(p) = p$. *If f has no fixed points, f is called* **fixed point free**. *P is said to have the* **fixed point property** *iff each order-preserving self map* $f : P \to P$ *has a fixed point. For any ordered set P and any order-preserving map* $f : P \to P$ *we set*

$$\mathrm{Fix}(f) := \{p \in P : f(p) = p\}.$$

An original motivation for working with fixed points in ordered sets is a proof of the Bernstein–Cantor–Schröder theorem, which we will give later in the text. Yet the property also became interesting in itself. For more background on the fixed point property, consider Remark 2 in the "Remarks and Open Problems" section of this chapter.

We will use the fixed point property as a vehicle to introduce the reader to new order-theoretical notions. The fixed point property is well-suited for this task, because it combines properties of the set with properties of the maps on the set. For every new structure we introduce, the fixed point property can provide a familiar setting in which to investigate the structure. The underlying problem is

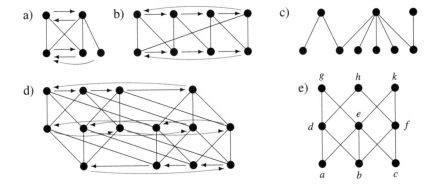

Figure 1.4: Examples of ordered sets with and without the fixed point property. Fixed point free maps are indicated for sets without the fixed point property.

> **Open Question 1.4.3** (Cf. [223].) *Characterize those (finite) ordered sets that have the fixed point property.*

We will frequently prove fixed point results in this text to show how one can work with a certain class of ordered sets or a certain structure. Ultimately (cf. Theorem 12.4.5) we will prove that the NP version of this problem turns out to be NP-complete.[3] This will lead into our discussion of methods to tackle "hard" algorithmic problems. At this stage we only give a few examples of sets with and without the fixed point property (cf. Figure 1.4).

For the ordered sets in Figure 1.4 that do not have the fixed point property, a fixed point free order-preserving self map is indicated. The question is how to prove that such a map does *not* exist. It is certainly instructive to try and do this for the remaining sets before reading on. We will soon have more sophisticated methods to prove that the ordered sets in Figure 1.4 c) and e) have the fixed point property. To get the reader more acquainted with the basic notions of order theory we give a quick proof that the ordered set in 1.4 e) has the fixed point property. The proof will be combinatorial with potentially many cases to be treated separately. To reduce the number of cases to be treated, one can often use symmetry or, formally, the notion of an automorphism.

Definition 1.4.4 *Let P be an ordered set. The self map $f : P \to P$ is called an* **(order-)automorphism** *iff f is an isomorphism.*

Example 1.4.5 For the ordered set in Figure 1.4 e) the map Φ that maps $a \mapsto c$, $b \mapsto b$, $c \mapsto a$, $d \mapsto f$, $e \mapsto e$, $f \mapsto d$, $g \mapsto k$, $h \mapsto h$, $k \mapsto g$ is an automorphism. We can see that this map "reflects the ordered set across an axis

---

[3]Formally, deciding if a finite ordered set has the fixed point property is co-NP-complete.

through the middle" if the set is drawn as in Figure 1.4. This illustrates the common interpretation that automorphisms reveal the symmetries of a combinatorial structure.

Proposition 1.4.6 *The ordered set in Figure 1.4 e), call it P, has the fixed point property.*

Proof. Assume to the contrary that there is an order-preserving map $F : P \to P$ such that F has no fixed point. Then $F(b)$ cannot be related to b. Indeed, otherwise $b < F(b)$ and applying F twice to this inequality leads to (since F has no fixed points) $F(b) < F^2(b)$ and $F^2(b) < F^3(b)$. This however is not possible, since there are no four distinct elements $w, x, y, z \in P$ such that $w < x < y < z$. Thus $F(b) \in \{a, c\}$. Since Φ in Example 1.4.5 is an automorphism, we can assume without loss of generality that $F(b) = a$. (Otherwise we would apply the whole following argument to $\Phi^{-1} \circ F \circ \Phi$, which is also a fixed-point-free order-preserving self map.)

Since $F(b) = a$ we have $F[\{d, e, g, h, k\}] \subseteq \{a, d, e, g, h, k\}$. Since $F(g)$ cannot be related to g (proved similar to $b \not\sim F(b)$) we must have $F(g) \in \{h, k\}$. If $F(g) = h$, then we must have that $a = F(b) \leq F(d) \leq F(g) = h$.

$F(d) = h$ would lead to $F(h) \geq F(d) = h$ and then $F(h) = h$, which is not possible. We exclude $F(d) = a$ in similar manner. This leaves $F(d) = d$, a contradiction.

Therefore we must have $F(g) = k$, which then can be lead to a contradiction in similar fashion. Thus P has no fixed-point-free order-preserving self maps and hence P has the fixed point property. ∎

With automorphisms being indicators of symmetry, ordered sets that have automorphisms without any fixed points should have a very high degree of symmetry, since there will be at least one way to move every point and still have the same order-theoretical structure. Existence of fixed point free automorphisms is also a problem similar to asking about the existence of a fixed point free order-preserving self map. We record

Definition 1.4.7 *Let P be an ordered set. P is called **automorphic** iff P has a fixed point free automorphism.*

> **Open Question 1.4.8** *Characterize the (finite) automorphic ordered sets.*

Automorphic ordered sets play a role in the investigation of the fixed point property as we will see in Theorem 4.2.2, part 2. The NP version of problem 1.4.8 turns out to be NP-complete as well (cf. [277]). We conclude this section by showing that while the fixed point property implies the ordered set is not automorphic, nonautomorphic ordered sets need not have the fixed point property.

Proposition 1.4.9 *The ordered set in Figure 1.4 a), call it Q, is not automorphic. However, it does not have the fixed point property.*

Proof. A fixed point free order-preserving map is indicated in the figure. To see that Q has no fixed point free automorphism, note that Q has exactly one point with exactly one upper cover. This point must be fixed by any automorphism. ∎

1.5   Ordered Subsets/The Reconstruction Problem

Having introduced the objects of our studies and their structure-preserving maps, the last basic notion to expose the reader to are the subobjects. These subobjects are the ordered subsets of an ordered set, which are defined analogous to subgroups in algebra or topological subspaces in topology.

Proposition 1.5.1 *Let $(P, \leq_P)$ be an ordered set and let $Q \subseteq P$. Then Q with the restriction $\leq_Q := \leq_P |_{Q \times Q}$ of the order on P to Q is also an ordered set.*

Proof. The proofs of reflexivity, antisymmetry and transitivity of $\leq_Q$ are trivial. Every property that holds for all elements of P will clearly hold for all elements of Q. ∎

Knowing that the order properties are not destroyed when restricting ourselves to a subset, the following definition is sound.

Definition 1.5.2 *Let $(P, \leq_P)$ be an ordered set and let $Q \subseteq P$. If $Q \subseteq P$ and $\leq_Q = \leq_P |_{Q \times Q}$ we will call $(Q, \leq_Q)$ an* **ordered subset** *(or* **subposet***) of P. Unless indicated otherwise we will always assume that subsets of ordered sets carry the order induced by the surrounding ordered set.*[4]

Example 1.5.3 Every set of sets $\mathcal{S}$ that is ordered by inclusion is an ordered subset of $\mathcal{P}\left(\bigcup \mathcal{S}\right)$.

Order-theoretical properties may or may not carry over to ordered subsets. As we have not explored many properties yet, all we can do is record a negative example.

Example 1.5.4 Not every subset of an ordered set with the fixed point property again has the fixed point property. Indeed the subset $\{a, c, g, k\}$ of the ordered set in Figure 1.4 e) does not have the fixed point property. A fixed point free order-preserving map on $\{a, c, g, k\}$ would be $a \mapsto b, b \mapsto a, g \mapsto k, k \mapsto g$.

Connecting ordered subsets with order-preserving functions is simple, since of course for each $f : P \to Q$ the set $f[P]$ is an ordered subset of Q. Moreover it

---

[4]For those readers acquainted with graph theory this means that the notion of ordered subset is similar to the notion of an *induced* subgraph.

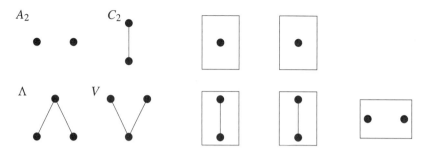

Figure 1.5: Two examples of nonisomorphic ordered sets with isomorphic decks. Are these the only ones?

is easy to see that, if P is finite and f is injective, then $f[P]$ has at least as many comparabilities as P. However, Figure 1.2 part c) shows that $f[P]$ need not be isomorphic to P, even if f is bijective. Order-preserving mappings for which P is isomorphic to $f[P]$ are called embeddings.

Definition 1.5.5 *Let P, Q be ordered sets. Then $f : P \to Q$ is called an* (**order**) **embedding** *iff*

1. *f is injective.*

2. *For all $p_1, p_2 \in P$ we have*

$$p_1 \leq p_2 \Leftrightarrow f(p_1) \leq f(p_2).$$

Proposition 1.5.6 *Let P be an ordered set and let $f : P \to Q$ be order-preserving. Then P is isomorphic to $f[P]$ iff f is an embedding. In this case f is an isomorphism between P and $f[P]$.*

Proof. This follows immediately from Definition 1.5.5 and Proposition 1.3.5. ∎

To further illustrate ordered subsets we are now ready to introduce our second main research problem (characterizing the fixed point property was the first). Like the fixed point property, this problem will provide us with a familiar setting in which to investigate many new structures we introduce. Draw each subset of an ordered set P that has one point less than P on a card (without labeling the points). Is it possible to take these pictures and reconstruct the original ordered set from them up to isomorphism? Examples that the reconstruction does not work in general are shown in Figure 1.5. However these are the only examples known so far. The natural question that arises is now if these are all such examples.

Definition 1.5.7 *For an ordered set P, we call the class of all ordered sets that are isomorphic to P the* **isomorphism class** *of P. We denote the isomorphism class of P by $[P]$.*

Definition 1.5.8 *Let P be a finite ordered set. For $x \in P$, the ordered subset $P \setminus \{x\}$ is called a* **card** *of P. Cards are generally considered unlabeled. That is, we have no way of determining which element of the card corresponds to which element of P.*

Let C be the set of all isomorphism classes of ordered sets with underlying set contained in $\mathbb{N}$. The **deck** *of P is the function $\mathcal{D}_P : C \to \mathbb{N}$ such that for each $[C] \in C$ we have that $\mathcal{D}_P([C])$ is the number of cards of P that are isomorphic to the elements of $[C]$.*

Open Question 1.5.9 The Reconstruction Problem. *Is every (finite) ordered set with more than three elements uniquely reconstructible from its deck? That is, is it true that if P, Q are ordered sets with more than three elements such that $\mathcal{D}_P = \mathcal{D}_Q$, then P and Q must be isomorphic?*

For more background on the reconstruction problem consider Remark 4 in the "Remarks and Open Problems" section of this chapter. We conclude this section with two simple-looking results which will be helpful when working on reconstruction problems. The first (Proposition 1.5.11) is a result about a partial success by proving a positive answer to the problem in a restricted class of ordered sets. The second (Proposition 1.5.14) shows that a certain parameter can be reconstructed from the deck of any ordered set. These results provide some (of the many) indications the answer might be positive in general. They will also be tools in later investigations. Both results are representative of possible approaches to this problem. One could prove reconstructibility for more and more special classes of ordered sets until every ordered set must belong to one class that has been proved to be reconstructible. Or, one could reconstruct more and more parameters of ordered sets until every ordered set is uniquely determined just by knowing a set of reconstructible parameters. So far, we are far from either of these goals.

Definition 1.5.10 *We will say an ordered set P is* **reconstructible** *from its deck $\mathcal{D}_P$, if all ordered sets Q with $\mathcal{D}_P = \mathcal{D}_Q$ are isomorphic to P. A class $\mathcal{K}$ of ordered sets is reconstructible iff each of its members $P \in \mathcal{K}$ is reconstructible. We will call a class $\mathcal{K}$ of ordered sets* **recognizable** *iff for each ordered set $P \in \mathcal{K}$ and any ordered set Q, $\mathcal{D}_P = \mathcal{D}_Q$ implies that $Q \in \mathcal{K}$.*

Proposition 1.5.11 *Let P be a finite ordered set with $|P| \geq 4$ and a smallest element s. That is, $s \in P$ is such that for all $p \in P$ we have $s \leq p$. Then P is reconstructible from its deck.*

Proof. We will first prove that ordered sets with a smallest element are recognizable. Indeed, if P has a smallest element s, then for any $x \in P \setminus \{s\}$ we have that s is the smallest element of $P \setminus \{x\}$. Thus there is at most one card of P that does not have a smallest element. On the other hand, if Q does not have a smallest element, then there are at least two incomparable elements in Q that do not have any strict lower bounds. Such sets Q have at most two cards that have a smallest element and hence at least two cards that do not. (Cards are counted with multiplicity here.) Thus the ordered sets with a smallest element are exactly those ordered sets whose deck has at most one card without a smallest element.

Now we are ready to prove reconstructibility. If P has a smallest element and the deck of P has a card C that does not have a smallest element, then this card must by the above be unique. Thus $C = P \setminus \{s\}$ and P is isomorphic to the ordered set obtained by attaching a smallest element to C. Formally, P is isomorphic to $C \cup \{s\}$ ordered by $\le \cup \{(s, c) : c \in C \cup \{s\}\}$, where $\le$ is the order on C.

If all cards of P have a smallest element, then removal of s must have introduced a new smallest element and we argue as follows. For every finite ordered set Q with a smallest element there is a number l_Q such that

1. All elements of Q with fewer than l_Q lower bounds are unique.

2. Q contains zero or at least two elements that have l_Q lower bounds.

Essentially, $l_Q - 1$ is the number of times a smallest element can be removed before we arrive at an ordered set without a smallest element. For each card C of P find the number l_C. The cards C with $l_C \le l_K$ for all cards K of P are all isomorphic. Moreover, $P \setminus \{s\}$ is isomorphic to one of them. Pick one and call it C. Then P is isomorphic to the ordered set obtained from C by attaching a new smallest element below the smallest element of C.   ∎

Definition 1.5.12 *An* (**order**) **invariant** $\alpha(\cdot)$ *of ordered sets is a function from the class of all ordered sets to another set or class, such that if P and Q are isomorphic, then $\alpha(P) = \alpha(Q)$.*

Invariants are mostly numerical, but the deck for example is also an invariant. The simplest invariant is probably the number of elements, closely followed by the number of comparabilities.

Definition 1.5.13 *An invariant α is called* **reconstructible** *iff for all ordered sets P and Q we have that $\mathcal{D}_P = \mathcal{D}_Q$ implies $\alpha(P) = \alpha(Q)$.*

The most easily reconstructed invariant is the number of elements. It is simply one more than the number of elements of any card. The number of comparabilities is to be reconstructed in Exercise 25a. If one could reconstruct a complete set of invariants, i.e., a set of invariants so that two ordered sets with the same invariants must be isomorphic, then the reconstruction problem would be solved. Unfortunately no such complete set of invariants has materialized yet. A helpful invariant is the number of subsets of a certain type.

Proposition 1.5.14 (A Kelly Lemma, cf. [136], Lemma 1 and [151], Lemma 4.1.) *Let P and Q be two finite ordered sets with $|P| > |Q|$ and $|P| > 3$. Then the number $s(Q, P)$ of ordered subsets of P that are isomorphic to Q is reconstructible from the deck.*

Proof. Let $d_Q := \sum_C |\{S \subseteq C : S \text{ isomorphic to } Q\}|$, where the sum runs over all cards C of P, with multiplicity. Clearly d_Q can be computed from the deck. Let $P_Q \subseteq P$ be a fixed subset of P that is isomorphic to Q. Then P_Q is contained in exactly $|P| - |Q|$ cards of P, namely in exactly those cards obtained by removing an element outside P_Q. Since this is true for any subset of P that is isomorphic to Q, we have that

$$d_Q = s(Q, P)(|P| - |Q|)$$

and we have reconstructed $s(Q, P)$ as

$$s(Q, P) = \frac{d_Q}{(|P| - |Q|)}.$$

■

Exercises

1. Prove that the following are ordered sets:

   (a) The set $\{a, b, c, d, e, f\}$ with the relation

   $$\begin{aligned} \leq \; := \quad &\{(a, a), (a, c), (a, d), (a, e), (a, f), (b, a), (b, b), (b, c), \\ &(b, d), (b, e), (b, f), (c, c), (c, e), (c, f), (d, d), (d, e), \\ &(d, f), (e, e), (f, f)\}. \end{aligned}$$

   (b) Let L and U be disjoint ordered sets. Construct $(P, \leq)$ as follows. $P := L \cup U$ and $a \leq b$ iff $a \in L$ and $b \in U$ or $a, b \in L$ and $a \leq_L b$ or $a, b \in U$ and $a \leq_U b$.

   (c) For a finite set X, consider the power set $\mathcal{P}(X)$ with the following relation. The set A is said to be **dominated** by the set B iff there is a k such that $|\{1, \ldots, k\} \cap A| < |\{1, \ldots, k\} \cap B|$ and for all $l < k$ we have $|\{1, \ldots, l\} \cap A| = |\{1, \ldots, l\} \cap B|$. Define $A \leq B$ iff $A = B$ or A is dominated by B.

   (d) Any tree $T = (V, E)$ ordered as follows. $x \leq y$ iff $x = y$ or there is a path from x to y that does not go through the root and the distance of x to the root is less than the distance of y to the root.

2. Let P be a set and let $\{\leq_i\}_{i \in I}$ be a family of order relations on P. Prove that $\bigcap_{i \in I} \leq_i$ is an order relation.

3. (Alternative definition of order.) Prove the following.

   (a) If $\leq$ is an order relation on P, then $<$ is an antireflexive (that is for all $p \in P$ we have $p \not< p$) and transitive relation.

(b) If $<$ is antireflexive and transitive, then $\leq := < \cup =$ is an order relation.

4. Prove that if $a_1 \leq a_2 \leq \cdots \leq a_n \leq a_1$ in an ordered set P, then $a_1 = a_2 = \cdots = a_n$.

5. Define the transitive closure of a relation on an arbitrary set. Then prove the analogue of Proposition 1.2.6.

6. Give an example of an infinite ordered set for which Proposition 1.2.7 fails. (Use the definition of the transitive closure from Exercise 5.)

7. Let $\prec$ be a relation that is contained in the transitive relation $\leq$. Prove that the transitive closure $\prec^t$ of $\prec$ is contained in $\leq$. Conclude that the transitive closure of a relation $\prec$ is the intersection of all transitive relations that contain $\prec$.

8. Draw the diagram of the ordered set.

   (a) The ordered set in Exercise 1a.

   (b) An ordered set as constructed in Exercise 1b with L and U isomorphic to ordered set as in Exercise 1a.

   (c) The set $\mathcal{P}(\{1, 2, 3, 4\})$ ordered as indicated in Exercise 1c.

9. Does the "spider" in Figure 1.1, part f) have a three-element subset C such that any two elements of C are comparable?

10. Prove that each of the following maps is order-preserving.

    (a) With the natural numbers $\mathbb{N}$ ordered as described in Example 1.1.2, part 6, define $f : \mathbb{N} \to \mathbb{N}$ by $f(x) := \begin{cases} \frac{x}{2}; & \text{if } x \text{ is even,} \\ x; & \text{if } x \text{ is odd.} \end{cases}$

    (b) For the ordered set $P2$ in Figure 1.1, part b) define $F : P2 \to P2$ by $F(a) = b$, $F(b) = a$, $F(c) = c$, $F(d) = e$, $F(e) = d$, $F(f) = g$, $F(g) = f$, $F(h) = h$, $F(i) = k$, $F(k) = i$.

    (c) In the power set $\mathcal{P}(X)$ of any topological space (X, τ), the map $f : \mathcal{P}(X) \to \mathcal{P}(\mathcal{P}(X))$ that maps A to its closure.

11. Let P and Q be ordered sets.

    (a) Prove that if P, Q are finite the following are equivalent for a map $f : P \to Q$.

       i. f is order-preserving.

       ii. For all $a, b \in P$ we have that $a < b$ implies $f(a) \leq f(b)$.

       iii. For all $a, b \in P$ we have that $a \prec b$ implies $f(a) \leq f(b)$.

    (b) Prove that parts 11(a)i and 11(a)ii are equivalent for infinite ordered sets also.

(c) Find an example that shows that 11(a)iii is in general not equivalent to 11(a)i and 11(a)ii.

12. (The relation between covers and isomorphisms.)

(a) Let $f : P \to Q$ be an isomorphism. Prove that $x \prec_P y$ implies $f(x) \prec_Q f(y)$.

(b) Prove that if P and Q are finite, $f : P \to Q$ is an isomorphism iff f is bijective and $x \prec_P y$ is equivalent to $f(x) \prec_Q f(y)$.

(c) Give an example of a bijective function $f : P \to Q$ such that $x \prec_P y$ is equivalent to $f(x) \prec_Q f(y)$ and yet f is not an isomorphism.

13. Let P be an ordered set and let $f : P \to P$ be an injective order-preserving self map.

(a) Prove that if P is finite, then f is an automorphism.

(b) Show that in general f need not be an automorphism.

14. Prove Proposition 1.3.6.

15. For two ordered sets $(P, \leq_P)$ and $(Q, \leq_Q)$ the ordered set $P \odot Q$ has the set $P \times Q$ as its points and $(p_1, q_1) \leq (p_2, q_2)$ iff $p_1 <_P p_2$ or $p_1 = p_2$ and $q_1 \leq_Q q_2$. Under what circumstances is $P \odot Q$ isomorphic to $Q \odot P$?

16. Let P be a finite ordered set and let $f : P \to P$ be order-preserving. Prove that if there is a $p \in P$ with $p \leq f(p)$, then f has a fixed point. Then find an infinite ordered set in which this result fails.

17. Prove that the ordered set $(\emptyset, \emptyset)$ does not have the fixed point property.

18. Prove that the following ordered sets have the fixed point property.

(a) The ordered set in Figure 1.4 c)

(b) The ordered set in Figure 1.1 b)

(c) The ordered set in Figure 1.3

19. Prove that the range of the map in Figure 1.2 a) does not have the fixed point property.

20. Let P be an ordered set and let $\Psi : P \to P$ be an automorphism.

(a) Prove that $f : P \to P$ has a fixed point iff $\Psi^{-1} \circ f \circ \Psi$ has a fixed point.

(b) What general hypotheses can be imposed on Ψ to assure that f has a fixed point iff $\Psi \circ f \circ \Psi$ has a fixed point?

21. Let P, Q be ordered sets, let $f : P \to Q$ be order-preserving and injective and let P be finite. Prove that if $f[P]$ contains as many comparabilities as P, then f is an embedding.

22. It is clear that for every ordered set the identity and the constant function are order-preserving self maps. Call two order relations $\leq_1$ and $\leq_2$ on the same ground set P **perpendicular** iff the only order-preserving self maps they have in common are the constant functions and the identity. Prove that if $\leq_1$ and $\leq_2$ are perpendicular, then their intersection is $\leq_1 \cap \leq_2 = \{(x, x) : x \in P\}$.

23. (a) Find all nonisomorphic ordered sets with up to five elements. (Hint. The numbers of nonisomorphic sets are given in Remark 3 in Chapter 11.)

    (b) Verify that the reconstruction problem 1.5.9 is solvable for sets with four and five elements.

    (c) Attempt a positive solution of the reconstruction problem by hand or with a computer for small sets with more than five elements.[5]

24. An **isolated point** is a point in an ordered set that is only comparable to itself. Prove that ordered sets with an isolated point are reconstructible.

25. (a) Prove that the number of comparabilities in an ordered set is recon-structible,

    (b) For $p \in P$, the **degree** $\deg(p)$ is the number of elements that are comparable to p. Prove that for every card $P \setminus \{x\}$ of P the degree of the missing element x can be reconstructed.

26. Prove that the ordered set in Figure 1.1 c) is reconstructible.

Remarks and Open Problems

The "Remarks and Open Problems" sections are intended to give the reader some more background on the material that is discussed in the main body of the text. For this first section it consists solely of remarks, while in later sections we will list more and more open problems. To keep things organized and easy to refer to, these sections are normally enumerated lists.

1. The reader interested in the abstract underpinnings on objects and mor-phisms should look at category theory. A good introductory text is [3].

2. The fixed point property originated in topology using topological spaces and continuous functions (for a survey on the topological fixed point prop-erty, cf. [33]). It was apparently first studied for ordered sets by Tarski and

---

[5]A graduate student (MS) of the author's is at eleven at press time.

Davis (cf. [51, 260]) and has since steadily gained in attention. The essentials of Tarski's result are already present in joint work with Knaster mentioned in [145]. A complete characterization of the sets with the fixed point property might be beyond the realm of possibilities. Problems 1.4.3 and 1.4.8 have been proved to be (co-)NP-complete, cf. [62, 277], co-NP-completeness of problem 1.4.3 will be proved here in Theorem 12.4.5. Yet there are interesting connections to other fields (cf. [12, 192, 278], explored here in Chapters 6, 12 and Appendix B). Moreover it is possible to produce nice insights into the fixed point property for certain classes of sets (cf. [2, 51, 192, 216, 235, 241, 260]), which in turn can be used in applications (cf. [113, 114]) or to provide deeper structural insights into the theory of ordered sets (cf. [159, 162, 163] or Theorem 4.5.1 here).

3. The graph-theoretical analogue of an ordered subset is not the concept of a subgraph, but that of an *induced* subgraph. Subgraphs in graph theory are obtained by removing some vertices and possibly also some edges from the original graph. This works well, since graph theory has no a priori assumptions on the edges. In order theory, removal of comparabilities is not easy, since it might affect transitivity. Thus generating a substructure by removing comparabilities is not a widely used notion.

4. The reconstruction problem has its origins in graph theory and was recently also posed for ordered sets (cf. [127, 150, 151, 240]). According to "reliable sources" in [23] the problem is originally due to P.J. Kelly and S. Ulam who discovered the problem in 1942. The visualization via pictures on cards was suggested by Harary in 1964. For more background on the graph-theoretical reconstruction problem the reader may check for example the survey article [23] by Bondy and Hemminger. For a thorough survey of order reconstruction, cf. [213].

5. Note that in this section we have frequently used a certain luxury theorists have. In Definitions 1.2.4 and 1.2.5 we defined the diagram of an ordered set and the transitive closure of a relation. We have then shown how the diagram and the original order relation are linked via the taking of subsets and the formation of the transitive closure. In this fashion we are able to use these tools interchangeably and build the theory with them. The question we did not address was how to specifically translate between diagrams and order relations. What step-by-step procedure can one follow to translate one to the other and back? Similarly whenever in a proof we say "one can find", we do not address the issue how to find the object we are looking for. These issues, together with the algorithmic ramifications of the fixed point property and order isomorphism, are discussed in Chapter 12. Until then, we will freely use the mathematician's prerogative, which is that if an object exists or a transformation can be made, we will assume we can pick the object or make the transformation as necessary without worrying how long it might take us to find or execute it.

2
Chains, Antichains and Fences

Chains and antichains are arguably the most common kinds of ordered sets in mathematics. The elementary number systems $\mathbb{N}$, $\mathbb{Z}$, $\mathbb{Q}$ and $\mathbb{R}$ (with the exception of course being $\mathbb{C}$) are chains. Chains are also at the heart of set theory. The Axiom of Choice is equivalent to Zorn's Lemma (which we will adopt as an axiom) and the Well-Ordering Theorem. Both latter results are results about chains.

Antichains on the other hand are common when people are not talking about order, as an antichain is essentially an "unordered" set. Dilworth's Chain Decomposition Theorem (cf. Theorem 2.5.7) shows how these two concepts are linked together. Fences are not as widely known as either chains or antichains. Yet they do play a fundamental role as they are the analogue of paths in graph theory.

2.1   Chains and Zorn's Lemma

Chains appear to be the most natural order, as any two elements in a chain are comparable. This means the hierarchy is total. When people talk about ranking objects, they typically are talking about a chain.

Definition 2.1.1 *An ordered set C is called a* **chain** *(or a* **totally ordered set** *or a* **linearly ordered set***) iff for all $p, q \in C$ we have $p \sim q$.*

Example 2.1.2 Examples of chains are

1. The natural numbers, integers, rational and real numbers with their natural orders,

2. Every set of ordinal numbers,

3. The set $\mathbb{R} \times \mathbb{R}$ with the order $(a, b) \sqsubseteq (c, d)$ iff $a < c$ or $(a = c$ and $b \leq d)$,

4. If $f : P \to P$ is order-preserving and $p \in P$ is such that $p \leq f(p)$, then $\{f^n(p) : n \in \mathbb{N}\}$ is a chain.

Proof. We shall only prove 4. First note that for all $n \in \mathbb{N}$ we have $p \leq f^n(p)$. The proof is a simple induction on n. The basis step $n = 0$ is trivial as $p = f^0(p)$. For the induction step $n \to (n + 1)$ assume $p \leq f^n(p)$. Then $f(p) \leq f(f^n(p)) = f^{n+1}(p)$ and by assumption $p \leq f(p)$. Hence $p \leq f^{n+1}(p)$.

Now let $n, m \in \mathbb{N}$ with $m \leq n$. Then we have $p \leq f^{n-m}(p)$ and hence $f^m(p) \leq f^m(f^{n-m}(p)) = f^n(p)$. ∎

Remark 2.1.3 In relation to problems 1.4.3, 1.4.8 and 1.5.9 we can record:

1. Not every chain has the fixed point property. Simply consider $\mathbb{N}$ with the map $f(x) = x + 1$, which is order-preserving and fixed point free. It is possible to classify exactly those chains that have the fixed point property, and we will do so in Theorem 5.2.1.

2. Finite chains have exactly one automorphism, which is the identity. Infinite chains however can have even fixed point free automorphisms. Consider $\mathbb{Z}$ with the map $f(x) = x + 1$.

3. Finite chains with more than two points are reconstructible. They are the only ordered sets whose deck has only chains, which means they are recognizable. Since there is (up to isomorphism) only one chain with n elements, this means finite chains are reconstructible.

Chains are the foundation upon which Zorn's Lemma is built. We record:

Definition 2.1.4 *Let P be an ordered set.*

1. *If $A \subseteq P$, then $u \in P$ is called an* **upper bound** *of A iff for all $a \in A$ we have $u \geq a$.*

2. *P is called* **inductively ordered** *iff every nonempty chain $C \subseteq P$ has an upper bound.*

If u is an upper bound of A we will also write $u \geq A$ and if U is a set of upper bounds of A we will write $U \geq A$.

Definition 2.1.5 *Let P be an ordered set. An element $m \in P$ is called* **maximal** *iff for all $p \in P$ with $p \sim m$ we have $p \leq m$. We denote the set of maximal elements of P by* $\mathrm{Max}(P)$.

Example 2.1.6 To illustrate bounds and maximal elements, consider the ordered set P in Figure 1.4, part e). The set $\{a, b, d\}$ has upper bounds d, g, and h. (So an upper bound of a set can be part of the set, too.) The elements g, h, and k are maximal elements in P and all other elements are not maximal. Finally note that $\{g, h, k\} \geq \{a, b, c\}$.

It should also be noted that not every ordered set has maximal elements. Just consider $\mathbb{N}$ with its natural order.

Zorn's Lemma connects the above concepts as follows.

Axiom 2.1.7 (Zorn's Lemma) *Let P be an inductively ordered set. Then P contains a maximal element M.*

Zorn's Lemma is in fact a standard tool for mathematicians who freely use the Axiom of Choice (to which Zorn's Lemma is equivalent). It is used for example in algebra to establish the existence of maximal ideals in rings with unity or in functional analysis in the proof of the Hahn–Banach theorem (Exercise 12 in this chapter). On the other hand the Axiom of Choice can be used to establish the existence of non-measurable sets and such counterintuitive things as the Banach–Tarski Paradox (cf. [269]). This is why some mathematicians decide to avoid its use. The philosophical issues that arise from using the infinitary generalization of a statement that is obvious in finite structures (cf. Exercise 5) are beyond the scope of this text. We will freely use Zorn's lemma and anything equivalent here.

The setup of a proof involving Zorn's Lemma is fairly standard. The idea is to design an appropriate nonempty set of objects that is inductively ordered and such that the desired element (if it exists) is maximal in it. The order one uses often is set inclusion. We give several examples in the following.

Our first example is the Axiom of Choice. The proof will feature all the characteristics of a proof using Zorn's Lemma without any additional order theory involved. Subsequent examples will (naturally) be examples that apply to order theory. These examples will often work with several orders at once and one will need to distinguish these orders carefully. Thus the Axiom of Choice is a good "warm-up" to the slightly more complex arguments that follow. The reader should note that we do not give an *absolute* proof of the Axiom of Choice. All that is shown in the following is that the Axiom of Choice is true *if Zorn's Lemma is true*.

Theorem 2.1.8 (Axiom of Choice) *If $\{P_\alpha\}_{\alpha \in I}$ is an indexed family of nonempty sets, then there is a function $f : I \to \bigcup_{\alpha \in I} P_\alpha$ (also called a **choice function**) such that for all $\alpha \in I$ we have $f(\alpha) \in P_\alpha$.*

Proof (using Zorn's Lemma) Let $\mathcal{P}$ be the set of all functions $f : D \to \bigcup_{\alpha \in I} P_\alpha$ such that $D \subseteq I$ and for all $\alpha \in D$ we have $f(\alpha) \in P_\alpha$. Clearly this set is not empty, since for any finite set $D \subseteq I$ such functions exist.

Since every function f is just the set $\{(\alpha, f(\alpha)) : \alpha \in \text{domain}(f)\}$, we have that $\mathcal{P}$ is ordered by set inclusion. (For those who prefer to think about restrictions to subsets we can say that if $f_i : D_i \to \bigcup_{\alpha \in I} P_\alpha$, $i = 1, 2$ are functions in P, then $f_1 \subseteq f_2$ iff $D_1 \subseteq D_2$ and $f_2|_{D_1} = f_1$. All arguments that we give in the following using unions can be re-written in language similar to the previous sentence. We choose not to do so, as we will see that the formation of unions is a much more compact way of arguing.)

To show $\mathcal{P}$ is inductively ordered, let $\mathcal{C} \subseteq \mathcal{P}$ be a nonempty chain. Define $g := \bigcup \mathcal{C}$. Then g is well-defined. Indeed let $\alpha \in \text{domain}(g)$. Then for all functions $f \in \mathcal{C}$ for which $f(\alpha)$ is defined we have $(\alpha, f(\alpha)) \in g$. If $f_1, f_2 \in \mathcal{C}$ with $\alpha \in \text{domain}(f_1) \cap \text{domain}(f_2)$, we can assume without loss of generality that $f_1 \subseteq f_2$. But that means $f_1(\alpha) = f_2(\alpha)$ and so g is well-defined.

Moreover g is a (partial) choice function. For each $\alpha \in \text{domain}(g)$, there is an $f \in \mathcal{C}$ with $\alpha \in \text{domain}(f)$. This means $g(\alpha) = f(\alpha) \in P_\alpha$.

By Zorn's Lemma we can now conclude that $\mathcal{P}$ has a maximal element F. What is left to be shown is that F is a choice function on I and not just on some subset $D \subset I$. To do so assume $\text{domain}(F) \neq I$. Then there is an $\alpha \in I \setminus \text{domain}(F)$. Let $p_\alpha \in P_\alpha$ and define $G := F \cup \{(\alpha, p_\alpha)\}$. Then $G \in \mathcal{C}$ and $G \supsetneq F$ is a strict upper bound of F, contradicting the maximality of F. Thus $\text{domain}(F) = I$ and F is the desired choice function. ∎

Having warmed up to Zorn's Lemma we now focus on chains that are as large as can be.

Proposition 2.1.9 *Let $(P, \leq)$ be a nonempty ordered set and let $C_0 \subseteq P$ be a chain. Then there is a chain $M \subseteq P$ which is maximal with respect to set inclusion and such that $C_0 \subseteq M$.*

Proof. Let $\mathcal{C}$ be the (nonempty) set of all chains $C \subseteq P$ with $C_0 \subseteq C$. This set is ordered in a natural way by set inclusion $\subseteq$.

To show that $\mathcal{C}$ is inductively ordered, let $\mathcal{K} \subseteq \mathcal{C}$ be a chain with respect to inclusion. Consider the set $K := \bigcup \mathcal{K}$ (with the induced order from P). To see that $K \in \mathcal{C}$ we must show that K is a chain. Let $x, y \in K$. Then there are $C_x, C_y \in \mathcal{K}$ such that $x \in C_x$ and $y \in C_y$. Since $\mathcal{K}$ is a chain, we can assume without loss of generality that $C_x \subseteq C_y$. Thus $x, y \in C_y$ and since C_y is a chain we have $x \sim_P y$. Thus K is a chain and hence K is in $\mathcal{C}$. Since for all $C \in \mathcal{K}$ we trivially have $K \supseteq C$, we conclude that K is an upper bound of $\mathcal{K}$ in $\mathcal{C}$. Thus $\mathcal{C}$ is inductively ordered.

Therefore by Zorn's Lemma $\mathcal{C}$ has a maximal element M. By choice of $\mathcal{C}$, M is a chain that contains C_0 and is maximal with respect to inclusion. ∎

Chains that are maximal with respect to inclusion are useful at times, so we define

Definition 2.1.10 *Let P be an ordered set. A chain C in P will be called a **maximal chain** iff for all chains $K \subseteq P$ with $C \subseteq K$ we have $C = K$.*

In the next section we will see another application of Zorn's Lemma when we prove that every set can be well-ordered.

2.2   Well-ordered Sets

Well-ordered sets are a particularly nice type of chain. Ordinal numbers in set theory are examples of well-ordered sets. In fact, they are up to isomorphism *all* well-ordered sets. Hence this class of ordered sets is sufficiently simple to be understood completely.

Definition 2.2.1 *Let P be an ordered set and let $S \subseteq P$. Then $s \in S$ is called the* **smallest element** *of S iff $s \leq S$.*

Definition 2.2.2 *An ordered set W is called* **well-ordered** *iff each non-empty subset $A \subseteq W$ has a smallest element.*

It is easy to see that well-ordered sets are chains and that not all chains are well-ordered. The next proposition shows that well-ordered sets are very closely related to the notion of counting. For every non-maximal element there will be a "next" element.

Proposition 2.2.3 *Let W be a well-ordered set. Then every non-maximal element $w \in W$ has an* **immediate successor**. *That is, there is an element w^+ such that $w < w^+$ and for all $p > w$ we have $p \geq w^+$.*

Proof. Let $w \in W$ be not maximal. Then $\{p \in W : p > w\}$ is not empty and it therefore has a smallest element. This element is w^+. ∎

Example 2.2.4 Every finite chain is well-ordered. The natural numbers $\mathbb{N}$ are the smallest infinite well-ordered set. It would be tempting to try to prove that (just as in $\mathbb{N}$) every element of a well-ordered set has an immediate predecessor also. An immediate predecessor of w would of course be an element w^- such that $w > w^-$ and for all $p < w$ we have $p \leq w^-$. The well-ordered set $\mathbb{N} \oplus \{\infty\}$ (consisting of $\mathbb{N}$ with a largest element ∞ attached) shows that not every element of a well-ordered set has an immediate predecessor. In this example ∞ does not have an immediate predecessor.

Informally speaking, well-ordered sets can be built as we just described. Pick a well-ordered set and attach a new largest element to get a new well-ordered set. This generates chains of well-ordered sets, which then can be united to form even bigger well-ordered sets. How far one can push this process will depend on how strong a version of the Axiom of Choice one is willing to accept. The standard class of examples of well-ordered sets (and as the reader will prove in Exercise 13 the only class of examples) is the class of ordinal numbers.

Example 2.2.5 (Cf. [103], p. 75.) An **ordinal number** is a well-ordered set α such that for each $\xi \in \alpha$ we have that $\xi^+ = \{\eta \in \alpha : \eta \leq \xi\}$. That is, the immediate successor of each ordinal number ξ is the set of all ordinal numbers up to and including ξ. As one can form sets of existing objects, this is a well-defined operation that allows formation of ordinal numbers. The set-theoretical problems start once one encounters infinite ordinals.

The **first infinite ordinal number** is (isomorphic to) the set of natural numbers. In ordinal arithmetic it is denoted by ω. By simply defining the successor via the above equation, we obtain ω^+, $(\omega^+)^+$ and so on. In ordinal arithmetic the nth successor of an ordinal α is called $\alpha + n$. Note that ω does not have an immediate predecessor. Indeed, any ordinal number that is less than ω is finite and thus its successor is also finite and not equal to ω. Ordinal numbers that do not have an immediate predecessor are also called **limit ordinals**.

The indicated counting process continues through all $\omega + n$ and the next limit ordinal is 2ω. Continuing in the above fashion we reach the limit ordinals 3ω, 4ω, ..., until we reach ω^2. What follows are $\omega^2 + 1$, $\omega^2 + 2$, ..., $\omega^2 + \omega$, ..., $\omega^2 + 2\omega$, ..., ω^3, ..., ω^4, ..., ω^ω. As in [275], p. 10 we will denote the **first uncountable ordinal number** by ω_1.

Proposition 2.2.3 and Examples 2.2.4 and 2.2.5 show that well-ordered sets are a natural generalization of the natural numbers. In a well-ordered set one has a natural notion of counting (the immediate successor of each element is the "next" element as we count) and we can count "past infinity" if the well-ordered set is big enough. On the other hand well-ordered sets can arise anywhere (if one believes Zorn's Lemma) as we will show that any set can be well-ordered.

Definition 2.2.6 *Let W be a well-ordered set. A well-ordered subset $V \subseteq W$ is called an* **initial segment** *of W iff $V \subseteq W$ and for all $w \in W \setminus V$ we have that w is an upper bound of V.*

Theorem 2.2.7 (Well-Ordering Theorem) *For every set S there is an order $\leq\,\subseteq S \times S$ such that $(S, \leq)$ is well-ordered.*

Proof (using Zorn's Lemma) Let $\mathcal{W}$ be the set of all pairs $(\leq, M)$ such that $M \subseteq S$ and $\leq\,\subseteq M \times M$ is a well-ordering. Then $\mathcal{W}$ is not empty as for each $s \in S$ the pair $(\{(s, s)\}, \{s\})$ is in $\mathcal{W}$. Order $\mathcal{W}$ as follows. We will say $(\leq_1, M_1) \sqsubseteq (\leq_2, M_2)$ iff

1. $M_1 \subseteq M_2$,

2. $\leq_1\,\subseteq\,\leq_2$ (as relations),

3. M_1 is an initial segment of M_2 ordered by $\leq_2$.

To prove that $\mathcal{W}$ is inductively ordered, let $\mathcal{K}$ be a $\sqsubseteq$-chain in $\mathcal{W}$. Let

$$M_t := \bigcup \{M : (\leq, M) \in \mathcal{K}\}$$

and equip it with the relation

$$\leq_t := \bigcup \{\leq : (\leq, M) \in \mathcal{K}\}.$$

We need to show that M_t is well-ordered by $\leq_t$ and that for all $(\leq, M) \in \mathcal{K}$ we have that M is an initial segment of M_t. We shall first show that M_t is totally ordered. Let $x, y, z \in M_t$. Then there are M_x, M_y, and M_z such that $x \in M_x$, $y \in M_y$ and $z \in M_z$. We can assume without loss of generality that $(\leq_x, M_x) \sqsubseteq (\leq_y, M_y) \sqsubseteq (\leq_z, M_z)$. Thus $x \leq_x x$, which implies $x \leq_t x$ for all $x \in M_t$. If $x \geq_t y$ and $x \leq_t y$, then $x \geq_y y$ and $x \leq_y y$ (there is a small argument here, which the reader shall produce as Exercise 7), which implies $x = y$. If $x \leq_t y$ and $y \leq_t z$, then $x \leq_z y$ and $y \leq_z z$, which implies $x \leq_z z$ and $x \leq_t z$. Finally for all $x, y \in M$ we have either $x \geq_y y$ or $x \leq_y y$ which means $x \geq_t y$ or $x \leq_t y$ and $\leq$ is a total order on M.

To show that $\leq_t$ is a well-ordering, let $A \subseteq M_t$ be nonempty. Then there is a $(\leq, M) \in \mathcal{K}$ such that $A \cap M \neq \emptyset$. Let $a \in A$ be the $\leq$-smallest element of A. Then for all $b \in A \cap M$ we have $b \geq a$, hence $b \geq_t a$. For $b \in A \setminus M$ find a $(\leq', M') \in \mathcal{K}$ such that $b \in M'$. Then by condition 3 we must have $(\leq, M) \sqsubseteq (\leq', M')$ (the other comparability is not possible since $b \in M' \setminus M$) and thus all elements $x \in M' \setminus M$ are upper bounds of M. Since $b \in M' \setminus M$, we have that b is an upper bound of M, hence $b \geq' a$, which implies $b \geq_t a$. Thus a is the $\leq_t$-smallest element of A. Hence $\leq_t$ is a well-ordering and $(\leq_t, M_t)$ is a $\sqsubseteq$-upper bound of $\mathcal{K}$. This means $(\mathcal{W}, \sqsubseteq)$ is inductively ordered.

Let $(\leq, M) \in \mathcal{W}$ be a $\sqsubseteq$-maximal element as guaranteed by Zorn's Lemma. If $M = S$, we are done. Assume there is an $s \in S$ that is not in M. Define

$$\leq' := \leq \cup \{(x, s) : x \in M \text{ or } x = s\} \subseteq (M \cup \{s\}) \times (M \cup \{s\}).$$

Then $\leq'$ is easily verified to be a well-ordering. But then $(\leq', M \cup \{s\}) \in \mathcal{W}$ is a $\sqsubseteq$-upper bound of $(\leq, M)$ that is not equal to $(\leq, M)$, which is a contradiction. Thus $\leq$ must be a well-ordering of $S = M$.   ∎

It is notable that in the proof of the well-ordering theorem we need to define $\sqsubseteq$ in as complicated a fashion as we did. Indeed, if we just used containment of the orders as $\sqsubseteq$, then the candidates for upper bounds of chains as defined in the proof need not be well-ordered. That is, they might not be in $\mathcal{W}$. For illustration, consider the set of chains

$$\mathcal{B} := \left\{ \left\{ \pm \frac{1}{k} : k = 1, \ldots, n \right\} : n \in \mathbb{N} \right\},$$

with each individual chain ordered with the order inherited from $\mathbb{Q}$. Set containment induces a total order on this set of chains and any two chains in this set are well-ordered (after all, all chains in $\mathcal{B}$ are finite). Yet the union of these chains is the set $\bigcup \mathcal{B} = \left\{ \pm \frac{1}{n} : n \in \mathbb{N} \right\}$, which is not well-ordered. So we need assumption 3 in the definition of $\sqsubseteq$ to prevent chains that we unify from "filling in holes" instead of "building upwards in a well-ordered fashion".

A final observation about well-ordered sets is that they have many of the properties that increasing sequences have. In the near future (cf. Theorem 4.2.11) we will encounter situations in which we are only interested in "where the tops of certain chains go". In such situations it will be helpful to replace the chains we have with chains that have the same "growth" or "convergence behavior" and are otherwise well-behaved. The result that allows one to do so is the fact that any chain has a cofinal (cf. Definition 2.2.8) well-ordered subchain (cf. Proposition 2.2.9).

Definition 2.2.8 *Let* P *be an ordered set and let* $A \subseteq B \subseteq P$. *Then* A *is called* **cofinal (coinitial)** *in* B *iff for every* $b \in B$ *there is an* $a \in A$ *such that* $a \geq b$ $(a \leq b)$.

Proposition 2.2.9 *Let* P *be an ordered set. Then for every chain* $C \subseteq P$ *there is a well-ordered cofinal subchain* $W \subseteq C$.

Proof. Left as Exercise 10. Hint: Order the well-ordered subchains of C with an order like in the proof of Theorem 2.2.7.   ∎

2.3 A Remark on Duality

Zorn's Lemma and the Well-Ordering Theorem have an "upwards bias". Indeed, all arguments are geared towards increasing in size and then finding a bound above. Since most people would agree that an increase in the size of a set means we are "going up", this is a natural visualization.

Of course, set-theoretically the notions of taking subsets and supersets have different properties. "Getting smaller" (taking of subsets) does not cause any set-theoretical difficulties. "Getting larger" (generating supersets with certain properties) can lead to problems such as Russell's paradox (if one is not careful). Thus when working with Zorn's Lemma most often one works with containment of sets in a fixed surrounding universe, say a power set (this avoids Russell's paradox). Visually, the increase in size (the "upward motion") is what is desired.

For the order-theorist, going up and going down however are closely related. Indeed if we were to re-designate in our minds that going to a superset is interpreted as "going down",[1] then the statements and proofs in Sections 2.1 and 2.2 would have a "downward bias". This re-designation is called dualizing.

In Example 1.1.2, part 8 we mentioned that for each ordered set P there is another ordered set P^d, called its dual ordered set, that is obtained by reversing all comparabilities. Reversing all comparabilities also changes definitions that depend on the order. If u is an upper bound of A in P, then with $\leq_d$ being the order of P^d we have that for all $a \in A$ the relation $u \leq_d a$ holds. Properties obtained from each other by reversing all comparabilities are called **dual properties**. The dual notion to being an upper bound is that of being a lower bound.

---

[1] This is done for example for search trees in computer science, so this notion is not far fetched at all. For more on search trees cf. Section 12.5.1 in this text.

Definition 2.3.1 *Let P be an ordered set. If $A \subseteq P$, then $l \in P$ is called a* **lower bound** *of A iff for all $a \in A$ we have $l \leq a$*

If l is a lower bound of A we will also write $l \leq A$ and if L is a set of lower bounds of A we will write $L \leq A$. Similar to the above we obtain the dual notion of a maximal element, which is called a minimal element.

Definition 2.3.2 *Let P be an ordered set. An element $m \in P$ is called* **minimal** *iff for all $p \in P$ with $p \sim m$ we have $p \geq m$. We denote the set of minimal elements of P by* $\mathrm{Min}(P)$.

Note that we were talking about minimal elements already in the proof of Proposition 1.5.11. Duality is a powerful tool when results are proved that are biased in one direction. Instead of re-stating and re-proving everything, one can simply invoke duality. For example, any set can also be dually well-ordered (that is, ordered in such a way that every nonempty subset has a maximal element).

There are notions that are their own duals such as the notion of being order-preserving. If $f : P \rightarrow Q$ is order-preserving, then so is $f : P^d \rightarrow Q^d$. Thus for example being an isomorphism is a self dual notion and any result proved about the relation between isomorphisms and, say, upper bounds is also a result on isomorphisms and lower bounds. The reader is invited to state (and briefly prove) the duals of the results we have proved so far and in the future. In this fashion when the need for dualization arises it will not hold any surprises. We shall conclude this section with one simple example on the use of duality, complete with formal proof. Future uses of duality will not be elaborated on as much.

Proposition 2.3.3 *Let P, Q be finite ordered sets and let $\Phi : P \rightarrow Q$ be an isomorphism. Then for each $p \in P$ the image $\Phi(p)$ has as many upper bounds in Q as p has upper bounds in P. The same statement holds for lower bounds.*

Proof. Let $p \in P$ and let $x \in P$ be such that $x \geq p$. Then $\Phi(x) \geq \Phi(p)$, since Φ is order-preserving. Moreover, since Φ is injective, no two upper bounds of p are mapped to the same point. Thus $\Phi(p)$ has at least as many upper bounds as p. Now suppose that $\Phi(p)$ has an upper bound q that is not the image of an upper bound of p. Then $\Phi^{-1}(q) \not\geq p$ even though $q \geq \Phi(p)$, a contradiction. This proves that the numbers of upper bounds of p and of $\Phi(p)$ are equal.

The reader should see that to prove the same statement for lower bounds we could simply follow the above argument with reversed comparabilities. A formal proof using duality could work as follows. Note that Φ also is an isomorphism between P^d and Q^d. Thus $\Phi(p)$ has as many Q^d-upper bounds in Q^d as p has P^d-upper bounds in P^d. For any ordered set R, an R^d-upper bound in R^d is an R-lower bound in R. Thus $\Phi(p)$ has as many lower bounds in Q as p has lower bounds in P. In the future such short arguments will be replaced with saying "By duality the statement for the lower bounds holds". ∎

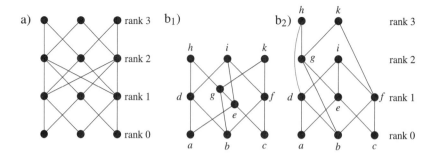

Figure 2.1: Two ordered sets for which drawing points with the same rank at the same height has different effects. The set in a) has a very orderly drawing revealing several symmetries. The second set is drawn twice. Once (in b_1) to reveal its vertical symmetry (it is isomorphic to its own dual) and once (in b_2) by drawing points of the same rank at the same height.

2.4   The Rank of an Element

The rank is a parameter that indicates the "height" at which one can find an element in an ordered set.

Definition 2.4.1 *Let P be a finite ordered set. For $p \in P$ we define the* **rank** $\mathrm{rank}(p)$ **of** *p recursively as follows. If p is minimal, let $\mathrm{rank}(p) := 0$. If the elements of rank $< n$ have been determined and p is minimal in the ordered set $P \setminus \{q \in P : \mathrm{rank}(q) < n\}$ we set $\mathrm{rank}(p) := n$.*

The rank can be used to somewhat standardize the drawing of a finite ordered set. As was stated in the drawing procedure on page 7, $p < q$ forces that p is drawn with a smaller y-coordinate than q in the diagram. Using the rank function, a drawing of a diagram can be standardized by drawing all points of the same rank at the same height. In fact this convention has been used in all drawings of diagrams of finite ordered sets so far. The ordered set in Figure 2.1 part a) shows once more that this convention can lead to good drawings of an ordered set. In part b) of Figure 2.1 we show that this convention should not be applied mechanically however. The down-up bias of ordering the points by rank can mask symmetries that are not compatible with the rank.

Example 2.4.2 It is possible for two points in an ordered set to be adjacent and yet have their ranks differ by more than 1. Just consider the ordered set in part b) of Figure 2.1.

Another good characterization of the rank is through the length of chains.

Definition 2.4.3 *Let C be a finite chain. The* **length of** C *is the number $l(C) := |C| - 1$.*

Proposition 2.4.4 *Let P be a finite ordered set. The rank of an element $p \in P$ is the length of the longest chain in P that has p as its largest element.*

Proof. Let $p \in P$. We will show by induction on the rank of p that rank(p) is the length of the longest chain from p to a minimal element. If rank$(p) = 0$, there is nothing to prove.

Let rank$(p) = k > 0$ and assume the statement is true for all elements of P of rank $< k$. Then p has a lower cover l of rank $k - 1$. By induction hypothesis there is a chain C of length $k - 1$ that has l as its largest element. Thus $\{p\} \cup C$ is a chain of length k with p as its largest element. Now suppose there is a chain $K \subseteq P$ of length $> k$ such that p is the largest element of K. Let c be the unique lower cover of p in K. Then c is of rank $< k$ and yet $K \setminus \{p\}$ is a chain of length $\geq k$ with largest element c, a contradiction. Thus the largest possible length of a chain in P that has p as its largest element is k, and we are done.   ∎

Proposition 2.4.5 *Let P, Q be finite ordered sets and let $\Phi : P \to Q$ be an isomorphism. Then for all $p \in P$ we have $\text{rank}_P(p) = \text{rank}_Q(\Phi(p))$.*

Proof. First note that the image of a chain under an order-preserving map is again a chain (it is left to Exercise 3 to spell out the details). Thus, since Φ is injective, the image of any k-element chain under Φ is again a k-element chain. By Proposition 2.4.4 this means that $\text{rank}_P(p) \leq \text{rank}_Q(\Phi(p))$.

Now Φ^{-1} is an isomorphism also, so the result of the previous paragraph also applies to Φ^{-1}. This means that

$$\text{rank}_P(p) \leq \text{rank}_Q(\Phi(p)) \leq \text{rank}_P(\Phi^{-1}(\Phi(p))) = \text{rank}_p(p),$$

which implies $\text{rank}_P(p) = \text{rank}_Q(\Phi(p))$.   ∎

As we have seen, the elements of largest rank in the ordered set determine how high the drawing of the order is. This motivates the following definition, which also makes sense for infinite ordered sets.

Definition 2.4.6 *Let P be an ordered set. The **height** of P is the length of the longest chain in P. If there are chains of arbitrary length in P, we will say that P is of infinite height.*

*The height of an ordered set is sometimes also called its **length**.*

Proposition 2.4.7 *The height of a finite ordered set with > 3 elements is reconstructible from the deck of the ordered set.*

Proof. Let P be a finite ordered set with > 3 elements. If P is a chain, then P is reconstructible by Proposition 1.5.11 and we are done. If not, the height of P is the maximum of the heights of the cards $P \setminus \{x\}$.   ∎

2.5   Antichains and Dilworth's Chain Decomposition Theorem

Logically, antichains are the simplest possible order, as we in fact impose no comparabilities at all on the points. If chains are totally ordered, one could say that antichains are totally unordered.

Definition 2.5.1 *An ordered set P is called an* **antichain** *iff for all $p, q \in P$ with $p \neq q$ we have $p \not\sim q$.*

Proposition 2.5.2 *Let P be an ordered set and let $A \subseteq P$ be an antichain. Then there is an antichain $B \subseteq P$ that is maximal with respect to set inclusion and contains A.*

  Proof. Left as Exercise 18.   ∎

  Antichains can be used to pictorially characterize chains via a "forbidden subset" (also compare with Theorem 8.1.5).

Proposition 2.5.3 *An ordered set C is a chain iff C does not contain any two-element antichains.*   ∎

  Just as chains lead to a natural notion of height in ordered sets, antichains lead to a notion of width.

Definition 2.5.4 *Let P be an ordered set. We define the* **width** *$w(P)$ of P to be the size of the largest antichain in P if such an antichain exists and to be ∞ otherwise.*

  Note that while chains allow us to define a natural vertical ranking of ordered sets, there is no possibility to define a horizontal ranking using antichains. We can, however, reconstruct the width just as easily as the height.

Proposition 2.5.5 *The width of a finite ordered set with > 3 elements is reconstructible from the deck.*

  Proof. We will need to distinguish the case in which P is an antichain and the case in which P is not. If P is an antichain, then all its cards are antichains. Since only antichains can have a deck consisting of antichains, this would mean that P is reconstructible as the unique (up to isomorphism) antichain with $|P|$ elements.
  If P is not an antichain, then the width of P will be the largest width of any card of P.   ∎

  A nice connection between chains and antichains is provided by Dilworth's Chain Decomposition Theorem. Dilworth's Chain Decomposition Theorem says that any ordered set of width k can be written as the union of k chains. This appears quite obvious. Call our ordered set P. Every chain in an ordered set is contained in a maximal chain by Proposition 2.1.9. Let C be a maximal chain. Then

Figure 2.2: An ordered set for which there is a maximal chain (marked) whose removal does not decrease the width.

$P \setminus C$ should have width $w(P) - 1$ and should thus be (if we argue inductively) the union of $w(P) - 1$ chains. Throw in C and P is the union of $w(P)$ chains. Unfortunately what we just gave was a "poof", not a proof. Figure 2.2 shows an example of a maximal chain whose removal does not decrease the width. Of course in this example we simply did not remove the "right" maximal chain. A formalization of what the "right" maximal chain may be appears quite hard. For our proof (which is very similar to the proofs that can be found in [22], Chapter III.4, Theorem 11, [263], Theorem 3.3 or [266]; for another proof consider [193]) it will be useful to have the following notation at hand.

Definition 2.5.6 *Let P be an ordered set. For $p \in P$ we define the* **filter** *or* **up-set** *of p to be $\uparrow p := \{q \in P : q \geq p\}$ and the* **ideal** *or* **down-set** *of p to be $\downarrow p := \{q \in P : q \leq p\}$.*

Theorem 2.5.7 (Dilworth's Chain Decomposition Theorem, cf. [57] or Exercise 24 for the proof for infinite ordered sets.) *Let P be a finite ordered set of width k. Then P is the union of k chains. That is, there are chains $C_1, \ldots, C_k \subseteq P$ such that $P = \bigcup_{i=1}^{k} C_i$.*

Proof. The proof is an induction on $n := |P|$. For $n = 1$ the result is obvious.

For the induction step let us assume Dilworth's Chain Decomposition Theorem holds for ordered sets with $\leq n$ elements and let P be an ordered set of width k with $(n + 1)$ elements. Let us first assume there is an antichain $A = \{a_1, \ldots, a_k\}$ with k elements in P such that at least one a_j is not maximal and at least one a_j is not minimal. Then the sets $L := \bigcup_{i=1}^{k} \downarrow a_i$ and $U := \bigcup_{i=1}^{k} \uparrow a_i$ both have width k and $\leq n$ elements. Thus U is the union of chains U_i and L is the union of chains L_i with i going from 1 to k, respectively. Without loss of generality we can assume that a_i is the largest element of L_i and the smallest element of U_i. However then each $L_i \cup U_i$ is a chain and $P = \bigcup_{i=1}^{k} (L_i \cup U_i)$.

If the only antichain(s) with k elements in P are the antichain of maximal elements or the antichain of minimal elements, let $C \subseteq P$ be a maximal chain. Then C contains a maximal and a minimal element. Thus we have $|P \setminus C| \leq n$ and

$w(P \setminus C) = k - 1$. By the induction hypothesis, there are chains $C_1, \ldots, C_{k-1}$ such that $P \setminus C = \bigcup_{i=1}^{k-1} C_i$. Now we set $C_k := C$ and we are done.   ∎

Dilworth's Chain Decomposition Theorem can be used to provide a surprisingly easy solution to the order-theoretical analogue of a hard graph-theoretical problem. The graph-theoretical analogue of a chain is a **complete graph**, that is, a graph in which any two vertices are connected by an edge. Complete graphs with s vertices are also denoted K^s. The natural analogue of antichains are graphs in which no two vertices are connected by an edge. These are called **discrete graphs**.

Definition 2.5.8 (For an introduction on Ramsey numbers cf. [22], Chapter VI; for an introduction to more sophisticated Ramsey Theory for ordered sets cf. [263], Chapter 10, Section 5.) *Let $s, t \in \mathbb{N}$, $s, t \geq 2$. Then the **Ramsey number** $R(s, t)$ is the smallest natural number n such that any graph G with $\geq n$ vertices contains a complete subgraph K^s or a discrete subgraph with t elements.*

One of Ramsey's theorems (cf. [212]) states that the Ramsey numbers actually are finite. Very few Ramsey numbers are known. In fact, the precise value for $R(5, 5)$ is still unknown. Ordered sets have a richer structure than graphs and the availability of Dilworth's Chain Decomposition Theorem allows us to easily solve the analogous problem for ordered sets.

Proposition 2.5.9 *Let $s, t \in \mathbb{N}$, $s, t \geq 2$ and let the **ordered set Ramsey number** $R_{\mathrm{ord}}(s, t)$ be the smallest natural number n such that every ordered set with $\geq n$ elements contains a chain with s elements or an antichain with t elements. Then*

$$R_{\mathrm{ord}}(s, t) = (s - 1)(t - 1) + 1.$$

Proof. To see that $R_{\mathrm{ord}}(s, t) > (s - 1)(t - 1)$ consider the ordered set that consists of $t - 1$ pairwise disjoint chains with $s - 1$ elements each and no further comparabilities involved. This is an ordered set with $(s - 1)(t - 1)$ elements and no s-element chain and no t-element antichain.

Now suppose P has $(s - 1)(t - 1) + 1$ elements. We need to show that P contains an s-element chain or a t-element antichain. If P contains an antichain with t elements, there is nothing to prove. If every antichain in P has at most $t - 1$ elements, then P has width at most $t - 1$. By Dilworth's Chain Decomposition Theorem P is the union of at most $t - 1$ chains. But then one of these chains must have more than $s - 1$ elements.   ∎

We can immediately conclude that infinite size means we have an infinite chain or an infinite antichain.

Corollary 2.5.10 *Every infinite ordered set contains an infinite chain or an infinite antichain.*

Proof. Let P be an infinite ordered set. From Proposition 2.5.9 we conclude that P must contain chains of arbitrary length or antichains of arbitrary length. Indeed, if the longest chain in P was of length c and the longest antichain in P was of length a, then P would have at most ca elements.

If P contains an infinite chain, then there is nothing to prove. If P does not contain any infinite chains, then for each element $p \in P$ we can define $\text{rank}_P(p)$ as the length of the longest chain that has p as its largest element. If for any k, P has infinitely many elements of rank k, we are done.

Finally, in case P has no infinite chains and for each k only finitely many elements of rank k, there must be an element of rank k for each $k \in \mathbb{N}$. Then there must be an infinite sequence $k_1, k_2, \ldots$ of natural numbers, such that for each i, there is a maximal element m_{k_i} of rank k_i. This sequence is constructed as follows. Start with a minimal element b_1. Since b_1 is not the bottom element of an infinite chain, there is a maximal element above b_1. The rank of this maximal element is k_1; we call the element m_{k_1}. Once k_i is found, let b_i be a non-maximal element of rank k_i. There is a maximal element above b_i. The rank of said maximal element is k_{i+1}; we call the element $m_{k_{i+1}}$.

Since no two maximal elements are comparable, the set $\{m_{k_i} : i \in \mathbb{N}\}$ is an infinite antichain. ∎

2.6 Dedekind Numbers

Antichains also give rise to a simple-looking, but still baffling counting problem.

Open Question 2.6.1 Dedekind's problem. (Cf. [203].) *What is the number of antichains in the power set $\mathcal{P}_n$ of an n-element set (ordered of course by inclusion)?*

In this section we will record two possible starts in working on such a problem. One is to break the problem into smaller subproblems, the other is to translate the problem into another venue. These ideas are intended to give the reader a flavor of the problem. We start with a possible breakup of the problem into smaller problems.

Definition 2.6.2 *Let $n \in \mathbb{N}$. We define*

1. *D_n to be the number of antichains in $\mathcal{P}(\{1, \ldots, n\})$; D_n is also called the nth **Dedekind number**.*

2. *T_n to be the number of antichains in $\mathcal{P}(\{1, \ldots, n\})$ such that the union of these antichains is $\{1, \ldots, n\}$.*

Clearly $D_0 = 1$, $T_0 = 1$ (the empty set is an antichain), $D_1 = 2$ and $T_1 = 1$. The connection between the D_n and the T_n is the following.

Proposition 2.6.3 $D_n = \sum_{k=0}^{n} \binom{n}{k} T_k.$

Proof. Let $\mathcal{A}_k$ be the set of all antichains in $\mathcal{P}(\{1, \ldots, n\})$ whose union is a k-element set. Clearly $D_n = \sum_{k=0}^{n} |\mathcal{A}_k|$. For $B \subseteq \{1, \ldots, n\}$ let $\mathcal{A}_B$ be the set of all antichains in $\mathcal{P}(\{1, \ldots, n\})$ whose union is B. Then

$$|\mathcal{A}_k| = \sum_{|B|=k} |\mathcal{A}_B|.$$

Now if $B = \{b_1, \ldots, b_j\}$, then $\mathcal{P}(B)$ is order-isomorphic to $\mathcal{P}(\{1, \ldots, j\})$ via $\varphi : \mathcal{P}(\{1, \ldots, j\}) \to \mathcal{P}(B), \{z_1, \ldots, z_i\} \mapsto \{b_{z_1}, \ldots, b_{z_i}\}$. Hence $|\mathcal{A}_B| = T_{|B|}$. Now since an n-element set has $\binom{n}{k}$ k-element subsets, we infer $|\mathcal{A}_k| = \binom{n}{k} T_k$, which directly implies the result. ∎

Thus the task of computing the D_n has been reduced to the task of computing the T_n, which is equally formidable. We can record the following.

Definition 2.6.4 *For $n \geq 1$ let T_n^j be the number of j-element antichains of $\mathcal{P}(\{1, \ldots, n\})$ whose union is $\{1, \ldots, n\}$.*

Proposition 2.6.5 *For $n \geq 1$ we have $T_n = \sum_{j=1}^{n} T_n^j.$*

Proof. This is trivial. ∎

We have, as is often done in a counting task, reduced our task to a number of more specific counting tasks. Unfortunately, such reductions do not always lead to success, and Dedekind's problem is one example in which this is the case. The only further advance the author can record is

Proposition 2.6.6 *For $n \geq 1$ we have $T_n^2 = \frac{1}{2} \sum_{k=1}^{n-1} \binom{n}{k} (2^k - 1).$*

Proof. We will count the number of all pairs

$$(F, S) \in \mathcal{P}(\{1, \ldots, n\}) \times \mathcal{P}(\{1, \ldots, n\})$$

such that $\{F, S\}$ is an antichain and $F \cup S = \{1, \ldots, n\}$. Clearly this number is $2T_n^2$. Our counting process will be that we first count the ways in which F can be chosen and then we multiply for each possible choice of F with the number of ways S can be chosen.

If the first set F has k elements, then $k \neq 0$ and $k \neq n$. Hence we must sum from $k = 1$ to $n - 1$ and for each k there are $\binom{n}{k}$ possible choices for F. Once

F is chosen, the second set S must contain $\{1, \ldots, n\} \setminus F$. For $F \cap S$ there are $2^k - 1$ possibilities (only $F \cap S = F$ cannot happen). This gives the indicated formula. ∎

It is now easy to see that a similar process will give us the values for T_n^3, T_n^4 and so on. Unfortunately, the formulas involved become so unwieldy as to not be of very much use any more. For example, the author was unable to express the T_n^j solely in terms of T_k^i with $k < n$ and $i < j$, which could lead to recursion formulas solvable with a sufficiently precise integer arithmetic on a computer. For a computational approach to the problem and a list of the known Dedekind numbers so far (also given in Remark 4 in Chapter 11), cf. [273].

Another time honored approach to difficult problems is to translate them into a different venue. The connection we present here is to order-preserving maps.

Proposition 2.6.7 *Let P be a finite ordered set. Then the number of antichains in P is equal to the number of order-preserving maps from P into the two-element chain.*

Proof. We need to show that there is a bijective map B from the set $\mathcal{A}(P)$ of all antichains in P, to the set $\text{Hom}(P, \{0, 1\})$ of order-preserving maps from P into the two element chain $\{0, 1\}$. Let $A = \{a_1, \ldots, a_m\} \in \mathcal{A}(P)$. Define

$$L(A) := \bigcup_{i=1}^{m} {\downarrow} a_i$$

(note that for $A = \emptyset$ we have that $L(A) = \emptyset$) and define

$$f_A(p) := \begin{cases} 1; & \text{if } p \notin L(A), \\ 0; & \text{if } p \in L(A). \end{cases}$$

Then $f_A : P \to \{0, 1\}$ is order-preserving. Indeed, if $p \le q$ and $f_A(q) = 1$, there is nothing to prove, while if $p \le q$ and $f_A(q) = 0$, then $q \in L(A)$, which implies $p \in L(A)$ and $f_A(p) = 0$.

Define the map B mentioned at the beginning by $B(A) := f_A$. Then B is injective, since for $A_1 \ne A_2$, we have that

$$(L(A_1) \setminus L(A_2)) \cup (L(A_2) \setminus L(A_1)) \ne \emptyset$$

and thus there will be a point

$$p \in (L(A_1) \setminus L(A_2)) \cup (L(A_2) \setminus L(A_1))$$

such that $f_{A_1}(p) \ne f_{A_2}(p)$. To prove that B is surjective, let $f : P \to \{0, 1\}$ be given. If f is the function that is identical to 1, then $f = B(\emptyset)$. Otherwise, let A be the set of maximal elements of $f^{-1}[0]$. Then $f^{-1}[0] = L(A)$ and $f = f_A = B(A)$. Thus B is the desired bijection. ∎

We will consider the problem of counting maps in more detail in Chapters 11 and 12. The translation in Proposition 2.6.7 reaches farther than Dedekind's problem. Proposition 2.6.7 applies to arbitrary finite ordered sets. Thus a natural generalization of Dedekind's problem is to ask for the number of antichains in an arbitrary ordered set. Beyond what is presented in this text, the author is

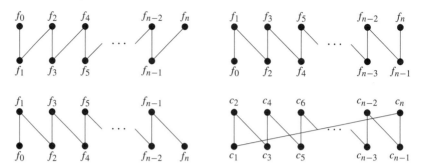

Figure 2.3: Pictures of fences and crowns. On the left are the two nonisomorphic fences with an odd number of elements, top right is the unique (up to isomorphism) fence with an even number of elements. The unique (up to isomorphism) crown with n elements (n even) is given on the bottom right.

unaware of any classes of ordered sets for which this question was asked or answered. However, it was proved in [202] that the computational enumeration must be considered "hard".

2.7   Fences and Crowns

The next-simplest ordered structures after chains and antichains appear to be fences and crowns. Fences can be considered analogous to paths connecting distinct points in topology or graph theory. Crowns are very much analogous to closed paths. We start with fences.

Definition 2.7.1 *Let P be an ordered set. An $(n + 1)$-**fence** (cf. Figure 2.3) is an ordered set $F = \{f_0, \ldots, f_n\}$ such that $f_0 > f_1, f_1 < f_2, f_2 > f_3, \ldots, f_{n-1} < f_n$ or $f_0 < f_1, f_1 > f_2, f_2 < f_3, \ldots, f_{n-1} > f_n$ if n is even, respectively $f_0 < f_1, f_1 > f_2, f_2 < f_3, \ldots, f_{n-1} < f_n$ or $f_0 > f_1, f_1 < f_2, f_2 > f_3, \ldots, f_{n-1} > f_n$ if n is odd, and such that these are all comparabilities between the points. The **length** of the fence is n. The points f_0 and f_n are called the **endpoints** of the fence.*

Remark 2.7.2 Note that the "spider" in Figure 1.1 f) is made up of countably many fences with their left endpoints "glued together".

Proposition 2.7.3 *Every fence has the fixed point property.*

Proof. We will soon have more sophisticated tools to prove this result. However the direct proof that follows nicely demonstrates the analogy between fences and intervals on the real line. It also provides some practice in working with order-preserving maps.

Let $P = \{p_0, p_1, p_2, \ldots, p_n\}$ be a fence and assume without loss of generality that $p_0 < p_1$. Suppose $f : P \to P$ was a fixed point free order-preserving map. For each $k \in \{0, \ldots, n\}$ let $g(k)$ be the number l such that $f(p_k) = p_l$. Then for $|k_1 - k_2| \leq 1$ we have $|g(k_1) - g(k_2)| \leq 1$, since points in P are comparable iff their indices are adjacent. Moreover if $|k - g(k)| \leq 1$, then $p_k \sim f(p_k)$ and $f(p_k)$ is a fixed point of f. Thus $|k - g(k)| \geq 2$ for all k.

Let m be the smallest number such that $g(m) \leq m$. Then $g(m) \leq m - 2$ and $g(m - 1) \geq m + 1$, a contradiction to $|g(m) - g(m - 1)| \leq 1$. Thus P cannot have any fixed point free order-preserving self-maps f.   ∎

Fences are a sufficiently simple structure to allow a solution to a variation on Dedekind's problem.

Definition 2.7.4 *For an ordered set P we define #A(P) to be the number of antichains in P.*

Proposition 2.7.5 ([155]) *Let F_n denote a fence with n elements. The two versions of the n-fence are isomorphic if n is even and they are duals of each other when n is odd. Thus #A(F_n) is independent of what kind of n-fence F_n is. Moreover #A$(F_1) = 2$, #A$(F_2) = 3$, #A$(F_3) = 5$ and*

$$\text{#A}(F_n) = 2\text{#A}(F_{n-2}) + \text{#A}(F_{n-3}).$$

Proof. The numbers for #A(F_1), #A(F_2), and #A(F_3) are easily verified. Just remember that the empty set is an antichain also. Now let F_n be an n-fence. Every antichain in F_n either contains f_{n-1}, f_n or neither.

1. There are #A(F_{n-2}) antichains in F_n that contain neither f_{n-1} nor f_n.

2. There are #A(F_{n-2}) antichains in F_n that contain f_n.

3. There are #A(F_{n-3}) antichains in F_n that contain f_{n-1}.

This proves the recursive formula.   ∎

A closed formula can now be obtained via standard methods for solving linear recurrence relations (assume the solution is of the form λ^n and find λ, then solve the initial value problem). Since the two-element chain is nothing but a 2-fence, a closed formula could also be obtained by using Proposition 2.6.7 and the high-powered results in [80]. The reader is invited to choose between a direct proof or this route for Exercise 35, which proves a similar result for crowns, our next topic.

Just as one turns a path from one point to another into a closed path by joining the endpoints, one turns a fence with an even number of points into a crown by making the endpoints comparable in the appropriate manner.

Definition 2.7.6 *Let $n \in \mathbb{N}$ be even and ≥ 4. An n-crown (cf. Figure 2.3) is an ordered set C_n with point set $\{c_1, \ldots, c_n\}$ such that $c_1 < c_2, c_2 > c_3, c_3 < c_4, \ldots, c_{n-1} > c_{n-1}, c_{n-1} < c_n, c_n > c_1$ are the only strict comparabilities.*

Just as moving from injective paths with distinct endpoints to closed paths destroys the fixed point property in topology, moving from fences to crowns also makes us lose the fixed point property.

Proposition 2.7.7 *Let $n \in \mathbb{N}$ be even and ≥ 4. Then C_n does not have the fixed point property.*

Proof. The map that maps $c_k \mapsto c_{k+2}$ for $k = 1, \ldots, n-2$, $c_{n-1} \mapsto c_1$ and $c_n \mapsto c_2$ is order-preserving and fixed point free. ∎

Pictorially it is easy to see that fences are parts of crowns and that crowns cannot be parts of fences. This observation can be formalized as follows.

Proposition 2.7.8 *Any fence of length k can be embedded into an l-crown with $l > k + 1$. However no n-crown can be embedded into a fence of any length.*

Proof. Let $F = \{f_0, \ldots, f_k\}$ be a fence of length k and let $C = \{c_1, \ldots, c_l\}$ be an l-crown with $l > k + 1$. Assume without loss of generality that $f_0 < f_1$. Then the map $f_i \mapsto c_{i+1}$ is an embedding.

Now suppose that $C = \{c_1, \ldots, c_n\}$ is an n-crown and that $f : C \to F$ is an embedding. Let $f_k := f(c_1)$ and assume without loss of generality that $f(c_2) = f_{k+1}$. Then for all $i \in \{1, \ldots, n\}$ we must have $f_{k+i-1} = f(c_i)$. In particular $f(c_n) = f_{k+n-1} \not> f_k = f(c_1)$ and f could not have been an embedding. ∎

2.8   Connectivity

The overall class of ordered sets includes many discrete objects, such as finite ordered sets. Therefore it would be futile to try to mimic topological notions such as openness in general for all ordered sets. Thus the appropriate notion of connectivity for ordered sets (as well as the analogous notion for graphs) is inspired by what is called pathwise connectivity in topology.

Definition 2.8.1 *Let P be an ordered set. P is called **connected** iff for all $a, b \in P$ there is a fence $F \subseteq P$ with endpoints a and b. An ordered set that is not connected is called **disconnected**.*

Similar to topology the fixed point property implies connectivity.

Proposition 2.8.2 *Let P be an ordered set with the fixed point property. Then P is connected.*

Proof. Assume that P is not connected. Let $a, b \in P$ be two points such that there is no fence with endpoints a, b. Define

$$f(p) := \begin{cases} a; & \text{if there is a fence with endpoints } b \text{ and } p, \\ b; & \text{otherwise.} \end{cases}$$

Then f is order-preserving and has no fixed points, contradiction. ∎

A related problem is the question what properties the set of fixed points of an order-preserving mapping has when it is nonempty. Though there are nice positive results (cf., e.g., Theorem 5.2.1) we present a negative result here.

Example 2.8.3 Even a finite ordered set P with the fixed point property can have a self-map $f : P \rightarrow P$ such that $Fix(f)$ is not connected. (Cf. Exercise 38.)

Analogous to topology we can define the connected components of an ordered set. Components and the metric notions defined thereafter will be useful for our proof that disconnected ordered sets are reconstructible.

Proposition 2.8.4 *Let P be an ordered set. If $S \subseteq P$ is a connected subset of P, then there is a maximal (with respect to inclusion) connected subset C of P such that $S \subseteq C$.*

Proof. This can be done with a standard argument using Zorn's Lemma. However one can also avoid using Zorn's Lemma here altogether. Let $\mathcal{C}$ be the set of all connected subsets $C \subseteq P$ that contain S. Then $C := \bigcup \mathcal{C}$ is a connected subset of P. Indeed if $a, b \in C$, let s be an arbitrary point in S. Let $C_a \in \mathcal{C}$ be such that $a \in C_a$ and let $C_b \in \mathcal{C}$ be such that $b \in C_b$. Then there is a fence F_a in C_a from a to s and there is a fence F_b in C_b from s to b. Thus (upon possibly removing some elements from $F_a \cup F_b$) there is a fence in C from a to b and C is connected.

The definition of C shows that C is the largest connected subset of P that contains S. ∎

Definition 2.8.5 *The maximal (with respect to inclusion) connected subsets of an ordered set are called the **(connected) components** of the ordered set.*

Metric notions such as distance and diameter arise in ordered sets through the lengths of fences.

Definition 2.8.6 *The **distance** dist(a, b) between two points $a, b \in P$ is the length of the shortest fence from a to b. If a and b are in different components of P, we will say the distance is infinite. The **diameter** diam(P) of an ordered set P is the largest distance between any two points in P. If P contains points that are arbitrarily far apart or if P is disconnected, we say the diameter of P is infinite.*

Example 2.8.7

1. The "one-way infinite fence" $F_\omega := \{f_0 < f_1 > f_2 < f_3 > \cdots\}$ is connected and has infinite diameter and infinite width.

2. The "spider" in Figure 1.1 is an ordered set with infinite diameter that contains no infinite fences.

3. Let A be an antichain. Obtain the set V by adding an element to A that is an upper cover of all elements of A. Then V has width $|A|$ but its diameter is 2.

Diameter and width appear to be related notions, as they both measure how far away an ordered set is from being a chain. However, both notions are distinct, and there are no globally tight inequalities that relate either to each other. Indeed, Example 2.8.7, part 3 shows that there is no possibility to bound the width of an ordered set by using the diameter. Proposition 2.8.8 gives an inequality in the opposite direction. While the inequality cannot be improved in general, Example 2.8.7, part 3 shows that it need not be very good in special cases.

Proposition 2.8.8 *Let P be an ordered set of finite diameter. Then the diameter of P is bounded by twice the width of P minus 1 and this inequality cannot be improved.*

Proof. A fence of length n contains an antichain of length $\left\lceil \dfrac{n+1}{2} \right\rceil$. Thus $\left\lceil \dfrac{\mathrm{diam}(P)+1}{2} \right\rceil \le w(P)$, which implies $\mathrm{diam}(P) \le 2w(P) - 1$.

The fences with $2k$ elements all have length $2k - 1$ and width k. Thus the above inequality cannot be improved. ∎

All the above notions merge nicely in the proof that disconnected ordered sets are reconstructible.

Proposition 2.8.9 *Let P be a disconnected finite ordered set with $|P| \ge 4$. Then P is reconstructible from its deck.*

Proof. First let us prove that disconnected ordered sets are recognizable. If P is disconnected, then all cards of P are disconnected unless P has a component with one point. In this case all but one card of P are disconnected.

On the other hand, if Q is a connected ordered set we can show that Q has at least two connected cards. Let $a, b \in P$ be such that $\mathrm{dist}(a, b) = \mathrm{diam}(P)$. Then $P \setminus \{a\}$ must be connected. Indeed, otherwise there is a $c \in P$ such that every fence from c to b goes through a, which means $\mathrm{dist}(c, b) > \mathrm{dist}(a, b)$, contradicting the choice of a and b. Similarly we prove that $P \setminus \{b\}$ is connected.

Thus disconnected ordered sets are recognizable. They are the ordered sets whose decks contain at most one connected card. For reconstruction note that if the deck of a disconnected ordered set D contains a connected card, then the components of D are the connected card and a singleton. This leaves the case in which all cards of D are disconnected. Find a card C of D such that the sum of the squares of the component sizes is minimal. This card must have been obtained by removing an element from a minimum-sized component. Thus C has a unique smallest component, which is the unique component of C that is not a component of D. Let s be the size of the unique smallest component of C. Then the last component of D is the unique ordered set S with $s + 1 < |D|$ elements such that D contains more isomorphic copies of S than C. Since the number of isomorphic copies of S in D can be determined via the Kelly Lemma (cf. Proposition 1.5.14), we have reconstructed D. ∎

Exercises

1. For the ordered set in Figure 2.1 part b), find

   (a) The upper bounds of the set $\{e\}$,

   (b) The upper bounds of the set $\{a, b, c\}$,

   (c) The upper bounds of the set $\{f, g\}$,

   (d) The upper bounds of the set $\{d, f, g\}$,

   (e) The lower bounds of the set $\{g, h, k\}$,

2. Find the maximal elements of the ordered sets in Figure 2.1.

3. Chains and order-preserving maps. Let P, Q be ordered sets, $f : P \to Q$ be order-preserving and let $C \subseteq P$ be a chain. Prove that $f[C]$ is a chain.

4. Prove that for every finite ordered set P there is an order-preserving map from P onto a $|P|$-element chain.

5. Without using Zorn's Lemma, prove that in a finite ordered set every element is below at least one maximal element. Use this to prove quickly that in a finite ordered set every element is above at least one minimal element.

6. Prove that a well-ordered set has the fixed point property iff it has a largest element.

7. In the proof of the Well-ordering Theorem, we claim that if $x \in M_x$ (ordered by $\leq_x$) and $y \in M_y$ (ordered by $\leq_y$) with $(\leq_x, M_x) \sqsubseteq (\leq_y, M_y)$ and $x \leq_t y$, then $x \leq_y y$. Formally we only know that $x \leq y$ for some $(\leq, M)$ where $x, y \in M$. Prove that $x \leq_y y$.

8. Prove that the Well-ordering Theorem implies Zorn's Lemma.

9. Prove that the Axiom of Choice is equivalent to Zorn's Lemma.

10. Prove Proposition 2.2.9.

11. Let V be a vector space. A **basis** of V is a subset $B \subseteq V$ such that any finite subset of B is linearly independent and such that any $v \in V$ has a (necessarily unique) representation as a finite linear combination of elements of B. Prove that every vector space has a basis.

12. (Hahn–Banach theorem) Let V be a normed vector space over $F = \mathbb{R}$ or $F = \mathbb{C}$. A **(continuous) linear functional** is a linear function $\Phi : V \to F$ such that there is a $c > 0$ such that for all $v \in V$ we have $|\Phi(v)| \leq c\|v\|$. Let $W \subset V$ be a linear subspace of V and let $v \in V \setminus W$. Prove that there is a continuous linear functional $\Phi : V \to F$ such that $\Phi|_W = 0$ and $\Phi(v) = 1$.

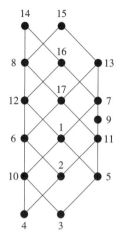

Figure 2.4: An ordered set (from [70]) without fibres of size $\leq \dfrac{|P|}{2}$.

13. Prove that every well-ordered set is isomorphic to one of the ordinal numbers in Example 2.2.5.

14. State the definition of a dual inductive order and state and prove the dual of Zorn's Lemma.

15. Prove Proposition 1.2.7 for ordered sets that have no infinite chains.

16. Prove that for any finite ordered set P and any distinct $x, r \in P$ we have that $\text{rank}_{P \setminus \{x\}}(r) \in \{\text{rank}_P(r), \text{rank}_P(r) - 1\}$. Give an example of an ordered set P and two points x and r such that the rank of r in $P \setminus \{x\}$ is less than the rank of r in P.

17. Determine the largest class of ordered sets in which a sensible rank function can be defined.

18. Prove Proposition 2.5.2.

19. Consider the ordered set P_0 in Figure 2.4 (also cf. Remark 9).

    (a) Prove that $\{1, 9, 17\}$ is a maximal antichain in P_0.

    (b) Find all two-element maximal antichains in P_0.

    (c) Prove that every set in P_0 that intersects all maximal antichains and that contains an odd-numbered element must have at least 9 elements.

    (d) Prove that every set in P_0 that intersects all maximal antichains must have at least nine elements.

20. Write the following ordered sets as a union of as few chains as possible.

(a) The ordered set in Figure 1.3,

(b) The ordered sets in Figure 1.4,

(c) The ordered sets in Figure 2.1.

21. Prove that for every finite ordered set P of height h there are antichains
$A_0, \ldots, A_h \subseteq P$ such that $A_i \cap A_j = \emptyset$ for $i \neq j$ and such that $P = \bigcup_{i=0}^{h} A_i$.

22. Prove Propositions 2.4.7 and 2.5.5 using the Kelly Lemma.

23. In the following we present an *incorrect proof of Dilworth's Chain Decomposition Theorem (cf. Theorem 2.5.7)*. Find the mistake.

Induction on $n := |P|$, $n = 1$ is obvious. For the induction step $n \to (n+1)$ let us assume Dilworth's Chain Decomposition Theorem holds for ordered sets with n elements and let P be an ordered set with $(n + 1)$ elements. Let $m \in P$ be a maximal element. Then $|P \setminus \{m\}| = n$. By the induction hypothesis, there are chains $C_1, \ldots, C_{w(P \setminus \{m\})}$ such that $P \setminus \{m\} = \bigcup_{i=1}^{w(P \setminus \{m\})} C_i$. If $w(P \setminus \{m\}) = k - 1$ we set $C_k := \{m\}$ and we are done. Otherwise, m has strict lower bounds and thus for some $i_0 \in \{1, \ldots, w(P \setminus \{m\})\}$ we have that m is an upper bound of C_{i_0}. Then $C_{i_0} \cup \{m\}$ is a chain and $C_1, \ldots, C_{i_0-1}$, $C_{i_0} \cup \{m\}$, $C_{i_0+1}, \ldots, C_{w(P \setminus \{m\})}$ are the desired chains. ∎??

24. Let us now finally prove the general version of Dilworth's Chain Decomposition Theorem. It states that *any* ordered set of width k can be written as the union of k chains. We follow Dilworth's original idea (cf. [57], p. 163).

The idea is of course an induction on k with $k = 1$ still being trivial, so assume the result holds for sets of width $k - 1$. Define a chain $C \subseteq P$ to be *strongly dependent* iff for every finite subset $S \subseteq P$ there is a representation of $S = K_1 \cup \cdots \cup K_l$ as a union of chains K_i such that $S \cap C \subseteq K_i$ for some $i \in \{1, \ldots, l\}$.

(a) Prove that there is a maximal strongly dependent chain C_1 in P.

(b) Prove that $P \setminus C_1$ has width $k - 1$. To do so, assume $\{a_1, \ldots, a_k\}$ is an antichain in $P \setminus C_1$.

  • For each i find a finite set $S_i \subseteq P$ such that no chain decomposition of S_i contains $S_i \cap (C_1 \cup \{a_i\})$ in one chain.

  • Apply the property of strong dependence of C_1 to $S = \bigcup_{i=1}^{k} S_i$.

  • Use the insight gained to find a j such that S_j has a chain decomposition such that $S_j \cap (C_1 \cup \{a_j\})$ is contained in exactly one chain.

25. A subset S of an ordered set P is called an **N** iff $S = \{a, b, c, d\}$ and $a \prec b \succ c \prec d$ with no further comparabilities. An ordered set that does not contain an N is called **N-free**. Prove that if a finite ordered set is N-free, then the removal of any maximal chain decreases the width. Give an example that shows that the converse is not true.

26. A characterization of finite N-free sets.

    (a) Prove that a finite ordered set is N-free iff every maximal chain intersects every maximal antichain.

    (b) Give an example of an infinite ordered set that is N-free and has a maximal chain that does not intersect all maximal antichains.

    (c) Is the other direction of part 26a true or false for infinite ordered sets?

27. (a) Find the number of all antichains in $\mathcal{P}_1$, $\mathcal{P}_2$, $\mathcal{P}_3$ and $\mathcal{P}_4$. (For a hint, consider Remark 4 in Chapter 11.)

    (b) Attempt to find the number of antichains in $\mathcal{P}_n$ for $n \geq 5$ via computer (cf. [273]). [Publish the result if you get past 8.]

28. Prove that the nth Dedekind number D_n is at least $2^{\binom{n}{\lfloor \frac{n}{2} \rfloor}}$.

29. Prove that a one-way infinite fence does not have the fixed point property.

30. Prove that fences are reconstructible.

31. For $j \in \{1, \ldots, m\}$ let $C_j := \{c_1^j, \ldots, c_{2n}^j\}$ be crowns with $2n$ elements such that for all $i \in \{1, \ldots, n\}$ and $j \in \{2, \ldots, m\}$ we have $c_{2i}^{j-1} = c_{2i-1}^j$. Let $\prec_j$ be the lower cover relation of C_j. Equip $T := \bigcup_{j=1}^{m} C_j$ with the order induced by the lower cover relation which is the union $\bigcup_{j=1}^{m} \prec_j$ of the individual lower cover relations. The ordered set T thus obtained is called a $2n$-**crown tower**. Prove that $2n$-crown towers have a fixed point free automorphism.

32. Prove that all fixed point free order-preserving self maps of a crown are automorphisms.

33. Prove that a six-crown-tower that is made up of three six-crowns (that is, it has twelve elements) has more than two fixed point free order-preserving maps.

34. Prove that crowns are reconstructible.

35. ([155]) Let #A(C_n) be the number of antichains in an n-crown ($n \geq 4$). Prove that #A(C_n) = #A(F_{n-1}) + #A(F_{n-3}).

36. Prove that if P is connected and $f : P \rightarrow Q$ is order-preserving, then $f[P]$ is connected.

37. Show that there are disconnected ordered sets P such that every automorphism of P has a fixed point (also cf. Exercise 20 in Chapter 10).

38.   (a) For the set in Figure 1.1 b) find an order-preserving self-map that has a disconnected set of fixed points.

    (b) Find an order-preserving map that has a disconnected set of fixed points for the set in Figure 1.3.

39. (The reconstruction problem has a negative answer for infinite ordered sets.) Let P be a connected ordered set of height 1 such that P has no crowns, every maximal element has countably many lower covers and every minimal element has countably many upper covers. Let Q be an ordered set with two connected components that are isomorphic to P. Show that P and Q have the same (infinite) deck. (Of course the deck would have to be a function that assigns every class $[C]$ of ordered sets the cardinality of $\{p \in P : P \setminus \{p\} \in [C]\}$.)

Remarks and Open Problems

This is the first time in this text that the remarks section also includes open problems. The open problems presented in this section are special cases of the main open questions or modifications of them.

1. The ordered set in Figure 2.2 also appears in [58], where R. P. Dilworth recounts the history of Theorem 2.5.7. For more on the work of R. P. Dilworth the reader should consult [19]. For connections between graph theory and Dilworth's theorem, consider Section 2.1 in [271].

2. Dilworth's theorem cannot be extended to ordered sets of infinite width. In [193] it is proved that for every infinite cardinal c there is an ordered set of cardinality c without infinite antichains, which cannot be decomposed into less than c chains. The reader will go through this construction in Exercise 11 in Chapter 10.

3. A conjecture on reconstruction of infinite ordered sets states that the example in Exercise 39 is characteristic for infinite ordered sets. The conjecture says that if P, Q are infinite ordered sets that are not isomorphic and have the same decks, then there is a $p \in P$ such that $P \setminus \{p\}$ is isomorphic to Q or there is a $q \in Q$ such that $Q \setminus \{q\}$ is isomorphic to P.

4. Find formulas for T_n^j for $j \geq 3$. (Success would of course lead to a solution of Dedekind's problem.)

5. Find formulas for the number of antichains in classes of ordered sets other than fences or crowns. This might give ideas for the solution of Dedekind's problem. We give a formula for interval ordered sets in Theorem 8.3.2.

6. Characterize the ordered sets of height 2 or those of width 3 that have the fixed point property. We will consider width 2 in Theorem 4.4.3 and height 1 in Theorem 4.4.6. The author conjectures that there is a polynomial algorithm to determine the fixed point property for ordered sets of width 3. In the light of the proof of Theorem 12.4.5, which says that it is NP-complete to decide if an ordered set of height 5 has a fixed point free order-preserving self map, the author has no intuition what might happen for height 2.

7. Prove that ordered sets of width 3 or of height 1 are reconstructible. We will consider width 2 in Exercise 8 of Chapter 3. The author believes that a proof of reconstructibility of ordered sets of width 3 should be possible with the reconstruction tools available today. Reconstruction of ordered sets of height 1 on the other hand appears almost as hard as the reconstruction problem in general.

8. For order-theoretical results beyond Proposition 2.5.9 that are in a Ramsey theoretical spirit, cf. [89, 187, 188, 198]. One of Ramsey's theorems (the one which guarantees the existence of finite Ramsey numbers, cf. [212]) essentially says that certain structures (complete graphs and discrete graphs) are so plentiful, that a representative of a certain size can be found in any graph. Embed the graph $G = (V, E)$ into a complete graph with $|V|$ vertices and color the edges of G red and the remaining edges blue. Then Ramsey's theorem says that any such coloring will always allow for a monochromatic complete subgraph of a certain size.

   The mentioned papers investigate and prove the following. Fix $r, s \in \mathbb{N}$. For every ordered set P it is possible to find an ordered set P' such that for any r-coloring χ of the s-chains of P', there is an embedding e of P into P' such that the s-chains of $e[P]$ are monochromatic under χ. So there is an ordered set that contains so many copies of P that even a partition through coloring will still allow us to find a copy in one of the elements of the partition.

9. Define a **fibre** of an ordered set to be a subset $F \subseteq P$ such that F intersects every maximal antichain with at least two elements.[2] Pictorially, the

---

[2]Note that sometimes one also refers to sets that intersect *every* maximal antichain as fibres. With this definition the overall upper bound on the size of a fibre is $|P|$ as can be seen considering chains. Interesting work can here be done in finding upper and lower bounds on the fibre size for individual ordered sets in given classes of ordered sets.

maximal antichains of an ordered set represent "horizontal separators" in the order. That is, every element of the ordered set will be either above or below some element of the maximal antichain. The notion of a fibre is then a subset that "stabs through all the separators".

It was conjectured in [170] that every finite ordered set has a fibre whose complement also is a fibre. Consequently, every ordered set would have a fibre of size at most $\dfrac{|P|}{2}$. This is not true, since by Exercise 19 the smallest size for a fibre of the ordered set P_0 in Figure 2.4 is $\dfrac{9}{17}|P_0|$.

The natural question that now arises is the following. What is the smallest λ such that any ordered set P is guaranteed to have a fibre of size at most $\lambda|P|$?

In [175] an iterative construction using the set in Figure 2.4 is used to show that $\lambda \geq \dfrac{8}{15}$. In [63], Theorem 1 it is shown that the elements of an ordered set can be 3-colored so that all nontrivial maximal antichains receive at least two colors. This means $\lambda \leq \dfrac{2}{3}$. The exact value of λ remains unknown.

3
Upper and Lower Bounds

Upper and lower bounds have already been defined in Definitions 2.1.4 and 2.3.1. From their use in Zorn's Lemma, as well as their occurrences in the proofs of Dilworth's Chain-Decomposition Theorem 2.5.7 and Proposition 2.6.7 (in both proofs, sets were defined in terms of their upper bounds), the reader can already infer that bounds of sets play an important role in ordered sets. In this chapter we consider various types of bounds and relate them to open problems as well as to each other.

3.1 Extremal Elements

Maximal elements and their duals, minimal elements, were defined in Definitions 2.1.5 and 2.3.2. We have seen their usefulness already through various arguments involving Zorn's Lemma. Maximal elements and minimal elements are also called **extremal elements**. Extremal elements, if they exist, are the absolute bounds of an ordered set. There will be no points strictly above a maximal element or strictly below a minimal element. On the other hand, in infinite ordered sets, extremal elements need not exist. For example the integers $\mathbb{Z}$ with their natural order have neither maximal nor minimal elements. In this section we study the relation between extremal elements and the fixed point property, isomorphism and the reconstruction problem, respectively. We start with some simple results on the fixed point property and on isomorphism. Then we will embark on the proof of a strong and useful result about the reconstruction of maximal cards (that is, cards obtained through the removal of a maximal element).

Figure 3.1: An ordered set without the fixed point property for which every rank-preserving map has a fixed point.

Proposition 3.1.1 *Let P be a finite ordered set. Then for every fixed point free order-preserving map $f : P \to P$ there is a fixed point free order-preserving map $g : P \to P$ such that $g \leq f$ and g maps minimal elements to minimal elements.*

Proof. Let $f : P \to P$ be a fixed point free order-preserving self map. For every minimal element $m \in P$ find a minimal element $x_m \leq f(m)$ in P. Define

$$g(p) := \begin{cases} f(p); & \text{if } p \text{ is not minimal,} \\ x_p; & \text{if } p \text{ is minimal.} \end{cases}$$

Clearly $g \leq f$ and g is order-preserving, since if $x < y$, then $g(x) \leq f(x) \leq f(y) = g(y)$. Finally, g has no fixed point. Indeed, if g had a fixed point, there would be an $m \in P$ such that $m = g(m) \leq f(m)$. By Exercise 16 in Chapter 1 this implies that f has a fixed point, a contradiction. ∎

Applying Proposition 3.1.1 and its dual, we see that if an ordered set has a fixed point free order-preserving map, then it must also have one such map that maps minimal elements to minimal elements and maximal elements to maximal elements. Especially for ordered sets of small height this insight can help reduce the number of candidates for fixed point free maps in a direct proof of the fixed point property, such as in Proposition 1.4.6 or in a brute-force computer search. Unfortunately, we cannot extend Proposition 3.1.1 to saying that there must even be a fixed point free map that preserves the rank of each element. Figure 3.1 shows an ordered set P such that every order-preserving map that preserves the rank of all elements has a fixed point (Rutkowski calls this the weak fixed point property) and yet P does not have the fixed point property. Indeed, P has only one element of rank 2 and thus every map that preserves the rank must fix that element. A fixed point free map is indicated in the picture.

Proposition 3.1.2 *Let P, Q be ordered sets and let $\Phi : P \to Q$ be an isomorphism. Then for each maximal element $m \in P$ we have that $\Phi(m)$ is a maximal element of Q.*

Proof. Let $m \in P$ be maximal and suppose $\Phi(m)$ is not maximal. Then there is a $q \in Q$ such that $q > \Phi(m)$. However then $\Phi^{-1}(q) > \Phi^{-1}(\Phi(m)) = m$, which contradicts the maximality of m. ∎

Example 3.1.3 The **largest (or greatest) element** l of an ordered set P is defined dually to the smallest element, that is, $l \geq P$. An ordered set can have a unique maximal element and yet not have a largest element. Consider the ordered set that consists of the natural numbers $\mathbb{N}$ united with a singleton set $\{p\}$ and no further comparabilities added. Then the element p is the unique maximal element of the ordered set, yet it is not the greatest element.

We will now proceed to prove some reconstruction results related to maximal elements. We start by reconstructing the decks of ideals and maximal ideals defined as follows.

Definition 3.1.4 *Let P be a finite ordered set. A **maximal ideal** is the ideal $\downarrow m$ of a maximal element $m \in P$.*

Definition 3.1.5 *Let P be an ordered set. For $[C] \in \mathcal{C}$ let $\mathcal{I}_P([C])$ denote the number of ideals $\downarrow_P p$ in P that are isomorphic to C and let $\mathcal{I}_P^M([C])$ denote the number of maximal ideals $\downarrow_P m$ in P that are isomorphic to C. $\mathcal{I}_P$ is called the **ideal deck** of P and $\mathcal{I}_P^M$ is called the **maximal ideal deck** of P.*

Theorem 3.1.6 *Let P be a finite connected ordered set with > 3 elements. Then $\mathcal{I}_P$ is reconstructible from the deck. That is, the ideal deck of P is reconstructible.*

Proof. (The proof in its present form is influenced by suggestions from J.-X. Rampon, M. Ellingham and an anonymous referee.) If P has a largest element, then by the dual of Proposition 1.5.11 there is nothing to prove. Otherwise we argue as follows.

First note that for given P all functions considered in this proof will have only finitely many arguments where they are not zero. Hence all sums and sets involved are finite. For any two ordered sets I, J that both have a largest element, let $s'(I, J)$ be the number of subsets of J that

- Contain the largest element of J and

- Are isomorphic to I.

Let I be a finite ordered subset of P with a largest element t. Then I is contained in the set $\downarrow t$ and I contains t. The number of isomorphic copies of I in P is the sum (over all ideals in P) of the number of copies of I that are contained in the ideal in such a way that the top point of the ideal is in the copy of I. Hence the number $s(I, P)$ can be computed from the numbers $\mathcal{I}_P([J])$ as follows (the sum runs over all nonisomorphic ideals J in P with more elements than I):

$$
\begin{aligned}
s(I, P) &= \sum_{|J| \geq |I|} \mathcal{I}_P([J]) s'(I, J) \\
&= \mathcal{I}_P([I]) + \sum_{|J| > |I|} \mathcal{I}_P([J]) s'(I, J),
\end{aligned}
$$

since if $|I| = |J|$, then $s'(I, J) = 1$ iff I is isomorphic to J and $s'(I, J) = 0$ otherwise. Solving for $\mathcal{I}_P([I])$ we obtain

$$\mathcal{I}_P([I]) = s(I, P) - \sum_{|J| > |I|} \mathcal{I}_P([J]) s'(I, J).$$

This means that we can reconstruct $\mathcal{I}_P$ recursively from $s(\cdot, \cdot)$. To see this, first note that essentially we are solving a (large) system of linear equations. In the following we prove that the system really is uniquely solvable. Let

$$k_0 := \max\{|I| : s(I, P) > 0 \text{ and } I \text{ has a largest element}\}.$$

Then by definition of k_0, there are no ideals of size $> k_0$. Moreover for $|I| = k_0$ we have $s(I, P) = \mathcal{I}_P([I])$ by the above equation.

Having found $\mathcal{I}_P([I])$ for $|I| \geq k_i$, let

$$k_{i+1} := \max\left\{|I| < k_i : s(I, P) - \sum_{|J| > |I|} \mathcal{I}_P([J]) s'(I, J) > 0 \right.$$

$$\left. \text{and } I \text{ has a largest element} \right\},$$

where the sum runs over all nonisomorphic ideals J in P with more elements than I. If the set on the right is empty, stop, $\mathcal{I}_P$ has been reconstructed. Otherwise, there are no ideals with size in $\{k_{i+1} + 1, \ldots, k_i - 1\}$ and for all I with $|I| = k_{i+1}$ we have

$$\mathcal{I}_P([I]) = s(I, P) - \sum_{|J| > |I|} \mathcal{I}_P([J]) s'(I, J),$$

where the sum runs over all nonisomorphic ideals J in P with more elements than I. Since the right side has already been reconstructed, we have reconstructed the left side. Continue to k_{i+2}. This process eventually terminates with the set used for computing k_{i+1} being empty. We have thus reconstructed $\mathcal{I}_P$.   ∎

Note that there are at least two ways to reconstruct a parameter of an ordered set. One is to prove equations that involve quantities available through the deck and obtain the parameter through these equations. This is what we have done above and what we will do mostly. Another way would be to consider two ordered sets with equal decks and show that the parameter must have the same value for either set.

In the proof of the following theorem we will need to identify a card with certain properties. Similar to the above, there are at least two ways to look at such a proof. One way is to consider an ordered set P and its deck and show that the card can be identified from the deck without using any knowledge about P. This is the way we choose. Another way is to start with a deck $\mathcal{D}$ of an ordered set and let P be *any* ordered set with $\mathcal{D}_P = \mathcal{D}$. Then one would need to show that a card with the desired properties can be found in $\mathcal{D}$. This approach looks formally a little cleaner, but the proofs do not change except for more cumbersome language. (For example, instead of saying "$C = P \setminus \{x\}$ is a card such that x satisfies $\cdots$", one would need to say "C is a card such that for any P with $\mathcal{D}_P = \mathcal{D}$ there is an $x \in P$ such that $\cdots$".) This is why we choose the first approach throughout.

Theorem 3.1.7 *Let P be a finite connected ordered set with > 3 elements. Then $\mathcal{I}_P^M$ is reconstructible from the deck. That is, the maximal ideal deck of P is reconstructible.*

Proof. To reconstruct $\mathcal{I}_P^M$, let $C = P \setminus \{x\}$ be a card such that there is a set I with a largest element such that

- $\mathcal{I}_C([J]) = \mathcal{I}_P([J])$ for all $[J] \neq [I]$,
- $\mathcal{I}_C([I]) = \mathcal{I}_P([I]) - 1$, and
- $|I|$ is as small as possible.

Since the ideal deck is reconstructible, such a card can be identified in the deck. First suppose x had been below at least two maximal elements. If one of these maximal elements had more lower bounds than the others, then removal of a maximal element with the fewest lower bounds would have produced a card with the same properties as demanded above, but with a smaller I. Thus all maximal elements above x must have the same number of lower bounds. But then for all maximal elements m above x the number $\mathcal{I}_C([\downarrow m])$ equals $\mathcal{I}_P([\downarrow m])$ minus the number of maximal elements above x whose ideal is isomorphic to $\downarrow m$. This contradicts the choice of C, which says that only one value of $\mathcal{I}_C$ is different from $\mathcal{I}_P$ and that difference is 1.

Thus x was below exactly one maximal element. By definition of C, x must have been below a maximal element m with as few lower bounds as possible and such that $[\downarrow_P m] = [I]$. This means that C contains all maximal ideals of P except for one copy of I. Moreover the maximal ideals of C with less than $|I|$ elements are exactly those maximal ideals of C that are not maximal ideals of P. We conclude that

$$\mathcal{I}_P^M([J]) = \begin{cases} \mathcal{I}_C^M([J]); & \text{for } |J| \geq |I| \text{ and } [J] \neq [I], \\ \mathcal{I}_C^M([I]) + 1; & \text{for } [J] = [I], \\ 0; & \text{for } |J| < |I|. \end{cases}$$

∎

Definition 3.1.8 *Let P be an ordered set. We denote the number of maximal elements of P by m_P.*

Corollary 3.1.9 *The number m_P of maximal elements of P is reconstructible.* ∎

Theorem 3.1.7 says we can reconstruct what is below single maximal elements. It seems natural to investigate what the set without the maximal element looks like. Maximal cards (cf. Definition 3.1.10) are essentially cards that were obtained

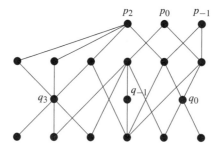

Figure 3.2: Illustration of how the removal of an element can increase or decrease the number of maximal elements.

by removing a maximal element. It is tempting to hope that a card is maximal iff it has fewer maximal elements than P. However this is not the case. The number of maximal elements does not behave in a nice predictable way when an element is removed from the ordered set. The number of maximal elements can increase, decrease or stay the same. Consider for example the ordered set in Figure 3.2. The elements p_i are maximal and removal of p_i changes the number of maximal elements by i. Fortunately the situation is not completely hopeless, as the most the number of maximal elements can decrease is 1.

Still, recognition of maximal cards, or even better, the reconstruction of the maximal deck (cf. Definition 3.1.12) remains an open problem. We will reconstruct one maximal card here. The original definition of a maximal card is given below. Note that an ordered set can have more maximal cards than it has maximal elements (cf. Exercise 5). This is because of possible isomorphisms between $P \setminus \{m\}$ and $P \setminus \{x\}$ for m maximal and x not maximal. We will address this problem in our Definition 3.1.12 of the maximal deck

Definition 3.1.10 *Let P be an ordered set and let C be an ordered set such that $\mathcal{D}_P([C]) > 0$. Then C is called a **maximal card** of P iff for any Q with $\mathcal{D}_Q = \mathcal{D}_P$ there is a maximal element m of Q such that $[C] = [Q \setminus \{m\}]$. (This definition is equivalent to Kratsch and Rampon's definition in [151].)*

In the following result the elements of the card that were not maximal in the ordered set turn out to be identifiable. This is because the removed element must have been maximal in *any* Q with $\mathcal{D}_Q = \mathcal{D}_P$.

Scholium 3.1.11 *(Cf. Theorem 6.3 in [151].) Let P be an ordered set with at least four elements. Then either P is reconstructible or from the deck we can identify one maximal card C that was obtained by removal of a maximal element with as few lower bounds as possible. Moreover, the elements of C that are not maximal in P can be identified.*

Proof. Let $C = P \setminus \{x\}$ be a card as in the proof of Theorem 3.1.7. Then x was below exactly one maximal element m and said maximal element has as few lower bounds as possible. Thus we are done if we identify a maximal card among these cards. We shall use notation from Theorem 3.1.7. If for the numbers of maximal elements we have $m_C \neq m_P$, x must have been a maximal element. Also, in case $m_C = m_P$, if C has a maximal ideal of size $\leq |I| - 2$, then x must have been maximal. Thus in either case C is a maximal card. Moreover, since x was maximal with as few lower bounds as possible in P, the maximal elements of C that are not maximal in P can be identified as the maximal elements of C with fewer than $|I|$ lower bounds.

This leaves the case in which for all the cards $C = P \setminus \{x\}$ as in the proof of Theorem 3.1.7 we have $m_C = m_P$ and C has a (necessarily unique) maximal ideal of size $|I| - 1$. In this case we will show that P is reconstructible.

Let m be the unique maximal element of P above x. Then m has a unique lower cover l. (Otherwise there would be a card C as in the proof of Theorem 3.1.7 with a maximal ideal of size $< |I| - 1$.) Moreover l has only m as its unique upper cover. (Otherwise there would be a card C as in the proof of Theorem 3.1.7 with $m_P - 1$ maximal elements.) This means that there are strict lower bounds y of m for which $\uparrow_P y$ is a chain and such that no strict upper bound of y has more than one lower cover.

Let the set J be obtained from I by removing the top element of I. We claim that if y is such that the card $K = P \setminus \{y\}$ as in the proof of Theorem 3.1.7 has a maximal ideal that is isomorphic to the set J, then $(\uparrow_P y) \setminus \{y\}$ is a chain such that no element has more than one lower cover. To see this claim, assume, for a contradiction, that $(\uparrow_P y) \setminus \{y\}$ was not a chain or that a strict upper bound of y has more than one lower cover. Let c be the largest point above y that has more than one lower cover. (Since $\uparrow_P y$ has a largest point, m, there is such a point $c > y$ in either case; moreover by the above, $c \leq l < m$.) Then J would contain an ideal isomorphic to $\downarrow_P c$, while $\downarrow_{P\setminus\{y\}} m$ does not. Since $\downarrow_{P\setminus\{y\}} m$ would have to be isomorphic to J on any such card $P \setminus \{y\}$, this is a contradiction. To finish the proof we use this fact to select one specific card now.

Let $K = P \setminus \{y\}$ be any card as in the proof of Theorem 3.1.7 so that the unique $(|I| - 1)$-element maximal ideal J' of K is isomorphic to J. Then, as shown above, $(\uparrow_P y) \setminus \{y\}$ must have been a chain such that no element has more than one lower cover. Moreover $y \leq m$ means that $\uparrow_P y$ is part of the chain at the top of J'. This however implies that P is isomorphic to K with a new top element attached to J' and P has been reconstructed. ∎

While Scholium 3.1.11 is an extremely useful tool for order reconstruction, we have to (unfortunately) conclude this section on a lower note. Even if we could reconstruct the maximal deck (which would be quite an advance in reconstruction), we would still not be able to reconstruct all ordered sets as Example 3.1.13 shows. Note that in the definition of the maximal deck below, the deck contains only as many cards as there are maximal elements. Thus even if we can find all maximal cards as defined in Definition 3.1.10, we might still not have the maximal deck, since several cards can be isomorphic and yet only some need be in the maximal deck.

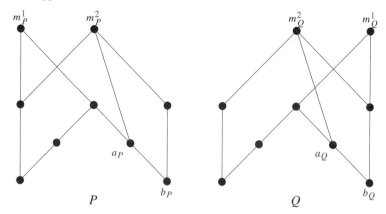

Figure 3.3: Two nonisomorphic ordered sets with the same maximal decks (due to Kratsch and Rampon, cf. [150]).

Definition 3.1.12 *Let P be an ordered set. The* **maximal deck** *of P is the function $\mathcal{M}_P : \mathcal{C} \to \mathbb{N}$ such that $\mathcal{M}_P([C])$ is the number of cards $P \setminus \{x\}$ of P that are isomorphic to C and that were obtained by removal of a maximal element.*

Example 3.1.13 (Cf. [150].) *Ordered sets are not reconstructible from their maximal decks.* An example of two nonisomorphic ordered sets with the same maximal decks is given in Figure 3.3. Indeed, it is easy to see that the maximal decks of these two ordered sets are the same. For $i = 1, 2$ the sets $P \setminus \{m^i_P\}$ and $Q \setminus \{m^i_Q\}$ are isomorphic. On the other hand the sets themselves are not isomorphic, since any isomorphism would have to map a_P to a_Q and hence b_P to b_Q and $\uparrow_P b_P$ to $\uparrow_Q b_Q$. This however is impossible since $\uparrow_P b_P$ is not isomorphic to $\uparrow_Q b_Q$.

Furthermore note that these two sets can be turned into infinite families of examples. Indeed, one can for example attach chains of the same length below each of the minimal elements of P and Q and thus obtain larger examples of nonisomorphic ordered sets with equal maximal decks.

3.2 Covers

Covers are a very local notion that we introduced in Definition 1.2.2 and investigated thereafter. We have seen in Proposition 1.2.7 that if we know all the covers in a finite ordered set, we also know all comparabilities. The relation between covers and isomorphisms was explored in Exercise 12 in Chapter 1. Also, in Section 3.1 we have seen the role of covers in reconstruction of maximal cards. In this section we will show that local knowledge about covers can have global consequences, such as the existence of a smallest element (cf. Theorem 3.2.3) and we will show that the number of covering relations is reconstructible (cf. Theorem 3.2.4).

Theorem 3.2.3 is inspired by the argument that is used in [79] to prove unique-ness of cores. (The reader will be asked to produce this very argument in Exercise 19 in Chapter 4.)

Lemma 3.2.1 *Let P be a finite ordered set such that any two distinct elements that have a common upper cover have a common lower cover. Then any two ele-ments of P that have a common upper bound have a common lower bound.*

Proof. Without loss of generality we can assume that P is connected. (Other-wise we work with the individual components.) For $x \leq y$ we define the cover-distance $c(x, y)$ from x to y to be the smallest number $n \in \mathbb{N}$ such that there is a chain $x = c_0 \prec c_1 \prec \cdots \prec c_n = y$ with each c_i being an upper cover of c_{i-1}. We claim that the following "parallelogram law" holds: If u is a common upper bound of $a, b \in P$, then there is a common lower bound l of a and b such that $c(l, b) \leq c(a, u)$ and $c(l, a) \leq c(b, u)$. (Clearly this implies the result.) The proof of the parallelogram law is by induction on $n = c(a, u) + c(b, u)$.

For $n = 0$ and $n = 1$ there is nothing to prove. In case $n = 2$, there is nothing to prove if $c(a, u) = 0$ or $c(b, u) = 0$, while if $c(a, u) = 1$ and $c(b, u) = 1$ the claim follows directly from the hypothesis.

For the induction step $\{0, \ldots, n-1\} \rightarrow n$ with $n > 2$ let $c(a, u) + c(b, u) = n$. We can assume without loss of generality that $c(b, u) > 1$. Since $c(b, u) > 1$, there is an upper cover b' of b such that $c(b', u) = c(b, u) - 1 > 0$. Thus there is a lower bound l' of a and b' such that $c(l', a) \leq c(b', u)$ and such that $c(l', b') \leq c(a, u)$. Now $c(l', b') \leq c(a, u)$ and $c(b, b') = 1 < c(b, u)$. Thus by induction hypothesis there is a common lower bound l of l' and b such that $c(l, l') \leq c(b, b')$ and $c(l, b) \leq c(l', b') \leq c(a, u)$. But then $c(l, a) \leq c(l, l') + c(l', a) \leq c(b, b') + c(b', u) = c(b, u)$ and we are done. ∎

Lemma 3.2.2 *Let P be a finite ordered set such that any two elements that have a common upper bound have a common lower bound. Then every component of P has a smallest element.*

Proof. Without loss of generality we can assume that P is connected. Let $m \in P$ be minimal and assume m is not the smallest element of P. Then there is an $x \in P$ such that $m \not\geq x$. Since P is connected there is a $y \in P$ such that $m \not\geq y$ and such that m and y have a common upper bound. However then m and y have a common lower bound b. Since $m \not\geq y$ we have $b < m$, a contradiction to the minimality of m. ∎

Theorem 3.2.3 *Let P be a finite ordered set such that any two elements that have a common upper cover have a common lower cover. Then every component of P has a smallest element.*

Proof. Simple combination of Lemmas 3.2.1 and 3.2.2. ∎

The above, as well as previous results on covers shows (unsurprisingly) that covers and bounds are tightly related to each other. It was straightforward to re-construct the number of comparability relations of an ordered set from the deck in Exercise 25a in Chapter 1. The short proof of the reconstruction of the number of covering relations should not lead the reader to believe that the reconstruction of the number of adjacencies is equally immediate. We are quoting a powerful lemma.

Theorem 3.2.4 (Cf. [151], Theorem 7.1.) *The number of covering relations (or adjacencies) in a finite ordered set is reconstructible from its deck.*

Proof. Let a_P be the number of adjacencies of P. For any ordered set I with a largest element, let s_I be the number of lower covers of the largest element. Then

$$a_P = \sum_{[I] \in \mathcal{C}} \mathcal{I}_P([I]) s_I,$$

which is reconstructible by Theorem 3.1.6.   ■

The number of covers is another tempting way to try to reconstruct the maximal deck. Removal of a maximal element decreases the number of covering relations. The temptation is to think that removal of a non-maximal element does not decrease the number of covering relations. To show this is not true consider once again Figure 3.2. Removal of any of the points q_k in Figure 3.2 changes the number of covering relations by k. Moreover, there is no bound on by how much the number of comparabilities can change by removal of one non-extremal point (cf. Exercise 12).

3.3   Lowest Upper and Greatest Lower Bounds

The existence of lowest upper and greatest lower bounds in an ordered set is normally a strong tool. For example, it is paramount to the development of analysis, as the completeness of the real numbers is equivalent to the existence of a lowest upper (greatest lower) bound for every subset that has an upper (a lower) bound. In fact our examples will show that in infinite ordered sets lowest upper and greatest lower bounds have a very similar feel as limits of monotone sequences. This strand of thought is followed in Section 3.4 where we investigate chain-completeness and in Proposition 5.1.7 in which chain-completeness is related to lattices.

Definition 3.3.1 *Let P be an ordered set and let $A \subseteq P$. Then*

1. *The point u is called the **lowest upper bound** or **supremum** or **join** of A iff $u \geq A$ and for all $p \in P$ with $p \geq A$ we have $p \geq u$.*

2. *The point l is called the **greatest lower bound** or **infimum** or **meet** of A iff $l \leq A$ and for all $p \in P$ with $p \leq A$ we have $p \leq l$.*

We will denote the supremum of a set A (if it exists) $\bigvee A$ *and the infimum (if it exists)* $\bigwedge A$. *For finite sets* $A = \{a_1, \ldots, a_n\}$ *we will also use the notation*

$$a_1 \vee a_2 \vee \cdots \vee a_n \quad := \quad \bigvee \{a_1, a_2, \ldots, a_n\},$$
$$a_1 \wedge a_2 \wedge \cdots \wedge a_n \quad := \quad \bigwedge \{a_1, a_2, \ldots, a_n\}.$$

We have already made the acquaintance of lowest upper bounds when we were working with Zorn's Lemma. Though the hypothesis of Zorn's Lemma only requires the existence of some upper bound for every nonempty chain, in all our proofs the bound that was constructed invariably was the lowest upper bound. The reader is invited to check over the proofs of Theorems 2.1.8 and 2.2.7 and Propositions 2.1.9 and 2.2.9 to verify this claim. To justify our talking about *the* lowest upper bound we note:

Proposition 3.3.2 *Let P be an ordered set and let $A \subseteq P$ be a set that has a lowest upper bound. Then the lowest upper bound of A is unique.*

Proof. Let $A \subseteq P$ be a subset and let u and u' be elements that are upper bounds of A and such that any upper bound of A has to be above u and above u'. Then in particular, since u is an upper bound of A, we have $u \geq u'$ and by symmetry we also have $u' \geq u$. But then $u = u'$ and we are done. ∎

There also is the possibility to iterate the formation of lowest upper bounds.

Proposition 3.3.3 *Let P be an ordered set and let A and B be subsets that have suprema. If $\bigvee A \vee \bigvee B$ exists, then it is the supremum of $A \cup B$.*

Proof. To abbreviate notation, let $s := \bigvee A \vee \bigvee B$. Then $s \geq A$ and $s \geq B$. Moreover, if $p \geq A \cup B$, then $p \geq \bigvee A$ and $p \geq \bigvee B$, which implies $p \geq s$. ∎

Example 3.3.4

1. In Figure 1.1 b) any two elements of rank 1 have a supremum or an infimum, but some sets of two elements of rank 1 do not have both. In Figure 1.1 c) any two minimal elements have a supremum and any two maximal elements have an infimum.

2. The ordered subset $\{x \in \mathbb{Q} : x \geq 0, x^2 \leq 2\}$ of $\mathbb{Q}$ has upper bounds but no lowest upper bound in $\mathbb{Q}$.

3. Every nonempty subset of $\mathbb{R}$ that has an upper bound has a lowest upper bound.

4. Let P be an arbitrary ordered set. The empty set $\emptyset$ as a subset of P has a supremum iff P has a smallest element. The empty set $\emptyset$ has an infimum iff P has a largest element.

5. If X is a set and $\mathcal{P}(X)$ is its power set ordered by inclusion, then for each $A \subseteq \mathcal{P}(X)$ we have $\bigvee A = \bigcup A$ and $\bigwedge A = \bigcap A$.

6. Let $A \subseteq C([0, 2], \mathbb{R})$ be finite. Then $f(x) := \bigvee \{g(x) : g \in A\}$ is the supremum of A.

7. For $n \in \mathbb{N}$ define $f_n \in C([0, 2], \mathbb{R})$ by

$$
f_n(x) := \begin{cases} 1; & \text{for } x \in \left[0, 1 - \frac{1}{n}\right), \\ -nx + n; & \text{for } x \in \left[1 - \frac{1}{n}, 1\right], \\ 0; & \text{for } x \in (1, 2]. \end{cases}
$$

Then $\{f_n : n \in \mathbb{N}\}$ has no supremum in $C([0, 2], \mathbb{R})$.

8. (Monotone Convergence Theorem) In the space $L^1(\Omega, \Sigma, \mu)$ let $\{f_n\}_{n \in \mathbb{N}}$ be a sequence of functions such that $f_n \leq f_{n+1}$ a.e.. Then $\{f_n : n \in \mathbb{N}\}$ has a lowest upper bound in $L^1(\Omega, \Sigma, \mu)$ iff it has an upper bound.

Proof. Part 1 allows a visualization of suprema and infima. Listing all sets that have suprema or infima and proving the claim is routine, if a little tedious. To see that all possibilities can occur, note that in the set in part b) the subset $\{c, d\}$ has a supremum and no infimum, the subset $\{c, f\}$ has a supremum and an infimum and the subset $\{i, k\}$ has neither a supremum, nor an infimum.

To prove part 2, assume that u is the lowest upper bound of $\{x \in \mathbb{Q} : x \geq 0, x^2 \leq 2\}$. Since there is no rational number q with $q^2 = 2$, and since there are rational numbers whose squares are arbitrarily close to 2 but less than 2, we must have $u^2 > 2$. Let $n \in \mathbb{N}$ be such that $2\frac{u}{n} < u^2 - 2$. Then $\left(u - \frac{1}{n}\right) < u$ and

$$
\left(u - \frac{1}{n}\right)^2 = u^2 - 2\frac{u}{n} + \frac{1}{n^2} > 2,
$$

contradicting the choice of u.

Note that part 2 normally is the motivation in elementary set theory or analysis to go from the rational number system to the real number system. While analysis is concerned more with limits than with suprema and infima, the above proof suggests how to recursively construct a monotone Cauchy sequence that does not have a limit in $\mathbb{Q}$.

Part 3 is either part of the axiomatic definition of the real numbers or (if the reals were constructed from smaller number systems) has to be proved from the construction of $\mathbb{R}$. Which it is depends on the author of the set theory chapter or text one is reading. Since we have not taken time to formally define the real numbers, we will not present a proof here. For the development of order theory this "hole" is not a problem. For examples that involve $\mathbb{R}$ the reader is advised to use any definition of $\mathbb{R}$ (s)he has seen in classes or another text.

The proof of part 4 goes back to the logical structure of the definition of suprema and upper bounds. The point $u \in P$ is an upper bound of $\emptyset$ iff u is above every element of $\emptyset$. Since $\emptyset$ has no elements, every $p \in P$ has this property. Therefore every element of P is an upper bound of $\emptyset$. This means if $\emptyset$ has a lowest upper bound, it would have to be below every element of P. Conversely, the smallest element of P, if P has one, would be the lowest upper bound of $\emptyset$. The infimum is treated dually.

To prove part 5 let $A \subseteq \mathcal{P}(X)$ be nonempty. Then $\bigcup A$ contains all sets in A and every set that contains all sets of A must contain $\bigcup A$. Since the union of an empty set of sets is the empty set, we have proved that unions are suprema. Infima are treated similarly.

To see part 6 first note that in any space of continuous functions comparability is normally defined pointwise, so the supremum of a set of continuous functions must be at least above the pointwise supremum. It is a standard result of analysis that the pointwise supremum of finitely many continuous functions is again continuous (cf. Theorem 7.22 in [120], for example; they prove it for lower semi-continuous functions). Thus we are done.

For part 7 first note that the pointwise supremum of the given set of functions is

$$f_{\mathrm{ptws}}(x) = \begin{cases} 1; & \text{for } x \in [0, 1), \\ 0; & \text{for } x \in [1, 2]. \end{cases}$$

Now for every continuous function $g \geq f_{\mathrm{ptws}}$, we have $g(y) \geq 1$ for $y \in [0, 1)$, and hence $g(1) \geq 1$. Therefore there is an $x_\varepsilon \in (1, 2]$ such that $g(x_\varepsilon) = \varepsilon > 0$. Let

$$h(x) := \begin{cases} 1; & \text{for } x \in [0, 1], \\ 2 - x; & \text{for } x \in (1, 2]. \end{cases}$$

Then the product gh is an upper bound of all the f_n, $gh \leq g$ and $gh(x_\varepsilon) < \varepsilon = g(x_\varepsilon)$. Thus no upper bound of the given set of functions can be its lowest upper bound. This means that $\{f_n : n \in \mathbb{N}\}$ does not have a supremum in $C([0, 2], \mathbb{R})$.

The monotone convergence theorem (part 8) is an important theorem in real analysis. In a sense it is what sets the spaces of Lebesgue integrable functions apart from spaces of continuous or Riemann integrable functions, where there is no analogue of the monotone convergence theorem. For continuous functions we have seen this in part 7. For Riemann integrable functions note that the function on $[0, 1]$ that is 1 for rational numbers and zero for irrational numbers (Dirichlet's function) is not Riemann integrable. Yet it can be represented as the supremum of functions that are 1 in only finitely many places. A proof of the monotone convergence theorem can be found for example in [40], Theorem 2.4.1.   ∎

Suprema and infima are of course linked together, since the supremum of a set is the infimum of its set of upper bounds, as we are to prove now.

Definition 3.3.5 *Let P be an ordered set and let $A \subseteq P$. Then we set*

$$\uparrow A \; := \; \{p \in P : p \geq A\},$$

$$\downarrow A \quad := \quad \{p \in P : p \leq A\}.$$

Lemma 3.3.6 *Let P be an ordered set and let $A \subseteq P$. Then*

1. *The equality $\bigvee A = \bigwedge \uparrow A$ holds in the sense that if either of the involved two quantities exists, then both exist and the equality holds.*

2. *If $\bigwedge A$ exists in P, then $\downarrow A = \downarrow \bigwedge A$.*

3. *If $\bigvee A$ exists in P, then $\downarrow\uparrow A = \downarrow \left(\bigvee A \right)$.*

Proof. For part 1, first suppose that $\bigvee A$ exists. Then $\bigvee A$ is a lower bound of $\uparrow A$. Since $\bigvee A \in \uparrow A$, any lower bound of $\uparrow A$ is below $\bigvee A$, so $\bigvee A$ is the infimum of $\uparrow A$.

Now suppose that $\bigwedge \uparrow A$ exists. Since all elements of A are lower bounds of $\uparrow A$, $\bigwedge \uparrow A$ is an upper bound of A. By definition it is below any upper bound of A, so it is the supremum of A.

In part 2 simply note that $x \leq A$ iff $x \leq \bigwedge A$.

To prove part 3 note that by part 1, the upper bounds of A have an infimum. Thus $\downarrow\uparrow A = \downarrow \left(\bigwedge \uparrow A \right) = \downarrow \left(\bigvee A \right)$. ∎

As we have seen, existence of suprema and infima is by no means guaranteed in arbitrary ordered sets. Yet power sets (part 5 of Example 3.3.4) show that in the perhaps most natural example of an ordered set that is neither a chain, nor an antichain, suprema and infima abound. In such a situation mathematicians often strive to at least identify the general structures as substructures of nicer structures. Examples of this procedure include the completion of metric spaces (if one cares about convergence of Cauchy sequences) or the various ways to compactify topological spaces (if one wants small coverings using open sets). Thus it is only natural to wish to embed an ordered set into some power set (or a subset of a power set). The following two results show how this can be done. We will later consider the Dedekind–MacNeille completion, which is another, nicer, way to embed an ordered set into an ordered set with suprema and infima.

Proposition 3.3.7 *Let P be an ordered set and let $\mathcal{P}(P)$ be the power set of P ordered by set inclusion. Then the map $\Phi : P \to \mathcal{P}(P)$, $p \mapsto \downarrow p$ is an embedding of P into $\mathcal{P}(P)$.*

Proof. This is a very simple consequence of the definition of an embedding and the definition of Φ. ∎

The embedding Φ is also nice in the sense that existing infima are mapped to intersections, so at least the lower bound operations in the two sets correspond to each other.

Proposition 3.3.8 (Cf. [49], Lemma 2.32.) *Let P be an ordered set and let $A \subseteq P$. If $\bigwedge A$ exists in P, then $\bigcap \{\downarrow a : a \in A\} = \downarrow \left(\bigwedge A \right)$.*

Proof. Note that

$$\bigcap \{\downarrow a : a \in A\} = \{p \in P : p \leq A\} = \left\{ p \in P : p \leq \bigwedge A \right\} = \downarrow \left(\bigwedge A \right).$$

∎

3.4 Chain-Completeness and the Abian–Brown Theorem

Clearly existence of suprema and infima is a strong tool. It may not occur as frequently as one wants, but it can still be seen as a fairly frequent occurrence in mathematics. Thus a lot of work has been devoted to the study of ordered sets in which any finite nonempty subset or any subset has a supremum and an infimum. Such sets are called lattices and complete lattices respectively. They are investigated in lattice theory, a fairly large branch of discrete mathematics. We will devote Chapter 5 to the study of lattices and Chapter 6 to truncated lattices in this text. In this section we will investigate a variation on the existence of suprema and infima that restricts its attention to nonempty chains. We will show how this condition can be used to prove an important fixed point result.

Definition 3.4.1 *Let P be an ordered set. Then P is called **chain-complete** iff each nonempty subchain $C \subseteq P$ has a supremum and an infimum.*

One has to be cautious with the use of the word "chain-complete". While the above definition appears to be the most common use of the word, some authors also demand that the empty chain has a supremum and an infimum. In this text we will not assume that a chain-complete ordered set has a largest or a smallest element unless we explicitly demand their existence.

Example 3.4.2

1. The power set of any set is chain-complete with respect to inclusion.

2. Every finite ordered set is chain-complete.

3. $C([0, 2], \mathbb{R})$ is not chain-complete.

Proof. By part 5 of Example 3.3.4, every subset of a power set has a supremum and an infimum, so clearly power sets must in particular be chain-complete. This proves part 1. For part 2 note that finite chains have in fact a largest and a smallest element. Finally, $C([0, 2], \mathbb{R})$ is not chain-complete, since in part 7 of Example 3.3.4 we have a nonempty chain of continuous functions on $[0, 2]$ that has no supremum. ∎

Proposition 3.4.3 *Let P be a chain-complete ordered set. For every $p \in P$ there is a maximal element $M \in P$ and a minimal element $m \in P$ such that $m \leq p \leq M$.*

Proof. This is a consequence of Zorn's Lemma and its dual.   ∎

Chain-completeness is a strong tool when working with the fixed point property. This is because of the Abian–Brown theorem, which we will prove in the following. The idea is very simple and was already used in the proof of Propositions 1.4.6 and 2.7.3. If $p \leq f(p)$, then all the $f^n(p)$ form a chain. In a finite ordered set, this chain eventually has to stop and the place where it stops has to be a fixed point. For infinite ordered sets a similar argument will work. We just need to have some way available to push past limit ordinal numbers. Chain-completeness provides a vehicle to do this. Whenever a chain goes up infinitely often, we will be able to take the supremum and continue. This is reflected in part 2 of the definition of f-chains below. Our proof follows along the lines of the proof in [1]. The methods developed here will be used for example in Exercise 24 in Chapter 10 to give an alternative proof for Roddy's product theorem (cf. Theorem 10.2.11) and in Appendix B. (It should not be surprising that a result this close to the monotone convergence theorem has applications in analysis. In fact, the motivation for the related work in [192] came from the desire to solve certain types of integral equations.)

Note that to allow for more versatility later on, we will not demand our underlying set to be chain-complete until Theorem 3.4.7.

Definition 3.4.4 (Cf. [1].) *Let P be an ordered set, let $f : P \to P$ be an order-preserving function and let $p \in P$ with $f(p) \geq p$. A well-ordered subset C of P is called an f-**chain starting at** p if and only if*

1. *p is the smallest element of C.*

2. *Using ordinal number terminology for elements of C, we have that for every $c \in C \setminus \{p\}$,*

$$
c = \begin{cases} f(c^-); & \text{if } c \text{ has an immediate predecessor } c^-, \\ \bigvee_P \left[((\downarrow c) \cap C) \setminus \{c\} \right]; & \text{if } c \text{ is a limit ordinal and} \\ & \text{the supremum exists.} \end{cases}
$$

Our first step is to find an f-chain that is as large as possible, that is, an f-chain maximal with respect to inclusion. This should remind the reader of Zorn's Lemma and indeed a simple Zorn's Lemma argument would prove the existence of maximal f-chains. However with just a little more effort we can show the same fact without using Zorn's Lemma. This approach has a theoretical and a practical aspect. From the theoretical angle, not needing Zorn's Lemma might allow us to get away with a somewhat smaller system of set-theoretical axioms.

Then again, demanding that our underlying universe (that is, our ordered set) is chain-complete establishes something stronger than Zorn's lemma at that level already. Thus with respect to the Abian–Brown theorem our argument "only" establishes that we do not need Zorn's lemma for a certain argument outside this universe. From a somewhat practical angle, showing that the f-chains are unique and only need to be unified to obtain the maximal f-chain opens the door for an (admittedly transfinite) algorithm to find fixed points in infinite sets. (For more on this idea, cf. Appendix B.)

Lemma 3.4.5 (Cf. [1], Lemma 1.) *Let P be an ordered set and let the map f : $P \to P$ be order-preserving. If $p \in P$ is such that $p \le f(p)$ and if C, K are two f-chains starting at p, then either C is an initial segment of K or K is an initial segment of C. Hence there is a unique maximal f-chain starting at p.*

Proof. Let C, K be f-chains starting at p and let

$$\mathcal{J} := \{I \subseteq C \cap K : I \text{ is an initial segment of } C \text{ and of } K\}.$$

Then $\mathcal{J} \neq \emptyset$, as $\{p\}$ is in $\mathcal{J}$. Consider $H := \bigcup \mathcal{J}$. Clearly $H \subseteq C \cap K$. Let $c \in C \setminus H = \bigcap_{I \in \mathcal{J}} C \setminus I$. Then c is an upper bound of all $I \in \mathcal{J}$ and thus an upper bound of H. Hence H is an initial segment of C and similarly it is an initial segment of K. Now assume $K \setminus H \neq \emptyset$ and $C \setminus H \neq \emptyset$. If H has a largest element h, then $f(h)$ is the immediate successor of h in C and in K. Hence $H \cup \{f(h)\}$ is an initial segment of C and of K, which is a contradiction. If H does not have a largest element, then $\bigvee H$ exists in P and it is the supremum of H in C and in K. Hence in this case $H \cup \left\{ \bigvee H \right\}$ is an initial segment of C and of K, which is a contradiction. Thus $H = C$ or $H = K$ and we are done. The unique maximal f-chain starting at p is thus the union of all f-chains starting at p. ∎

Lemma 3.4.6 *Let P be an ordered set and let $f : P \to P$ be an order-preserving self-map. If $p \in P$ is such that $f(p) \ge p$, then one of the following two holds.*

1. *The maximal f-chain starting at p has no supremum, or*

2. *There is a fixed point $q \in P$ of f such that $q \ge p$ and for all fixed points $x \ge p$ of f we have $x \ge q$.*

Proof. If the maximal f-chain C starting at p has no supremum, we are done, so let us assume $q = \bigvee_P C$ does exist. By the definition of f-chains, $q \in C$. Moreover if $f(q) > q$, then $C \cup \{f(q)\}$ would be an f-chain that contains C, so $f(q) \not> q$.

On the other hand we have $f(q) \ge q$ as the following shows. If q has an immediate predecessor q^-, then $f(q) = f(f(q^-)) \ge f(q^-) = q$. If

$$q = \bigvee_P \left[((\downarrow q) \cap C) \setminus \{q\} \right],$$

then since the image of every element of C is again in C, we have

$$f(q) = f\left(\bigvee_P\left[((\downarrow q) \cap C) \setminus \{q\}\right]\right) \geq \bigvee_P\left[((\downarrow q) \cap C) \setminus \{q\}\right] = q.$$

Therefore $f(q) = q$.

Now let $x \in \uparrow_P p$ be a fixed point of f and assume $x \not\geq q$. Let $c \in C$ be the smallest element of C such that $x \not\geq c$. Then c cannot have an immediate predecessor c^-, since then we would have $x = f(x) \geq f(c^-) = c$. Thus $c = \bigvee_P \downarrow_C c$. Via $x \geq (\downarrow_C c) \setminus \{c\}$ we again infer $x \geq c$, a contradiction. Therefore we must have that $x \geq q$. ∎

Note that the two possibilities in Lemma 3.4.6 are not mutually exclusive (cf. Exercise 27).

Theorem 3.4.7 *(The Abian–Brown theorem, cf. [2].) Let P be a chain-complete ordered set and let $f : P \to P$ be order-preserving. If there is a $p \in P$ with $p \leq f(p)$, then f has a smallest fixed point above p.*

Proof. Easy consequence of Lemma 3.4.6. Part 1 cannot occur since P is chain-complete. ∎

Corollary 3.4.8 *Every chain-complete ordered set with a smallest element has the fixed point property.* ∎

Exercises

1. Prove that in a finite ordered set a unique maximal element must be the greatest element of the set.

   Then give an example of an infinite ordered set that has a unique maximal element and no greatest element.

2. An ordered set satisfies the **descending chain condition** iff it contains no copies of the dual of $\mathbb{N}$. Prove that in an ordered set with the descending chain condition every element is above a minimal element.

3. Investigating the number of maximal elements cards can have.

   (a) Construct an ordered set for which all cards have the same number of maximal elements.

   (b) Prove that an ordered set cannot have one card with one maximal element, one card with three maximal elements and one card with five maximal elements.

(c) Prove that for every $n \in \mathbb{N}$ and every $(n + 1)$-element subset H of $\mathbb{N} \setminus \{1, \ldots, n-1\}$, there is an ordered set P_H such that for each $h \in H$, P_H has a card with h maximal elements.

4. Prove that for every $k \in \mathbb{N}$ and every finite ordered set with at least four elements the number of elements of rank k is reconstructible.

5. Prove that every card of a chain is a maximal card according to Definition 3.1.10.

6. (Extension of Scholium 3.1.11.) Let $s = \min\{k \in \mathbb{N} : I_P^M(k) \neq 0\}$. Prove that for every finite ordered set P with at least four elements *all* cards obtained by removing a maximal element with at most s lower bounds are reconstructible.

7. Let P be a finite ordered set with > 3 elements and let $C = P \setminus \{a\}$ be a maximal card that can be identified from the deck. A down-set is a set $H \subseteq P$ such that for all $h \in H$ we have $\downarrow h \subseteq H$. Prove that

   (a) It is possible to find a set A that is isomorphic to $\downarrow_P a$.

   (b) Let $H_1, \ldots, H_k \subseteq P \setminus \{a\}$ be the distinct down-sets in $P \setminus \{a\}$ that are isomorphic to $A \setminus \{\bigvee A\}$. Let P_j be the ordered set obtained by attaching a to C as an upper bound of H_j with no further new comparabilities. Then $[P] \in \{[P_1], \ldots, [P_k]\}$. In particular, if $k = 1$, then P is reconstructible.

   (c) Suppose the elements $l_1, \ldots, l_m$ in $P \setminus \{a\}$ have been identified as lower covers of a. Let $H_1, \ldots, H_k \subseteq P \setminus \{a\}$ be the distinct down-sets in $P \setminus \{a\}$ that are isomorphic to $A \setminus \{\bigvee A\}$ and contain $(\downarrow_C l_1) \cup \cdots \cup (\downarrow_C l_m)$. Let P_j be the ordered set obtained by attaching a to C as an upper bound of H_j with no further new comparabilities. Then $[P] \in \{[P_1], \ldots, [P_k]\}$. In particular, if $k = 1$, then P is reconstructible.

8. Prove that ordered sets of width 2 are reconstructible. (Hint: Use Exercise 7. The original solution is in [152].)

9. Call a maximal element m of an ordered set P **dominating** iff m is above all non-maximal elements of P. For the following we assume the sets in question have at least four elements.

   (a) Prove that ordered sets with a dominating maximal element are recognizable.

   (b) Prove that cards obtained by removal of a dominating maximal element can be identified from the deck.

   (c) Prove that ordered sets in which every element has at least two upper covers and which have a dominating maximal element are reconstructible.

(d) Prove that ordered sets of width 3 in which each element of rank 1 has at least two lower covers are reconstructible.

10. For an ordered set P, let $N_P(r, d, f, i, u, l)$ be the number of elements of rank r, of dual rank d (defined like the rank, only "dually"), with filter size f, ideal size i, u upper covers and l lower covers. Give an example that shows that $N_P = N_Q$ does not imply that P is isomorphic to Q. (Hint. Figure 3.3.)

11. Prove that the sets obtained from the sets in Figure 3.3 by deleting the element between a_P and m_P^1 in P and by deleting the element between a_Q and m_Q^1 in Q are also nonisomorphic with equal maximal decks.

12. (The relation between removal of a point and the number of covering relations.)

    (a) For each $n \in \mathbb{N}$ construct an ordered set such that removal of a certain point of the ordered set decreases the number of covering relations by exactly n.

    (b) For each $n \in \mathbb{N}$ construct an ordered set such that removal of a certain non-extremal point of the ordered set decreases the number of covering relations by exactly n.

    (c) For each $n \in \mathbb{N}$ find an ordered set such that removal of a certain point of the ordered set increases the number of covering relations by exactly n.

13. Find an ordered set that does not have the fixed point property and that is such that every cover-preserving map (that is, every map such that $x \prec y$ implies $f(x) \prec f(y)$) has a fixed point.

14. Recall from Exercise 25 in Chapter 2 that an ordered set P is called **N-free** iff P does not contain a subset $\{a, b, c, d\}$ such that $a < b > c < d$ and there are no further comparabilities between these elements. Let P be an N-free ordered set. Prove that

    (a) If $a, b \in P$ have a common upper cover, then a and b have the same upper covers.

    (b) If $a \in P$ has an upper cover u such that $u > b$, then $(\uparrow a) \setminus \{a\} \subseteq \uparrow b$.

15. Find a set S of rational numbers s with $s^2 > 2$, such that any rational number q with $q^2 > 2$ is above some $s \in S$.

16. Let P be an ordered set and let $A \subseteq B \subseteq P$. Prove that $\uparrow A \supseteq \uparrow B$ and $\downarrow A \supseteq \downarrow B$

17. Let P be an ordered set. Prove that $\uparrow \downarrow \uparrow A = \uparrow A$ for all $A \subseteq P$.

18. Let P, Q be ordered sets and let $f : P \to Q$ be order-preserving. Prove that for any set A such that A and $f[A]$ have a supremum we have the inequality $f\left[\bigvee A\right] \geq \bigvee f[A]$. Then give an example that shows that strict inequality can occur.

19. Prove that in the space $C([-1, 1], \mathbb{R})$ of continuous functions on $[-1, 1]$, the set of monomials $\{x^n : n \in \mathbb{N}\}$ has a greatest lower bound. Subsequently prove that it does not have a greatest lower bound in the space $C([-2, 2], \mathbb{R})$ of continuous functions on $[-2, 2]$.

20. Prove that if $c = a \vee b$ in P, then $c = a \wedge b$ in P^d, the dual of P.

21. (On distributivity.) Find an ordered set P and three points $a, b, c \in P$ such that $a \vee (b \wedge c) \neq (a \vee b) \wedge (a \vee c)$. That is, all involved quantities in the inequality exist and the inequality holds.

    This means that in general suprema do not distribute over infima. Use the above to conclude that in general infima do not distribute over suprema either.

    This abstract distributivity is motivated by the fact that distributivity holds for unions and intersection. We shall investigate distributivity in more detail in Section 5.5.

22. On $\mathcal{P}(\{1, \ldots, n\})$ define the mapping

$$f(A) := A \cup \{1, \ldots, \min(|A| + 1, n)\}.$$

    (a) Show that $f : \mathcal{P}(\{1, \ldots, n\}) \to \mathcal{P}(\{1, \ldots, n\})$ is order-preserving,
    (b) Find the fixed points of f,
    (c) Find the largest number k such that there is an $A \in \mathcal{P}(\{1, \ldots, n\})$ such that $f^k(A)$ is not a fixed point of f.

23. The mapping $f : \mathbb{R} \to \mathbb{R}, x \mapsto x + 1$ is order-preserving and has no fixed points. Is this a contradiction to Theorem 3.4.7?

24. Let $f : P \to P$ be an order-preserving map. Show that if $p \leq q$, $F(p)$ is the smallest fixed point of f above p and $F(q)$ is the smallest fixed point of f above q, then $F(p) \leq F(q)$.

25. Use Zorn's Lemma to prove that every f-chain is contained in a maximal f-chain.

26. Though we will ultimately have the stronger Theorem 5.2.1, it is a good exercise to prove that a chain has the fixed point property iff it is chain-complete.

27. Give an example of an ordered set P, an order-preserving map $f : P \to P$ and a $p \in P$ such that $p \leq f(p)$, the maximal f-chain starting at p has no supremum and f has a smallest fixed point $q \geq p$.

28. Let P and Q be ordered sets. Then the set of order-preserving maps f : $P \to Q$ can be ordered with the pointwise order. That is, $f \leq g$ iff $f(p) \leq g(p)$ for all $p \in P$. Show that if Q is chain-complete, then so is the set of all order-preserving maps from P to Q. Does the reverse implication hold, too?

Remarks and Open Problems

1. Theorem 3.2.3 is useful when showing that a removal procedure such as dismantling (cf. Section 4.3) or a consistency enforcing algorithm for constraint networks (cf. Exercise 23 in Chapter 12) yields the same end result independent of the order in which the removal steps are being taken.

2. A natural generalization of Sands' question if ordered sets can be reconstructed from their maximal decks (cf. [240]) is if ordered sets are reconstructible from their maximal and minimal decks. That is, does equality of the maximal *and* the minimal decks of two finite ordered sets with at least four elements force isomorphism of the underlying sets? Example 3.1.13 does not give an answer to this question, since the two ordered sets in the example have different minimal decks.

   The author has recently found examples of nonisomorphic sets with the same minimal and maximal decks (cf. [249]). Moreover, there appear to be examples of nonisomorphic sets with the same minimal and maximal decks for which there is a rank such that all cards obtained by removing an element of rank k are equal also (cf. [250]). Thus (pending review for [250]) even considerable partial information on extremal and other cards is not sufficient for reconstruction.

3. Are ordered sets reconstructible from their "ranked decks"? That is, does a deck such that for each card the rank of the removed element is known uniquely determine the ordered set?

   This problem is an extension/modification of Problem 2. More generally one could ask "What additional information besides the deck is needed to reconstruct ordered sets?". This amounts essentially to a more focused quest for a complete set of invariants. The motivation for these questions lies in the reconstruction of some further invariants such as the rank k neighborhood decks (that is, the decks that show all neighborhoods of points of rank k) and some maximal cards, which is recorded in [247].

4. Which of the many finitary results available for the fixed point property can be translated to infinite ordered sets and then brought to bear on, for example, analysis (cf. Appendix B)?

   This question is motivated by the fact that the Abian–Brown theorem still is the core for all nontrivial fixed point results on infinite ordered sets that the author is aware of.

4
Retractions

Retractions are an important tool when working with the fixed point property and when exploring the structure of ordered sets. They can be viewed as a way to reduce a problem to a problem on smaller, easier to handle structures. They also can be viewed (in reverse) as a tool to build larger examples of ordered sets with certain properties. (Similar statements certainly are also true for lexicographic sum decompositions, products and sets P^Q, which we will investigate in later chapters.) Sometimes investigations via retractions that start related to the fixed point property can have surprising consequences (cf. Theorem 4.5.1). For an excellent survey on retractions cf. [217].

4.1 Definition and Examples

Definition 4.1.1 *Let P be an ordered set. Then an order-preserving map $r : P \to P$ is called a **retraction** iff $r^2 = r$ (that is, iff r is **idempotent**). We will say that $R \subseteq P$ is a **retract** of P iff there is a retraction $r : P \to P$ with $r[P] = R$.*

Another way to describe retractions, which is less algebraic than idempotency, is to say that after the first application of r every point stays stationary. While the two conditions are easily seen to be equivalent, the latter provides a more pictorial way of thinking about retractions.

Proposition 4.1.2 *Let P be an ordered set and let $f : P \to P$ be an order-preserving map. f is a retraction iff $f|_{f[P]} = id_{f[P]}$.*

Proof. If $f^2 = f$, then for all $q = f(p) \in f[P]$ we have $f(q) = f(f(p)) = f(p) = q$. Conversely, if $f|_{f[P]} = id_{f[P]}$, then $f^2(p) = f|_{f[P]}(f(p)) = f(p)$ for all $p \in P$. ∎

Example 4.1.3 Many of the examples that follow could have been formulated as theorems. Parts 3, 4 and 5 will be important tools later on when retractions are used to prove other results. Part 8 is the starting point for proving the Li–Milner Structure Theorem.

1. For every ordered set P the identity id_P is a retraction, the trivial retraction on P.

2. The set $\{a, c, e, g, k\}$ in the ordered set e) in Figure 1.4 is a retract of the ordered set.

3. (Cf. [216, 241].) Let P be an ordered set and let $a, b \in P$. If

$$(\uparrow a) \setminus \{a\} \subseteq (\uparrow b) \text{ and } (\downarrow a) \setminus \{a\} \subseteq (\downarrow b),$$

    then

$$r(x) := \begin{cases} x; & \text{if } x \neq a, \\ b; & \text{if } x = a \end{cases}$$

    is a retraction. In this situation a is called **retractable** to b and r is a **retraction that removes a retractable point**.

4. (Cf. [216].) Let P be an ordered set and let $x \in P$ be such that $(\uparrow x) \setminus \{x\}$ has a smallest element u. Then x is retractable to u. In this situation (or its dual) we will say that x is **irreducible**[1] in P. The retraction r is referred to as a **retraction that removes an irreducible point**.

5. Let P be an ordered set and let $W \subseteq P$ be a well-ordered subchain such that there is no $p \in P \setminus W$ with $p > W$. Then W is a retract of P.

6. (Duffus, Rival and Simonovits, cf. [68]) Let P be an ordered set and let $C \subseteq P$ be a maximal (with respect to set inclusion) chain in P. Then C is a retract of P.

7. Let P be an ordered set and let $p \in P$. Then $\downarrow p$ is a retract of P.

8. For an ordered set P the point $p \in P$ is said to be **funneled through** $q \in P$ iff all maximal chains through p also go through q. The subset Q of P is called a **good subset** iff for all $p \in P$, there is exactly one $q \in Q$ such that p is funneled through q. Every good subset of P is a retract of P.

---

[1] This choice of language comes from lattice theory and will be motivated in Proposition 5.4.2.

Proof. Since the identity map trivially is idempotent, there is nothing to prove in part 1.

For part 2 note that the map that maps $h \mapsto k, d, f \mapsto e, b \mapsto a$ and leaves all other points fixed is a retraction. Also note that this is not the only retraction onto this set.

In part 3 it is clear that the proposed map is idempotent. To show r is order-preserving, let $x < y$. If $a \notin \{x, y\}$ there is nothing to prove, so let us assume $x = a$ (the case $y = a$ is handled dually). Then $y \in (\uparrow a) \setminus \{a\} \subseteq (\uparrow b)$ and hence $r(a) = b < y = r(y)$. Thus r is order-preserving.

Part 4 follows directly from the definition of retractable points.

To prove 5, let

$$r(x) := \bigvee_W \{w \in W : w \le x\}.$$

The map r is well-defined, since every initial segment of W that is not equal to W has a supremum and since if W has an upper bound, it must be contained in W. (Also recall that the supremum of the empty set is the smallest element of W, so suprema of empty sets are accounted for.) Idempotency is again trivial. Finally if $x \le y$, then

$$\{w \in W : w \le x\} \subseteq \{w \in W : w \le y\}.$$

Therefore we have

$$r(x) = \bigvee_W \{w \in W : w \le x\} \le \bigvee_W \{w \in W : w \le y\} = r(y).$$

The first step in proving part 6 is to put another order on C. Let $\sqsubseteq$ be any well-ordering of C. We define (cf. [217], p. 104)

$$r(x) := \min_{\sqsubseteq} \{c \in C : c = x \text{ or } c \not> x\}$$

Note that the minimum in the definition is the minimum with respect to the well-ordering, while the incomparability is with respect to the order C inherits from P. Maximality of C with respect to inclusion guarantees that r is well-defined. It is clear that r is idempotent. To see that r is order-preserving, let $x < y$. We need to prove that $r(x) \le r(y)$. If $x \in C$, this is trivial, as $r(y)$ is an element of the set $\{c \in C : c \not> y\} \cup \{y\}$, which consists entirely of upper bounds of x. The case $y \in C$ is treated similarly. This leaves the case in which $x, y \notin C$. If $r(x)$ is a $\le$-lower bound of $\{c \in C : c \not> y\}$ or $r(y)$ is a $\le$-upper bound of $\{c \in C : c \not> x\}$ we are done. Now $r(x)$ cannot be a $\le$-upper bound of $\{c \in C : c \not> y\}$, since this would imply that $x < y \le r(x)$, which is not possible for $x \notin C$. Similarly $r(y)$ is not a $\le$-lower bound of $\{c \in C : c \not> x\}$. This leaves the case in which $r(x), r(y) \in \{c \in C : c \not> x, y\}$. In this case $r(x)$ and $r(y)$ are the $\sqsubseteq$-minimal element of $\{c \in C : c \not> x, y\}$ and hence they are equal. Thus r is order-preserving.

The retraction needed in part 7 is

$$r(x) := \begin{cases} x; & \text{if } x \le p, \\ p; & \text{otherwise.} \end{cases}$$

The retraction for part 8 follows easily from the definition of the good subset Q. (The following adapted from [162], Lemma 2.5.) For each $p \in P$, we let $r(p)$ be the unique $q \in Q$ such that p is funneled through q. Clearly r is idempotent, since trivially for each $q \in Q$ the point q itself is the unique element of Q through which q is funneled.

To see that r is order-preserving, first note that if $x < y \le r(x)$, then $r(x) = r(y)$. Indeed, let C be a maximal chain that contains y. Then the chain $C' := [(\uparrow y) \cap C] \cup \{x\}$ is contained in a maximal chain K, which must contain $r(x)$. Thus $[(\uparrow y) \cap C] \cup \{r(x)\}$ is a chain and since $r(x) \ge y$, we have that $r(x) \cup C$ is a chain. By maximality of C we infer that $r(x) \in C$. Thus, since C was an arbitrary maximal chain containing y, y is funneled through $r(x) \in Q$. By definition of good subsets, this means $r(x) = r(y)$.

Now let $a \le b$ in P. If $r(a) = r(b)$ there is nothing to prove, so we can assume that $r(a) \ne r(b)$. Since $\{a, b\}$ is a chain that is contained in a maximal chain, $\{a, b, r(a), r(b)\}$ must be a chain. By the above we cannot have $a < b \le r(a)$. Thus $r(a) < b$. Similarly, since $r(a) \ne r(b)$ by the dual of the previous paragraph applied to $\{r(a), b\}$ we cannot have

$$r(b) \le r(a) = r(r(a)) < b.$$

Thus the only remaining possibilities are $r(a) < r(b) \le b$ or $r(a) < b < r(b)$. In either case we have proved that r is order-preserving.   ∎

Most of our examples above focused on the image as the central object of the retraction. While there are good reasons for that (retracts of ordered sets are interesting subsets), one should not forget that retractions are functions, which are in fact abundant. On a finite ordered set any order-preserving map can be turned into a retraction via the following proposition.

Proposition 4.1.4 *Let P be a finite ordered set and let $f : P \to P$ be an order-preserving map. Then $f^{|P|!}$ is a retraction on P.*

Proof. To abbreviate notation, let $n := |P|$. Let $p \in P$. Since P has only n elements, for the set $\{p, f(p), \ldots, f^n(p)\}$ there must be $0 \le k_p < l_p \le n$ such that $f^{k_p}(p) = f^{l_p}(p)$. However then for all $j \ge l_p$, we have

$$f^j(p) = f^{k_p + [(j - l_p) \bmod (l_p - k_p)]}(p) \in \{f^{k_p}(p), \ldots, f^{l_p - 1}(p)\}.$$

In particular, this means that every point $f^j(p)$ with $j \ge k_p$ is $(l_p - k_p)$-periodic. (That is, $f^{l_p - k_p}(f^j(p)) = f^j(p)$.)

Now let $q \in f^{n!}[P]$. We need to show that $f^{n!}(q) = q$. Let $p \in P$ be such that $f^{n!}(p) = q$. Then by the above, there is an $i \le n$ such that $f^i(p) = q$ and such that $q = f^i(p)$ is periodic with a period $t = l_p - k_p \leq n$. Thus

$$f^{n!}(q) = \left(f^t\right)^{\frac{n!}{t}}(q) = q$$

and we are done. ∎

Having some examples at hand, let us now investigate some properties of retractions. Retractions behave "friendly" with respect to compositions and with respect to suprema and infima as evidenced by the next two propositions.

Proposition 4.1.5 *Let P be an ordered set and let $r : P \to P$ and $s : r[P] \to r[P]$ be retractions. Then $s \circ r : P \to P$ is a retraction onto $s \circ r[P]$.*

Proof. Clearly $s \circ r$ is order-preserving. Now let $x \in s \circ r[P]$. Then $x \in s[r[P]] \subseteq r[P]$ and thus $s(r(x)) = s(x) = x$. ∎

Proposition 4.1.6 *Let P be an ordered set and let $r : P \to P$ be a retraction. If $A \subseteq r[P]$ has a supremum in P, then A has a supremum in $r[P]$ and*

$$r\left[\bigvee_P A\right] = \bigvee_{r[P]} A.$$

Proof. Suppose $A \subseteq r[P]$ has a supremum a in P. Then $r(a)$ is an upper bound of $A = r[A]$. Moreover if $x \in r[P]$ is an upper bound of A, then $x \geq a$ and hence $x = r(x) \geq r(a)$. ∎

Note however that retractions do not preserve suprema of arbitrary subsets of P (cf. Exercise 10).

4.2   Fixed Point Theorems

As mentioned earlier, retractions are valuable tools when working with the fixed point property. Here we present two approaches. Section 4.2.1 is geared towards removing points to reduce the task of determining whether an ordered set has the fixed point property to the (hopefully easier) task of answering the same question for a smaller set (or sets). The natural idea of iteration of this process will be discussed in Section 4.3. Section 4.2.2 presents an extension of the Abian–Brown theorem (cf. Theorem 3.4.7) which shows that finding a fixed point is closely related to finding a point comparable to its image in any ordered set (not just in chain-complete ones).

We start with an elementary result that shows the close relation of the Open Questions 1.4.3 and 1.4.8. In fact, one might want to also consider a modification of Open Question 1.4.8 which is to just characterize the minimal automorphic sets, defined as follows. (More about minimal automorphic sets in Section 4.4.3.)

Definition 4.2.1 *An ordered set P is called **minimal automorphic** iff P is automorphic and the only retract of P that does not have the fixed point property is P itself.*

Theorem 4.2.2 shows that retractions preserve the fixed point property and that minimal automorphic sets are omnipresent in the class of finite fixed point free ordered sets.

Theorem 4.2.2 *Let P be an ordered set.*

1. *If P has the fixed point property, then every retract of P has the fixed point property.*

2. *If P is finite, then P does not have the fixed point property iff P has a minimal automorphic retract.*

Proof. To prove 1 let $r : P \to P$ be a retraction and suppose P has the fixed point property. Let $f : r[P] \to r[P]$ be an order-preserving map. Then $f \circ r : P \to P$ is an order-preserving map and thus has a fixed point p. Since p must be in $r[P]$ we have $r(p) = p$ and thus $f(p) = f(r(p)) = p$.

For part 2 first note that "$\Leftarrow$" is clear. To prove "$\Rightarrow$" let $n := |P|$. Then by Proposition 4.1.4 for all order-preserving $f : P \to P$ we have that $f^{n!}$ is a retraction.

Now note that for all order-preserving $f : P \to P$ the map $f|_{f^{n!}[P]}$ is an automorphism. Indeed, if $p \in f^{n!}[P]$, then

$$f(p) = f(f^{n!}(p)) = f^{n!}(f(p)) \in f^{n!}[P],$$

so $f|_{f^{n!}[P]}$ is an order-preserving self map of $f^{n!}[P]$. The function $(f|_{f^{n!}[P]})^{n!-1}$ is an order-preserving inverse of $f|_{f^{n!}[P]}$.

The proof of 2 is now an induction on n. For $n = 1$ there is nothing to prove. For the induction step $\{1, \ldots, n-1\} \to n$, assume the result holds for all ordered sets with up to $n - 1$ elements. First note that if $|P| = n$ and all fixed point free maps of P are automorphisms, then there is nothing to prove (use $Q = P$). Otherwise let $f : P \to P$ be an order-preserving map that is not an automorphism and apply the induction hypothesis to $f^{n!}[P]$, which has fewer elements than P. Let Q be the minimal automorphic retract obtained via the induction hypothesis and let $r : f^{n!}[P] \to Q$ be a retraction. Then Q is a retract of P via $r \circ f^{n!}$ and we are done. ∎

4.2.1 Removing Points

The initial work in this direction is probably found in the paper [216] by Rival, which has since been extended in various directions. The idea is that when fixed point theory is concerned, any map that maps points to their upper or lower bounds should not affect the fixed point property, since we have the Abian–Brown theorem. (Hence the strong interest in comparative retractions as defined below.) We will present the best folklore version of Rival's original result in Theorem 4.2.5 and a modification in Theorem 4.2.6. Notice that chain-completeness is not needed in Theorem 4.2.6 and Scholium 4.2.7.

Definition 4.2.3 *Let P be an ordered set. A retraction $r : P \to P$ is called*

1. *A **comparative retraction** iff for all $p \in P$ we have that $r(p) \sim p$.*

2. *An **up-retraction** or **closure operator** iff for all $p \in P$ we have that $r(p) \geq p$.*

3. *A **down-retraction** or **interior operator** iff for all $p \in P$ we have that $r(p) \leq p$.*

Example 4.2.4 Retractions that remove one irreducible point are always up- or down-retractions. Retractions onto good subsets are always comparative. Retractions that remove a retractable point are comparative iff the retractable point is in fact irreducible (cf. Exercise 22).

Theorem 4.2.5 (Cf. [236], Theorem 2.) *Let P be a chain-complete ordered set and let $r : P \to P$ be a comparative retraction. Then P has the fixed point property iff $r[P]$ has the fixed point property.*

Proof. The direction "$\Rightarrow$" follows from part 1 of Theorem 4.2.2. To prove "$\Leftarrow$" let $r : P \to P$ be a comparative retraction and let $f : P \to P$ be order-preserving. Then $r \circ f|_{r[P]}$ has a fixed point p and since r is a comparative retraction we have

$$p = r \circ f|_{r[P]}(p) = r \circ f(p) \sim f(p).$$

Now the Abian–Brown Theorem implies that f has a fixed point. ∎

When we restrict ourselves to only moving one point in the retraction we obtain a similar result that holds for all ordered sets.

Theorem 4.2.6 (Cf. [241], Theorem 3.3.) *Let P be an ordered set and let $a \in P$ be retractable to $b \in P$. Then P has the fixed point property iff*

1. *$P \setminus \{a\}$ has the fixed point property, **and***

2. *One of $(\uparrow a) \setminus \{a\}$ and $(\downarrow a) \setminus \{a\}$ has the fixed point property.*

Proof. "$\Rightarrow$": Assume that P has the fixed point property. Condition 1 is a consequence of Example 4.1.3, part 3 and Theorem 4.2.2. To see 2 assume that $(\uparrow a) \setminus \{a\}$ and $(\downarrow a) \setminus \{a\}$ do not have the fixed point property. Let $g : (\uparrow a) \setminus \{a\} \to (\uparrow a) \setminus \{a\}$ and $h : (\downarrow a) \setminus \{a\} \to (\downarrow a) \setminus \{a\}$ be order-preserving maps without fixed points. Define

$$f(q) := \begin{cases} g(q); & \text{if } q \in (\uparrow a) \setminus \{a\}, \\ h(q); & \text{if } q \in (\downarrow a) \setminus \{a\}, \\ a; & \text{if } q \text{ is not comparable to } a, \\ b; & \text{if } q = a. \end{cases}$$

Clearly $f : P \to P$ has no fixed point. It is left to the reader as a good exercise to verify that f is order-preserving. We have arrived at a contradiction.

"$\Leftarrow$": Let $f : P \to P$ be order-preserving. We will show that f has a fixed point. Let $r : P \to P \setminus \{a\}$ be as in Example 4.1.3, part 3. Then

$$r \circ (f|_{P \setminus \{a\}}) : P \setminus \{a\} \to P \setminus \{a\}$$

is order-preserving and thus has a fixed point q. In case $q \neq b$ we have that

$$q = r \circ (f|_{P \setminus \{a\}})(q) = f|_{P \setminus \{a\}}(q) = f(q)$$

and we are done. In case $q = b$ we either have $f(b) = b$ or $f(b) = a$. In the first case we are done. In the second case assume without loss of generality that $(\uparrow a) \setminus \{a\}$ has the fixed point property. Then $(\uparrow a) \setminus \{a\} \neq \emptyset$. Since $f(b) = a$ we have that

$$f[(\uparrow a) \setminus \{a\}] \subseteq f[\uparrow b] \subseteq \uparrow a.$$

In case $f[(\uparrow a) \setminus \{a\}] \subseteq (\uparrow a) \setminus \{a\}$ we have by assumption that f has a fixed point in $(\uparrow a) \setminus \{a\}$ and are done. Otherwise there is a $q \in (\uparrow a) \setminus \{a\}$ such that $f(q) = a < q$. Now if $f(a) = a$ we are done. This leaves the case $d := f(a) < a$. In this case f maps $\downarrow d$ to itself. Yet $\downarrow d$ is a retract of $P \setminus \{a\}$ by part 7 of Example 4.1.3. Thus $\downarrow d$ has the fixed point property, which implies that f has a fixed point in $\downarrow d$. ∎

The difference between Theorem 4.2.6 and Scholium 4.2.7 is small but notable. When our retractable point is in fact irreducible, then we do not need condition 2.

Scholium 4.2.7 (Cf. [216], Proposition 1.) *Let P be an ordered set and let $a \in P$ be irreducible. Then P has the fixed point property iff $P \setminus \{a\}$ has the fixed point property.*

Proof. As the direction "$\Rightarrow$" is trivial, we only need to consider "$\Leftarrow$". Suppose without loss of generality that b is the unique lower cover of a and let $f : P \to P$ be order-preserving. As seen in the proof of Theorem 4.2.6 the only interesting case is the case $f(b) = a > b$. If $f(a) = a$, we are done. Otherwise $f(a) > f(b) = a$ and f maps $\uparrow f(a)$ to itself. However $\uparrow f(a)$ is a retract of $P \setminus \{a\}$ and has thus the fixed point property. Hence f has a fixed point in $\uparrow f(a)$. ∎

4.2.2 The Comparable Point Property

The Abian–Brown theorem showed that in chain-complete ordered sets it is enough to find a point that is comparable to its image to establish existence of a fixed point. On the other hand we have seen above that some fixed point arguments can be written in a way that avoids chain-completeness. This line of thought can be extended to show that understanding the fixed point property is essentially equivalent to understanding the (equally difficult) comparable point property.

Definition 4.2.8 *An ordered set P is said to have the **comparable point property** iff for each order-preserving map $f : P \to P$ there is a $p \in P$ with $f(p) \sim p$.*

Definition 4.2.9 *Let P be an ordered set and let $C_l, C_u \subseteq P$ be chains in P such that $C_l \leq C_u$. Then the set*

$$K := \{p \in P : C_l \leq p \leq C_u\}$$

*will be called the (C_l, C_u)-**core**. (Note that C_l, C_u or both could be empty).*

A word of caution to the reader. The word "core" is also used for certain entities defined via retractions (cf. Definition 4.3.5), which are in fact the more common entities in fixed point theory. We will use (C_l, C_u)-cores only in this section and in the proof of Theorems 5.2.1 and 5.2.6.

Lemma 4.2.10 *Let P be an ordered set and let $C_l \leq C_u$ be chains in P. If C_l is well-ordered and C_u is dually well-ordered, then*

$$C_l \cup \{p \in P : C_l \leq p \leq C_u\} \cup C_u$$

is a retract of P.

Proof. We define

$$r(x) := \begin{cases} x; & \text{if } C_l \leq x \leq C_u, \\ \max\{c \in C_l : c \leq x\}; & \text{if } x \not\geq C_l, \\ \min\{c \in C_u : c \geq x\}; & \text{if } x \geq C_l \text{ and } x \not\leq C_u. \end{cases}$$

The proof that r is a retraction is left to the reader as Exercise 11. ∎

Theorem 4.2.11 (Cf. [234], Theorem 1.) *The following are equivalent for an ordered set P.*

1. *P has the fixed point property.*

2. *Every (C, K)-core of P has the fixed point property.*

3. *Every (C, K)-core of P has the comparable point property.*

Proof. To prove that 1 implies 2 let $H \subseteq P$ be a (C, K)-core of P. By Proposition 2.2.9 and its dual we can assume without loss of generality that C is well-ordered and that K is dually well-ordered. By definition of (C, K)-cores we know that the supremum of C (if it exists) is in H and the same goes for the infimum of K. Let $f : H \to H$ be order-preserving. For $c \in C \setminus H$ let c^+ denote the immediate successor of c in the well-ordering of C and for $k \in K \setminus H$, let k^- denote the immediate predecessor of k in the dual well-ordering of K. With r denoting the retraction from Lemma 4.2.10 we define

$$F(x) := \begin{cases} f(x); & \text{if } x \in H, \\ r(x)^+; & \text{if } x \notin H \text{ and } r(x) \in C, \\ r(x)^-; & \text{if } x \notin H \text{ and } r(x) \in K. \end{cases}$$

Then F is order-preserving (Exercise 12). Thus F has a fixed point, which by definition of F must be in H and must hence be a fixed point of f.

The implication "2⇒3" naturally is trivial.

To prove "3⇒1" assume that every (C, K)-core of P has the comparable point property and that $f : P \to P$ is order-preserving. The set $\mathcal{H}$ of all (C, K)-cores H such that f maps H to itself is nonempty (P is the $(\emptyset, \emptyset)$-core) and inductively ordered by reverse inclusion $\supseteq$ (the supremum of a chain of (C, K)-cores being its intersection, cf. Exercise 12). Thus by Zorn's lemma there is a maximal (C, K)-core H that is fixed by f. Note that since (C, K)-cores have the comparable point property, H is not empty.

Assume that H is not a singleton. Since H has the comparable point property, there is a $p \in H$ with (without loss of generality) $f(p) \geq p$. If $f(p) = p$ we are done. If $f(p) > p$, then f maps $(\uparrow f(p)) \cap H$ (which is a (C, K)-core) to itself and $(\uparrow f(p)) \cap H$ is a proper subset of H. This is a contradiction to the maximality of H. Thus if H is a not singleton, f must have a fixed point.

If H is a singleton, then the element of H is a fixed point of f.   ∎

While the rest of this chapter primarily focuses on finite ordered sets, we will be using Theorem 4.2.11 to prove Theorem 5.2.1, which is one of the early big theorems of order theory.

4.3   Dismantlability

Theorems 4.2.5, 4.2.6 and Scholium 4.2.7 suggest a reduction procedure when determining the fixed point property for a set. Find a suitable retraction on the set and then decide if the retract has the fixed point property instead. Iterate this procedure until no suitable retraction can be found. In finite sets this procedure is well understood (cf. e.g., [12, 65, 79, 216]). In infinite sets one sometimes can iterate the above idea infinitely often, thus being faced with the problem how to get past the limit ordinal. This problem has been addressed successfully by Li and Milner (cf. [159, 160, 161, 162, 163, 165, 167]) for a large class of ordered sets.

The language surrounding dismantlability is a little confusing, as essentially the same idea (which is invariably referred to as "dismantlability") can be applied

- To finite and to infinite ordered sets,

- Using different types of retractions,

- With finitely as well as infinitely many iterations.

In this text we will apply dismantlability to finite as well as infinite ordered sets, but we will always assume that the number of iterations is finite. (For ways to work with infinitely many iterations cf. [159], or for a simpler approach, do Exercise 24 here.) The class of retractions used in the dismantling procedure will be indicated in the name.

Example 4.3.1 Some useful classes of retractions are:

1. $\mathcal{U}$: the class of all up-retractions,

2. $\mathcal{D}$: the class of all down-retractions,

3. $\mathcal{C}$: the class of all comparative retractions,

4. $\mathcal{G}$: the class of all retractions onto good subsets (cf. [162, 163]),

5. $\mathcal{I}$: the class of retractions that remove an irreducible point,

6. $\mathcal{R}_k$: the class of all retractions $r : P \to P$ such that $|P \setminus r[P]| \leq k$,

7. $\mathcal{F} := \bigcup_{k \in \mathbb{N}} \mathcal{R}_k$.

Definition 4.3.2 *Let P be an ordered set, let $\mathcal{R}$ be a class of retractions and let $Q \subseteq P$. Then P is called $\mathcal{R}$-**dismantlable to** Q iff there is a sequence*

$$P = P_0 \supseteq P_1 \supseteq \cdots \supseteq P_n$$

of subsets of P and a sequence

$$r_i : P_{i-1} \to P_{i-1}$$

of retractions in $\mathcal{R}$ such that $r_i[P_{i-1}] = P_i$.

Proposition 4.3.3 *Let P be an ordered set that is $\mathcal{R}$-dismantlable to Q. Then Q is a retract of P.*

Proof. The composition $r := r_n \circ r_{n-1} \circ \cdots \circ r_1$ is a retraction of P onto Q. ∎

Some notions of dismantlability are actually equivalent for finite ordered sets. This is one of the reasons why the language is somewhat mixed in the literature. In Exercise 25 we will explore the differences between these notions when considering infinite ordered sets.

Proposition 4.3.4 *Let P be a finite ordered set. Then the following are equivalent.*

1. *P is $\mathcal{C}$-dismantlable to Q.*

2. *P is $(\mathcal{U} \cup \mathcal{D})$-dismantlable to Q.*

3. *P is $\mathcal{I}$-dismantlable to Q.*

4. *P is $\mathcal{G}$-dismantlable to Q.*

Proof. Since $\mathcal{C} \supseteq (\mathcal{U} \cup \mathcal{D}) \supseteq \mathcal{I}$ the implications "3⇒2⇒1" are trivial. Similarly, since $\mathcal{C} \supseteq \mathcal{G} \supseteq \mathcal{I}$ the implications "3⇒4⇒1" are trivial.

To finish the proof we need to prove that part 1 implies part 3 for finite ordered sets. Let P be $\mathcal{C}$-dismantlable to Q. We proceed by induction on $|P|$. The case $|P| = 1$ is of course trivial. Now assume that the result has been proved for all sets $\tilde{P}$ with $|\tilde{P}| < m$, let $|P| = m$ and let $c_i : P_{i-1} \to P_{i-1}$ be comparative retractions with $c_i[P_{i-1}] = P_i$, $P_0 = P$ and $P_n = Q$.

Let $U \subseteq P$ be the set of all points $p \in P$ such that $c_1(p) > p$. Since U is finite, we can let u be a maximal element of U. Let $b > u$ be a strict upper bound of u. Then $b \geq c_1(b) \geq c_1(u) > u$, so $c_1(u)$ is the unique upper cover of u and hence u is irreducible. The map $c_1|_{P \setminus \{u\}}$ is a comparative retraction on $P \setminus \{u\}$ and $c_1|_{P \setminus \{u\}}[P \setminus \{u\}] = P_1$. Thus by induction hypothesis $P \setminus \{u\}$ is $\mathcal{I}$-dismantlable to Q. Since $P \setminus \{u\}$ is obtained from P by removal of an irreducible point, this finishes our proof.   ∎

Of course the dismantling procedures would be quite uninteresting if all we could prove about them were their mutual equivalence (or lack thereof). The idea is to use dismantlings to shrink the sets we are considering. This reduces the difficulty of problems such as deciding if an ordered set has the fixed point property. Since we are working in finite sets it is clear that the dismantling procedure has to stop at a certain point. The sets at which the dismantlings must stop are called cores.

Definition 4.3.5 *Let P be an ordered set and let $\mathcal{R}$ be a class of retractions. Then P is called an $\mathcal{R}$-**core** iff the only retraction $r : P \to P$ that is in $\mathcal{R}$ is id_P.*

Easy examples show that a set can have many cores (every singleton subset of a finite fence is a $\mathcal{C}$-core of that fence), but we can show, for example, that the $\mathcal{C}$-core of a finite set is unique up to isomorphism. For fixed point investigations this is exactly the result one needs.

Definition 4.3.6 *Let P be a finite ordered set and let $\mathcal{R}$ be a class of retractions. We say **the $\mathcal{R}$-core of P is unique** iff any two $\mathcal{R}$-cores C_1, C_2 to which P is $\mathcal{R}$-dismantlable are isomorphic.*

Lemma 4.3.7 *Let P be a chain-complete ordered set and let the order preserving map $f \neq id_P$ be such that $f(p) \sim p$ for all $p \in P$. Then there is a retraction $r \neq id_P$ such that $r(p) \sim p$ for all $p \in P$.*

Proof. First note that $f \vee id_P$ and $f \wedge id_P$ (suprema are taken pointwise) are both well-defined and order-preserving and that one of them (say $f \vee id_P$) is not equal to id_P. For all $p \in P$ we have $f(p) \vee p \geq p$ and so by the Abian–Brown theorem $f \vee id_P$ has a smallest fixed point above p. Let $r(p)$ be the smallest fixed point of $f \vee id_P$ above p. Then $r : P \to P$ clearly is idempotent, not equal to id_P and $r(p) \geq p$ for all $p \in P$. What remains to be proved is that r is order-preserving. Let $p < q$ be given. Then $r(p)$ is the smallest fixed point of $f \vee id_P$

above p and $r(q)$ is a fixed point of $f \vee \mathrm{id}_P$ above $q > p$. Thus $r(p) \le r(q)$ and we are done. ∎

We are now ready to prove uniqueness of some types of cores in finite sets. The main idea for the proof is that for some classes $\mathcal{R}$ of retractions, $\mathcal{R}$-cores have a certain resilience against $\mathcal{R}$-retractions. For these classes, $\mathcal{R}$-retractions will map any $\mathcal{R}$-core to which P is dismantlable to an isomorphic image of itself.

Theorem 4.3.8 (Cf. [65, 79].) *Let P be a finite ordered set. Then the $\mathcal{C}$-, $\mathcal{U}$-, $\mathcal{D}$-, and $\mathcal{R}_k$-cores of P are unique.*

Proof. To prove uniqueness of $\mathcal{C}$-cores, let $K \subseteq P$ be a $\mathcal{C}$-core such that P is $\mathcal{C}$-dismantlable to K. Let $S : P \to K$ be a composition of comparative retractions as in Definition 4.3.2 and the proof of Proposition 4.3.3. Let $P = P_0 \supset P_1 \supset \cdots \supset P_n$ and for $i = 1, \ldots, n$, let $c_i : P_{i-1} \to P_{i-1}$ be any sequence of comparative retractions with $c_i[P_{i-1}] = P_i$. For $j = 1, \ldots, n$, let $C_j := c_j \circ \cdots \circ c_1$ and let $C_0 = \mathrm{id}_P$.

We will prove inductively that for all j we have $SC_j|_K = id_K$. For $j = 0$ this is trivial, so let $j > 0$. First note that $SC_{j-1}|_K = id_K$. For all $p \in K$ we have that $c_j(C_{j-1}(p)) \sim C_{j-1}(p)$ and thus $S(c_j(C_{j-1}(p))) \sim S(C_{j-1}(p)) = p$. Thus since K is a $\mathcal{C}$-core by Lemma 4.3.7 we infer that $id_K = Sc_jC_{j-1}|_K = SC_j|_K$.

Thus we have that $SC_n|_K = id_K$. Now suppose K_1, K_2 are $\mathcal{C}$-cores of P with $S_i : P \to K_i$ being compositions of the retractions in the definition of dismantlability. Then by the above $S_1|_{K_2}S_2|_{K_1} = id_{K_1}$ and $S_2|_{K_1}S_1|_{K_2} = id_{K_2}$, and hence K_1 and K_2 are isomorphic.

To prove uniqueness of $\mathcal{U}$-cores, denote by F_P the set of all $p \in P$ that are funneled through a $u_p > p$. Note that a finite ordered set is a $\mathcal{U}$-core iff it does not have any points p that are funneled through a $u_p > p$. Moreover if $r : P \to P$ is a $\mathcal{U}$-retraction and p is funneled through u_p, then $r(p)$ is funneled through $r(u_p)$ in $r[P]$ (proved similar to the proof of Example 4.1.3, part 8).

Now let P be a finite ordered set. The above implies that no $\mathcal{U}$-core can contain any points $p \in F_P$. Thus all $\mathcal{U}$-cores are contained in $P \setminus F_P$. However $P \setminus F_P$ is itself a $\mathcal{U}$-core (cf. Exercise 15). Thus the $\mathcal{U}$-core of finite ordered sets is unique itself and not just up to isomorphism.

Proving uniqueness of $\mathcal{D}$-cores is the dual of the above.

Proving uniqueness of $\mathcal{R}_k$-cores is left as Exercise 17. ∎

We can thus talk about *the* $\mathcal{R}$-core of an ordered set.

Definition 4.3.9 *Whenever the $\mathcal{R}$-core of an ordered set P is unique we denote the set of all isomorphic $\mathcal{R}$-cores of P (or sometimes by abuse of notation one of its representatives) as $\mathcal{R}$-core(P).*

Theorem 4.3.10 *A finite ordered set P has the fixed point property iff $\mathcal{C}$-core(P) has the fixed point property.* ∎

4.3.1   (Connectedly) Collapsible Ordered Sets

When using retractable points to prove fixed point results, by Theorem 4.2.6 at every stage we not only have to contend with the set $P \setminus \{a\}$, but also with the sets $(\uparrow a) \setminus \{a\}$ and $(\downarrow a) \setminus \{a\}$. Thus $\mathcal{R}_1$-dismantlability, while an interesting structural property, is a priori not sufficient to make a determination whether an ordered set has the fixed point property or not. To make this decision we would also need to know if the sets $[(\uparrow a) \cup (\downarrow a)] \setminus \{a\}$ we encounter have the fixed point property. One way to ease this decision is to consider a recursively defined class of ordered sets for which any such decision can be made through successive $\mathcal{R}_1$-dismantlings.

Definition 4.3.11 *A finite ordered set P with n points is called* **collapsible** *iff $n \in \{0, 1\}$ or there is a point x such that*

1. *$P \setminus \{x\}$ is a retract of P,*

2. *$P \setminus \{x\}$ is collapsible,*

3. *$(\uparrow x \cup \downarrow x) \setminus \{x\}$ is collapsible.*

Definition 4.3.12 *A finite ordered set P with n points is called* **connectedly collapsible** *iff $n = 1$ or there is a point x such that*

1. *$P \setminus \{x\}$ is a retract of P,*

2. *$P \setminus \{x\}$ is connectedly collapsible,*

3. *$(\uparrow x \cup \downarrow x) \setminus \{x\}$ is connectedly collapsible.*

The above definitions of collapsibility and connected collapsibility appear almost equal. Yet there is a distinct difference between collapsibility and connected collapsibility as we are about to see.

Theorem 4.3.13 *A collapsible ordered set has the fixed point property iff it is connectedly collapsible.*

   Proof. The proof is an induction on $n = |P|$, with the base cases 0 and 1 being trivial. Now assume the result holds for all ordered sets of size less than n, $n > 1$ and let P be a collapsible ordered set of size $|P| = n$. Since P is collapsible, P has a retractable point a. By Theorem 4.2.6, P has the fixed point property iff $P \setminus \{a\}$ and one of $(\uparrow a) \setminus \{a\}$ and $(\downarrow a) \setminus \{a\}$ have the fixed point property. Moreover it is easy to see that this is the case iff $P \setminus \{a\}$ and $[(\uparrow a) \cup (\downarrow a)] \setminus \{a\}$ have the fixed point property. Since P is collapsible, $P \setminus \{a\}$ and $[(\uparrow a) \cup (\downarrow a)] \setminus \{a\}$ are collapsible. By induction hypothesis this means that both sets have the fixed point property iff they are connectedly collapsible. Thus P has the fixed point property iff P is connectedly collapsible. ∎

4.4   The Fixed Point Property for Ordered Sets of Width 2 or Height 1

Knowing what we know now about dismantlings we can conclude that the presence or absence of irreducible points does not affect the fixed point property. This realization allows us to characterize the fixed point property in the classes of ordered sets of width 2 or height 1.

4.4.1   Width 2

The characterization of the fixed point property for ordered sets of width 2 is quickly proven, as ordered sets of width 2 have only two possible kinds of $\mathcal{I}$-cores.

Definition 4.4.1 *A* **four-crown-tower** *is an ordered set* $\{a_0, b_0, a_1, b_1, \ldots, a_t, b_t\}$ *such that for all* $k \in 1, \ldots, t$ *the points* a_{k-1} *and* b_{k-1} *are the lower covers of* a_k *and* b_k.

Proposition 4.4.2 *Let* P *be a finite ordered set of width 2 with no irreducible points. Then* P *is either a singleton, an antichain or a four-crown-tower.*

Proof. If P is a singleton, there is nothing to prove, so assume that P is not a singleton. Let $t \in \mathbb{N}$ be the height of P. If $t = 0$, then P is the 2-antichain and we are done. Hence we can assume that $t > 0$. We shall prove by induction on k that for all $k \in \{0, \ldots, t\}$ P has exactly two points of rank k. Moreover we will prove that for $k > 0$ all points of rank k are upper bounds of all points of rank less than k.

For $k = 0$ the claims are trivial, since if P had only one minimal element, the elements of rank 1 would be irreducible. Assume the statement is proved for all natural numbers less than $k \in \{0, \ldots, t\}$. Let p be an element of rank k. Then p has at least one lower cover of rank $k - 1$. By induction hypothesis this means that p is not adjacent to any elements of rank less than $k - 1$. Therefore, since p cannot have a unique lower cover, p must have both elements of rank $k - 1$ as lower covers.

Now first assume that $k < t$. By the above, every element of rank k is above all elements of rank less than k. Since the elements of rank $k + 1$ must have at least two lower covers, this means that there must be two elements of rank k. In case $k = t$, note that the two elements of rank $t - 1$ cannot have a unique upper cover. Thus P must have two elements of rank t.

Therefore P must be a four-crown-tower.   ∎

Theorem 4.4.3 (Cf. Theorem 1 in [87] for the version for chain-complete ordered sets.) *Let* P *be a finite ordered set of width 2. Then* P *has the fixed point property iff* P *is* $\mathcal{I}$*-dismantlable to a singleton.*

Proof. Every ordered set of width 2 is $\mathcal{I}$-dismantlable to an $\mathcal{I}$-core. By Proposition 4.4.2 this $\mathcal{I}$-core is either a singleton, an antichain or a four-crown-tower and by Scholium 4.2.7, P has the fixed point property iff its $\mathcal{I}$-core does. Since four-crown-towers do not have the fixed point property (simply switch the a_i and b_i at every level for a fixed point free map), this proves the result.   ∎

4.4.2 Height 1

For ordered sets of height 1 the situation is not quite as easy. Still intuition dictates that crowns have to be part of the picture. Knowing this, one now needs the right types of crowns and the right idea what to do with them.

Lemma 4.4.4 *Let P be an ordered set of height 1 and let $C = \{c_0, \ldots, c_{n-1}\}$ be a crown of minimal size in P. Then C is a retract of P.*

Proof. (Compare with [217], p.118,119.) Assume without loss of generality that c_0 is minimal. Let P' be the ordered set obtained from P by erasing the cover (c_0, c_{n-1}). Define

$$r(x) := \begin{cases} c_{n-1}; & \text{if } d_{P'}(x, c_0) \geq n, \\ c_i; & \text{if } d_{P'}(x, c_0) = i < n. \end{cases}$$

Since C is a crown of smallest possible size in P, we have that $d_{P'}(c_i, c_0) = i$ for all i and the map r is idempotent. Now let $x < y$ (in P). Since r clearly preserves the comparability $c_0 < c_{n-1}$, we can assume $x \neq c_0$ or $y \neq c_{n-1}$. Then $d_{P'}(x, c_0)$ is even and $d_{P'}(y, c_0) \in \{d_{P'}(x, c_0) - 1, d_{P'}(x, c_0) + 1\}$. Thus if $d_{P'}(x, c_0) \geq n$, then $r(x) = r(y) = c_{n-1}$, while if $d_{P'}(x, c_0) < n$, then $r(x)$ is a minimal element of C and $r(y)$ is one of its maximal upper covers in C. Thus r is order-preserving and we are done.   ∎

Now we know that the presence of a crown destroys the fixed point property in sets of height 1. The next proposition considers the case of absence of crowns.

Proposition 4.4.5 *Let P be a finite connected ordered set of height 1 that does not contain any crowns. Then P must contain an irreducible point.*

Proof. We prove the counterpositive. Suppose P has no irreducible points and let $p_0 \in P$ be minimal. Let p_1 be an upper cover of p_0. Then p_1 is maximal and has at least two lower covers. Thus p_1 has a lower cover $p_2 \neq p_0$. Suppose mutually distinct $p_0, \ldots, p_n$ such that $p_0, \ldots, p_n$ is a fence have been constructed already and (without loss of generality) assume that p_n is minimal. Then p_n has at least two upper covers, so p_n has a maximal upper cover $p_{n+1} \neq p_{n-1}$.

Since P is finite, this construction must ultimately produce a p_{n+1} that is comparable to an earlier p_j with $j < n$. Let j_m be the largest $j < n$ such that p_{n+1} is comparable to p_j. Then $\{p_{j_m}, \ldots, p_{n+1}\}$ is a crown.   ∎

Theorem 4.4.6 (Cf. [191, 216].) *Let P be a connected finite ordered set of height 1. Then the following are equivalent.*

1. P has the fixed point property.

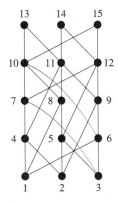

Figure 4.1: A minimal automorphic set of width 3.

2. *P contains no crowns.*

3. *P is C-dismantlable to a singleton.*

Proof. "1⇒2" follows directly from Lemma 4.4.4, "2⇒3" follows from Proposition 4.4.5 and "3⇒1" follows from Theorem 4.3.10. ∎

4.4.3   Minimal Automorphic Ordered Sets

For finite ordered sets of width 2 as well as for ordered sets of height 1, we have seen that $\mathcal{I}$-dismantlability to a singleton was equivalent to the fixed point property. Proposition 1.4.6 shows that there are ordered sets with the fixed point property that are not $\mathcal{I}$-dismantlable. Another way of looking at our results for width 2 and height 1 is to concentrate on minimal automorphic sets, as also is suggested by Theorem 4.2.2, part 2. Minimal automorphic sets for width 2 are the four-crown-towers (and the antichain with two elements) and for height 1 they are the crowns. Unfortunately for other classes of ordered sets it is hard to find the minimal automorphic sets. Graded minimal automorphic sets of width ≤ 4 have been characterized in [82] and we will see a large number of examples of minimal automorphic sets in Section 6.4. (We will consider graded ordered sets in Section 11.1.) The example that follows is one of the examples mentioned in [82]. It is not graded and seems to indicate that a characterization of all minimal automorphic sets even for width 3 would be quite difficult.

Proposition 4.4.7 *The ordered set M^{33333} in Figure 4.1 is minimal automorphic.*

Proof. Clearly the map Φ with cycles $(1, 2, 3)$, $(4, 5, 6)$, $(7, 8, 9)$, $(10, 11, 12)$, $(13, 14, 15)$ is an automorphism of M^{33333}.

To show that M^{33333} is minimal automorphic, let $f : M^{33333} \rightarrow M^{33333}$ be a fixed point free order-preserving map.

First note that $M^{33333} \setminus \{1\}$ is connectedly collapsible and hence has the fixed point property. Indeed, 4, 6 and 8 have unique lower covers in $M^{33333} \setminus \{1\}$. After removing them, 7 and 9 are retractable to 5 in $M^{33333} \setminus \{1, 4, 6, 8\}$ and their upper sets are 5-fences. The resulting set is such that 5 is comparable to all other elements. Thus, $M^{33333} \setminus \{1\}$ is connectedly collapsible and has the fixed point property. This means that we must have $1 \in f[M^{33333}]$. Repeated application of Φ shows that we must also have $2 \in f[M^{33333}]$ and $3 \in f[M^{33333}]$.

Now note that $(\uparrow 1) \cap (\uparrow 2) \cap (\uparrow 3) = \{7, \ldots, 15\}$. By the above

$$(\uparrow f(1)) \cap (\uparrow f(2)) \cap (\uparrow f(3)) = (\uparrow 1) \cap (\uparrow 2) \cap (\uparrow 3) = \{7, \ldots, 15\},$$

that is, $f[\{7, \ldots, 15\}] \subseteq \{7, \ldots, 15\}$. But $\{7, \ldots, 15\}$ is a stack of two 6-crowns, which is minimal automorphic (cf. [82] or Exercise 27). Thus $\{7, \ldots, 15\} \subseteq f[M^{33333}]$.

Finally none of $f(4)$, $f(5)$ and $f(6)$ is minimal or in $\{7, \ldots, 15\}$, since otherwise f would not be surjective on the minimal elements or on $\{7, \ldots, 15\}$. Thus $f(4)$ will be the unique lower cover of $f(7)$ that has rank 1, $f(5)$ will be the unique lower cover of $f(8)$ that has rank 1 and $f(6)$ will be the unique lower cover of $f(9)$ that has rank 1. This means that f must be surjective, hence injective and hence an automorphism. ∎

4.4.4   A Fixed Point Property for Graphs?

A property such as the fixed point property for ordered sets always invites the question as to whether this property can be investigated in a different setting. For us this setting would be graph theory. Graph homomorphisms, as the natural analogues of order-preserving maps, will take the place of order-preserving maps.

Definition 4.4.8 *Let $G = (V, E)$ and $H = (W, F)$ be graphs. A function $f : V \to W$ is called a* **graph homomorphism** *iff for all vertices x, y with $\{x, y\} \in E$ we have $f(x) = f(y)$ or $\{f(x), f(y)\} \in F$.[2] A homomorphism from a graph to itself is called a* **graph endomorphism**.

Unfortunately a fixed point property for graphs would be quite trivial, as no graph that has an edge would have the fixed point property. Indeed, if $\{a, b\}$ was an edge, then the map that maps all points not equal to a to a and a to b would be a fixed point free graph endomorphism. The following corollary resolves the question for when all maps fix a point or an edge. For further steps in this direction cf. Section 6.3.

Definition 4.4.9 *Let $G = (V, E)$ be a graph. Then G has a* **cycle** *iff there is a subset $\{c_1, \ldots, c_n\} \subseteq V$ such that c_k is adjacent to c_{k+1} for $k = 1, \ldots, n-1$, c_n is adjacent to c_1 and there are no further adjacencies.*

---

[2] Another definition of graph homomorphisms is that f is a homomorphism iff it preserves adjacencies. Readers who prefer to work with this definition only need to add loops at every vertex of the involved graphs to obtain the homomorphisms we want to consider.

Corollary 4.4.10 *(Cf. [191].) Let $G = (V, E)$ be a finite graph. Then every graph homomorphism of G has a fixed edge or a fixed vertex iff G is connected and has no cycles with ≥ 3 vertices.*

Proof. Let $G = (V, E)$ be a graph. We define an ordered set of height 1 with underlying set $P := \{\{v\} : v \in V\} \cup E$ and with the order being containment of sets. (This set is also sometimes referred to as the **split** of the graph.) Let $f : V \to V$ be a graph endomorphism of G. Then

$$F(x) := \begin{cases} \{f(v)\}; & \text{if } x = \{v\}, \\ \{f(v), f(w)\} & \text{if } x = \{v, w\} \end{cases}$$

is an order-preserving map of P. Moreover if F has a fixed point $\{v\}$ that is minimal in P, then f fixes v. If F fixes no singleton, but has a fixed point $\{v, w\}$ that is maximal in P, then f fixes $\{v, w\}$. Thus if P has the fixed point property, then every endomorphism of G fixes an edge or a vertex.

Let $h : P \to P$ be an order-preserving map of P that maps the minimal elements to minimal elements. If for $x \in V$ we let $H(x)$ be the unique element of the singleton set $h(x)$, then H is a graph endomorphism of G. Moreover if H has a fixed vertex v, then clearly h has a fixed point $\{v\}$, and if H has a fixed edge $\{v, w\}$, then $h[\{\{v\}, \{w\}\}] = \{\{v\}, \{w\}\}$ and hence $h(\{v, w\}) = \{v, w\}$. Thus if every endomorphism of G has a fixed vertex or a fixed edge, then every order-preserving map of P that maps minimal elements to minimal elements has a fixed point. Consequently by Proposition 3.1.1, in this case P has the fixed point property.

Hence P has the fixed point property iff every endomorphism of G has a fixed edge or a fixed vertex. However by Theorem 4.4.6, P has the fixed point property iff P is connected and has no crowns. This is the case exactly when G is connected and has no cycles with ≥ 3 elements. ∎

4.5   Li and Milner's Structure Theorem

Li and Milner's theorem (Theorem 4.5.1) is a surprising result. It says that the class of chain-complete ordered sets without infinite antichains is in fact very close to the class of finite ordered sets. Each of its members can be $\mathcal{G}$-dismantled to a finite set in finitely many steps. We thus have a family of infinite sets for which the fixed point property can be reduced to arguments on finite sets. The following proof of the Li–Milner theorem on chain-complete sets with no infinite antichains is due to J. D. Farley. It is a nice exhibition of many of the concepts we have introduced so far in this text. We start by stating the theorem, which we will be finished proving on p. 99.

Theorem 4.5.1 (Li–Milner Structure Theorem, cf. [163].) *Any chain-complete ordered set with no infinite antichain is $\mathcal{G}$-dismantlable to a finite core which is unique up to isomorphism.*

The proof of Theorem 4.5.1 has several stages, which we will formulate as lemmas and propositions before finally putting them together to finish the proof. What we need to prove is that we can reach finite sets in finitely many steps. To do this we embark on an argument by contradiction that assumes we cannot. Both published proofs of Theorem 4.5.1 (Li and Milner's original argument in [163] as well as Farley's proof in [81]) go this route. We start with some notation.

Definition 4.5.2 A **generalized perfect sequence** *of the ordered set P is a sequence $\{P_\alpha\}_{\alpha<\lambda}$ indexed by the ordinal numbers before λ such that*

1. *$P_0 = P$,*

2. *If $\alpha < \lambda$, then $P_{\alpha+1}$ is a good subset of P_α,*

3. *For all limit ordinals $\mu < \lambda$, we have $P_\mu = \bigcap\limits_{\alpha<\mu} P_\alpha$.*

A **perfect sequence** *is a strictly decreasing generalized perfect sequence such that the last set in the sequence has no nontrivial good subset.*

Note that perfect sequences are defined for arbitrary ordered sets. However in arbitrary ordered sets, they can be quite trivial and need not end with a finite set. For example a four-crown-tower of infinite height has no nontrivial good subsets, so its perfect sequence has only one element. Perfect sequences can also have infinite length as can be seen by considering the perfect sequence of the "spider" in Figure 1.1, part f).

What we need to show is of course that no perfect sequence of a chain-complete ordered set without infinite antichains is infinite and that the last element of the perfect sequence is finite. To do so we will *fix a chain-complete ordered set P without infinite antichains and a generalized perfect sequence $\{P_\alpha\}_{\alpha\leq\omega}$ of P.* (Recall that ω is the first infinite ordinal number.) Since no P_α has infinite antichains, we would (by Corollary 2.5.10) be done if we could show that some P_α with $\alpha < \omega$ had no infinite chains. We start with some general lemmas on chains.

Lemma 4.5.3 (Cf. [81], Lemma 6.7.) *Let $\{p_n\}_{n<\omega}$ be a sequence in an ordered set without infinite antichains. Then there is an infinite sequence $n_0 < n_1 < n_2 < \cdots < \omega$ such that $p_{n_0} \leq p_{n_1} \leq p_{n_2} \leq \cdots$ or $p_{n_0} \geq p_{n_1} \geq p_{n_2} \geq \cdots$.*

Proof. First note that by Corollary 2.5.10 we can assume without loss of generality that $\{p_n : n < \omega\}$ is a chain.

We shall assume that $\{p_n\}_{n<\omega}$ has no infinite increasing or decreasing subsequences. Find a maximal increasing subsequence of $\{p_n : n < \omega\}$ that starts at p_1 and call its top element p_{k_1}. Then no element with index greater than k_1 is $\geq p_{k_1}$. Now suppose we have already found $k_1 < k_2 < \cdots < k_m$ such that

1. If i is odd, no element with index greater than k_i is $\geq p_{k_i}$,

2. If i is even, no element with index greater than k_i is $\leq p_{k_i}$.

Without loss of generality assume m is odd. Find a maximal decreasing subsequence of $\{p_n : k_m \leq n < \omega\}$ that starts at p_{k_m} and call its smallest element $p_{k_{m+1}}$. Then no element with index greater than k_{m+1} is $\leq p_{k_{m+1}}$.

The above inductive procedure produces an infinite sequence of k_i satisfying parts 1 and 2 above. However this means that $\{p_{k_{2i}} : i \in \mathbb{N}\}$ is an infinite increasing sequence and $\{p_{k_{2i+1}} : i \in \mathbb{N}\}$ is an infinite decreasing sequence, a contradiction. ∎

Definition 4.5.4 (Cf. [162], p.328.) *Let P be an ordered set and let $X, Y \subseteq P$. Then $y \in Y$ is called the **supremum of X in Y**, denoted $\sup_Y(X)$ if y is an upper bound of X and for all $z \in Y$ with $z \geq X$ we have $z \geq y$.*

Lemma 4.5.5 (Special case of [162], Lemma 4.1 (1).) *Let $\{P_n\}_{n<\omega}$ be a decreasing sequence of chain-complete subsets of P and let $P_\omega := \bigcap_{n<\omega} P_n$. If $X \subseteq P$ is such that $\sup_{P_n} X$ exists for all $n < \omega$, then $\sup_{P_\omega} X$ also exists (and in particular, P_ω is chain-complete).*

Proof. Let $s_n := \sup_{P_n} X$. Since $P_n \supseteq P_{n+1}$ we have that $s_{n+1} \geq s_n$. If $s_m = s_n$ for some n and all $m > n$, then $s_n = \sup_{P_\omega} X$ and we are done. Otherwise every upper bound of X in P_ω is also an upper bound of the s_n. Thus in P_ω the chain of elements s_n has the same upper bounds as X. Therefore we can assume that X is a well-ordered chain C without a largest element.

Now for every well-ordered chain K with $C \subseteq K \subseteq P$ without a largest element, either $\sup_{P_\omega} K$ exists, or as above we can construct a countable chain C' such that $K \cup C'$ is well-ordered, has no largest element and all upper bounds of K in P_ω are also upper bounds of $K \cup C'$ (and hence they have the same upper bounds in P_ω). Consider the set of all well-ordered chains K with $C \subseteq K \subseteq P$ such that K and C have the same upper bounds in P_ω. Order this set by $K \sqsubseteq K'$ iff K is an initial segment of K'. Then this set is inductively ordered and must thus have a maximal element K_M. By maximality and the above, K_M must have a largest element, which is $\sup_{P_\omega}(C)$. ∎

Lemma 4.5.6 *If $x < p$ is funneled through p and $x < y < p$, then y is funneled through p.*

Proof. Exercise 5. ∎

For the rest of this proof we shall focus on a special kind of chain.

Definition 4.5.7 *A nonempty chain $C \subseteq P$ is called an **approaching chain** iff*

1. *C has no greatest element,*

2. *For all $n < \omega$ the set $C \cap P_n$ is cofinal in C.*

Now note that by definition of approaching chains and by Lemma 4.5.5 every approaching chain has a P_ω-supremum. We will also say that an approaching chain C **approaches** its P_ω-supremum $\sup_{P_\omega} C$.

Definition 4.5.8 *For the following we define*

$$\mathcal{A} := \left\{ \sup_{P_\omega} C : C \text{ is an approaching chain} \right\}.$$

Lemma 4.5.9 (Cf. [81], Lemma 6.8.) *Let C be a chain approaching $p \in P_\omega$. Then for each $c \in C$ there is a $p^c \in \mathcal{A}$ such that $c < p^c$ and $p^c \not\leq p$.*

Proof. Let C be a chain that approaches p. For each $c \in C$ and $n \in \mathbb{N}$ there is a $p_n^c \in P_n$ such that $c < p_n^c$ and $p_n^c \not> p$. Indeed, otherwise all P_n-maximal chains through c (if $c \notin P_n$, simply use an upper bound of c in C that is in P_n) would go through p, meaning all elements of $C \cap (\uparrow c) \cap P_n$ would be funneled through p and thus would not be in P_{n+1}, a contradiction.

If the sequence $\{p_n^c\}_{n \in \mathbb{N}}$ has an increasing subsequence $\{p_{n_k}^c\}_{k \in \mathbb{N}}$, then the desired element is $p^c := \sup_{P_\omega}\{p_{n_k}^c : k \in \mathbb{N}\}$. Since p^c can serve as the p^d for all $d \in C \cap (\downarrow c)$, we are done unless there is a $b \in C$ such that for all $c \geq b$ in C the sequence $\{p_n^c\}_{n \in \mathbb{N}}$ has no increasing subsequence. For the remainder of this proof we shall assume this is the case.

In this case for all $c \in C \cap (\uparrow b)$ the sequence $\{p_n^c\}_{n \in \mathbb{N}}$ has a decreasing subsequence $\{p_{n_k}^c\}_{k \in \mathbb{N}}$. Now for some k_c we must have that $p_{n_k} \not\geq C$ for $k \geq k_c$, since otherwise $\inf_{P_\omega}\{p_{n_k}^c : k \in \mathbb{N}\} \geq p$, which would imply $p_{n_{k_c}}^c \geq p$, a contradiction.

Therefore for any $c \in C \cap (\uparrow b)$ we can construct strictly increasing sequences $c = c_1 < c_2 < \cdots$ in $C \cap (\uparrow b)$ and $k_1 < k_2 < \cdots$ in $\mathbb{N}$ such that $p_{k_n}^{c_n} \not\geq c_{n+1}$. Now if $m < n$, then $p_{k_m}^{c_m} \not\geq p_{k_n}^{c_n}$, since otherwise $c_{m+1} \leq c_n \leq p_{k_n}^{c_n} \leq p_{k_m}^{c_m}$, a contradiction. Thus $\{p_{k_n}^{c_n}\}_{n \in \mathbb{N}}$ has an increasing subsequence whose P_ω-supremum p^c is not less than or equal to p. Since this construction can be started at any $c \in C \cap (\uparrow b)$, we are done. ∎

Before we continue we need to insert a proof of the Dushnik–Miller–Erdős theorem for graphs, which will allow us to finish the proof of the Li–Milner theorem.

Definition 4.5.10 *For every set S we define the **cardinality** of S to be the smallest ordinal number α such that there is a bijection between S and α. We define $|S| = \alpha$. Ordinal numbers α such that there is no bijection between α and any ordinal number strictly less than α are called **cardinal numbers**.*

Theorem 4.5.11 (Dushnik–Miller–Erdős, cf. [72], Theorem 5.23) *Let $G = (V, E)$ be a graph of infinite cardinality $|V| = \alpha$ such that every infinite subset contains two adjacent elements. Then G contains a complete subgraph of cardinality α.*

Proof. We shall give a proof by contradiction. Assume that G has no complete subgraph of cardinality α and let $M \subseteq V$ be a maximal complete subgraph. Then $|M| < \alpha$ and hence $|V \setminus M| = \alpha$. Now for each $v \in V \setminus M$ there is an $f(v) \in M$ such that v is not adjacent to $f(v)$. Thus $V \setminus M = \bigcup_{m \in M} f^{-1}(m)$, which means that for some $m \in M$ we have $|f^{-1}(m)| = \alpha$. This means that m is not adjacent to any element of $f^{-1}(m)$ and the induced subgraph $(f^{-1}(m), \{e \in E : e \subseteq f^{-1}(m)\})$ of G satisfies the above assumption. Iteration of this step produces a countable set of vertices in G such that no two are adjacent. This contradicts the hypothesis that any infinite set of vertices contains two adjacent vertices.   ∎

Corollary 4.5.12 *Let P be an ordered set without any infinite antichains, let α be a cardinal and let $X := \{x_\beta : \beta < \alpha\}$ be a family of elements of P. Then there exists a chain $C \subseteq X$ such that $\{\beta < \alpha : x_\beta \in C\}$ is cofinal in α.*

Proof. If α is finite, there is nothing to prove. Define the graph $G = (V, E)$ by

$$
\begin{aligned}
V &= \{(x_\beta, \beta) : \beta < \alpha\}, \\
E &= \{\{(x_\beta, \beta), (x_\gamma, \gamma)\} : \beta, \gamma < \alpha \text{ and } x_\beta \sim x_\gamma\}.
\end{aligned}
$$

By the Dushnik–Miller–Erdös theorem G contains a complete subgraph $H = (W, F)$ of cardinality α. Define $I := \{\beta < \alpha : (x_\beta, \beta) \in W\}$. Then I has cardinality α. Since no ordinal number before α has cardinality α, I must be cofinal in α. Thus $C = \{x_\beta : (x_\beta, \beta) \in W\}$ is a chain as desired.   ∎

Lemma 4.5.13 (Cf. [81], Lemma 6.9 or [163], Lemma 3.2.) *For all $p \in \mathcal{A}$ there is a $q \in \mathcal{A}$ such that $p < q$.*

Proof. Let C be a chain that approaches p. Without loss of generality assume that C is isomorphic to a cardinal. (This can be achieved by going to an appropriate subchain as necessary. The reader can prove it as Exercise 30.) By Lemma 4.5.9 for each $c \in C$ there is a $p^c \in \mathcal{A}$ such that $c < p^c$ and $p^c \not\leq p$. By Corollary 4.5.12 there is a nonempty chain $D \subseteq \{p^c : c \in C\}$ such that $\{c \in C : p^c \in D\}$ is cofinal in C. Since $D \subseteq \mathcal{A}$, D is an approaching chain that approaches, say $q \in P_\omega$. Since $p^c \not\leq p$ for all $c \in C$ we must have $p < q$.   ∎

Proof of Theorem 4.5.1. (Cf. proof of [81], Theorem 6.10 or [163], Corollary 3.4.) Since every chain in $\mathcal{A}$ is a chain in P_ω, every infinite chain $C \subseteq \mathcal{A}$ without a largest element is an approaching chain, whose P_ω-supremum is an upper bound of C in $\mathcal{A}$. Thus $\mathcal{A}$ is inductively ordered. (The reader should note that the only reason we needed to work with uncountable approaching chains was to show $\mathcal{A}$ is inductively ordered.) However by Lemma 4.5.13, $\mathcal{A}$ has no maximal elements in contradiction to Zorn's Lemma. Thus $\mathcal{A}$ is empty.

This implies that for some finite $n \in \mathbb{N}$, the set P_n must be finite. Indeed, otherwise we could find a sequence of mutually distinct $p_n \in P_n$ that must by Corollary 2.5.10 contain an infinite chain, which in turn would contain an approaching chain or a dual approaching chain, contradiction.

Therefore chain-complete ordered sets that do not contain infinite antichains are $\mathcal{G}$-dismantlable to a finite $\mathcal{G}$-core. Uniqueness of the core now follows from Theorem 4.3.8.   ∎

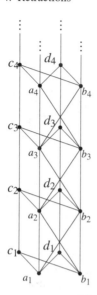

Figure 4.2: The "Tower of Doom"

We conclude this section by showing that the chain-completeness assumption in the Li–Milner theorem is vital. Indeed, the following example shows that even though every infinite ordered set without infinite antichains has a comparative retraction (cf. Exercise 32), the chain-completeness assumption cannot be relaxed.

Example 4.5.14 There is an ordered set of finite width that cannot be dismantled to a finite set in finitely many steps (cf. Figure 4.2). Let $A := \{a_n : n \in \mathbb{N}\}$, $B := \{b_n : n \in \mathbb{N}\}$, $C := \{c_n : n \in \mathbb{N}\}$, and $D := \{d_n : n \in \mathbb{N}\}$ be countable sets and let $T := A \cup B \cup C \cup D$ be ordered via $a_n \leq a_k, b_{k+1}, c_k, d_k$ for all $k \geq n$, $b_n \leq a_{k+1}, b_k, c_k, d_k$ for all $k \geq n$, $c_n \leq c_k$ for all $k \geq n$, and $d_n \leq d_k$ for all $k \geq n$. Let

$$R := A \cup B \cup \{c_{n_k} : k \in \mathbb{N}\} \cup \{d_{m_j} : j \in \mathbb{N}\}$$

be a subset of T where n_k and m_j are increasing sequences of natural numbers. We will show that for every retraction $r : R \to R$ such that $r(p) \sim p$ for all $p \in R$ the image $R[T]$ must contain $A \cup B$ and infinite subsequences of C and of D.

Let $r : R \to R$ be such a retraction. Then there is no point $p \in R$ with $r(p) < p$. Indeed otherwise let p be a minimal point with $r(p) < p$. Then $r(p)$ is the unique lower cover of p. However no point in R has a unique lower cover.

Moreover $r[A \cup B] = A \cup B$. First notice that $r(a_n) = c_{n_k}$ for any $n \leq n_k$ would mean that for all $m_j \geq n$ we have $r(d_{m_j}) \geq c_{n_k}$, that is, $r(d_{m_j}) \in C$ and hence $r(d_{m_j}) \not\sim d_{m_j}$. Similarly we show $r(a_n) \notin D$ and $r(b_n) \notin C \cup D$. Thus $r[A \cup B] \subseteq A \cup B$. Now if $r(a_n) > a_n$ for some $n \in \mathbb{N}$, then an easy inductive argument shows that for each $l > n$ we would have $r(a_l) > a_l$ or $r(b_l) > b_l$. This implies that for all $k, j \in \mathbb{N}$ we would have $r(c_{n_k}) > c_{n_k}$ and $r(d_{m_j}) > d_{m_j}$. Thus since $r[\{c_{n_k} : k \in \mathbb{N}\}] \subseteq \{c_{n_k} : k \in \mathbb{N}\}$ and $r[\{d_{m_j} : j \in \mathbb{N}\}] \subseteq \{d_{m_j} : j \in \mathbb{N}\}, r$ would not be idempotent. Hence $r[A \cup B] = A \cup B$ and r is the identity on $A \cup B$.

Finally since $r(p) \geq p$ for all p and $r[\{c_{n_k} : k \in \mathbb{N}\}] \subseteq \{c_{n_k} : k \in \mathbb{N}\}$ and $r[\{d_{m_j} : j \in \mathbb{N}\}] \subseteq \{d_{m_j} : j \in \mathbb{N}\}, r[T]$ must contain infinite subsequences of C and of D.

Thus if $T_0 := T$ and $r_k : T_k \to T_{k+1}, k = 0, \ldots, n - 1$ are retractions such that every point is comparable to its image under r_k, then T_n is of the same form as the set R above. Thus T cannot be dismantled to a finite set in finitely many steps. ∎

4.6 Isotone Relations

Isotone relations are a variation on the order-preserving mapping theme that has a complete solution to its analogue of the fixed point problem in the finite case (cf. Theorem 4.6.8). Most results in this section are due to or inspired by Walker (cf. [270]). The idea behind isotone relations is to define an order-preserving multi-function, that is, a function for which the images are sets instead of single points and for which some notion of preservation of order can be defined. The notion of ordering for the images that turns out to be the right one is the relation $\sqsubseteq$ of Proposition 1.1.3. Note that while this relation does not define an order relation for sets, it nonetheless does induce a hierarchy, which is all that is needed for us in the following.

Definition 4.6.1 *Let P be an ordered set and let $\mathcal{P}(P)$ be its power set. Then $\phi : P \to \mathcal{P}(P) \setminus \{\emptyset\}$ is called an* **isotone relation** *iff $p \leq q$ implies $\phi(p) \sqsubseteq \phi(q)$ in the sense of Proposition 1.1.3. That is*

1. *For all $a \in \Phi(p)$ there is a $b \in \Phi(q)$ with $a \leq b$ and*

2. *For all $d \in \Phi(q)$ there is a $c \in \Phi(p)$ with $d \geq c$.*

Definition 4.6.2 *An ordered set P is said to have the* **relational fixed point property** *iff for each isotone relation $\phi : P \to \mathcal{P}(P) \setminus \{\emptyset\}$ there is a fixed point $p \in P$, that is, a $p \in P$ such that $p \in \phi(p)$.*

We start our exposition of this property with a negative result (cf. Example 4.6.3). This example (together with Example 4.1.3, part 6 and Proposition 4.6.4, part 2) shows that a meaningful analysis of the relational fixed point property can only be carried out in ordered sets that do not contain a bi-infinite chain. Hence our later restriction to finite ordered sets is not as severe as it may first look.

Example 4.6.3 The set $\mathbb{Z} \cup \{\pm\infty\}$ with its natural order does not have the relational fixed point property.

Proof. Let $E \subseteq \mathbb{Z}$ be the set of even integers and let $O \subseteq \mathbb{Z}$ be the set of odd integers. The mapping $f : \mathbb{Z} \cup \{\pm\infty\} \to \mathcal{P}(\mathbb{Z} \cup \{\pm\infty\}) \setminus \{\emptyset\}$ defined by

$$f(x) := \begin{cases} E; & \text{if } x \in O, \\ O; & \text{if } x \in E, \\ \mathbb{Z}; & \text{if } x \in \{\pm\infty\} \end{cases}$$

clearly is an isotone relation with no fixed point.   ∎

Proposition 4.6.4 *Let P be an ordered set with the relational fixed point property. Then*

1. *P has the fixed point property.*

2. *Every retract of P has the relational fixed point property.*

Proof. For part 1 note that if $f : P \to P$ is an order-preserving map, then $\phi_f : P \to \mathcal{P}(P) \setminus \{\emptyset\}$ defined by $\phi_f(p) := \{f(p)\}$ is an isotone relation.

Part 2 is analogous to part 1 of Theorem 4.2.2. Its proof is left to the reader as Exercise 34.   ∎

As for order-preserving maps, there is a version of the Abian–Brown theorem for isotone relations.

Lemma 4.6.5 (Cf. [270], Proposition 5.2, compare with the Abian–Brown Theorem 3.4.7.) *Let P be an ordered set with no infinite chains and let $\phi : P \to \mathcal{P}(P) \setminus \{\emptyset\}$ be an isotone relation. If there are $p, q \in P$ such that $q \in \phi(p)$ and $p \leq q$, then ϕ has a fixed point.*

Proof. Exercise 35.   ∎

Since the Abian–Brown theorem was the key ingredient in proving Theorem 4.2.5, it should now be easy to prove

Theorem 4.6.6 *Every finite ordered set P that is $\mathcal{I}$-dismantlable to a singleton has the relational fixed point property.*

Proof. Let P be a finite ordered set and let $p \in P$ be irreducible. We will show that if $P \setminus \{p\}$ has the relational fixed point property, then P has the relational fixed point property, which clearly implies the result.

To do so, let $\phi : P \to \mathcal{P}(P) \setminus \{\emptyset\}$ be an isotone relation. Assume without loss of generality that $p \in P$ has a unique upper cover u and let $r : P \to P \setminus \{p\}$ be the retraction that maps p to u. Then $x \mapsto r[f(x)]$ is an isotone relation on $P \setminus \{p\}$, which must have a fixed point y. If $y \in f(y)$, then we are done. Otherwise $y = u$ and $u \notin f(u)$. However, since $u \in r[f(u)]$, we must have that $p \in f(u)$. By Lemma 4.6.5, f has a fixed point.   ∎

The most natural candidate for an isotone relation without a fixed point is obtained by mapping every point to the set of all points that are not comparable to it. If this candidate turns out not to be an isotone relation, the underlying set must already have a very special structure.

Lemma 4.6.7 (Compare [270], Theorem 5.6.) *Let P be a finite ordered set with more than one element. If*

$$\Phi(p) := \{x \in P : x \not> p\}$$

is not an isotone relation, then P has an irreducible point.

Proof. If Φ is not an isotone relation, then either there is a $p \in P$ such that $\Phi(p) = \emptyset$ or there are $a, b \in P$ such that $a \leq b$ and $\Phi(a) \not\subseteq \Phi(b)$.

In the first case p is comparable to all elements of P. Without loss of generality assume p has upper covers. Then for any upper cover u of p, p is the unique lower cover of u.

In the second case we can assume without loss of generality that there is an $x \in \Phi(a)$ such that there is no $y \in \Phi(b)$ that is above x. Then $x \sim b$ and since $x \not> a$, we have $x < b$. Let z be an upper bound of x such that $z \prec b$. Then no upper bound of z is in $\Phi(b)$, so all upper bounds of z are comparable to b. Since $z \prec b$, this means that all strict upper bounds of z are upper bounds of b and b is the unique upper cover of z. ∎

Lemma 4.6.7 already was the last ingredient needed to put us in a position to characterize the finite ordered sets that have the relational fixed point property.

Theorem 4.6.8 (Compare with [270], Theorem 5.7.) *A finite ordered set P has the relational fixed point property iff P is $\mathcal{I}$-dismantlable to a singleton.*

Proof. By Theorem 4.6.6, the direction "$\Leftarrow$" holds. To prove "$\Rightarrow$", we proceed by induction on $n = |P|$. The base case $n = 1$ is of course trivial. Now assume that $|P| = n$, that the result holds for ordered sets with less than n elements and that P has the relational fixed point property. Then Φ from Lemma 4.6.7 is not an isotone relation on P. Therefore P has an irreducible point p. By part 2 of Proposition 4.6.4 this means that $P \setminus \{p\}$ has the relational fixed point property. By induction hypothesis, $P \setminus \{p\}$ is $\mathcal{I}$-dismantlable to a singleton, and therefore so is P. ∎

Exercises

1. Another definition for retractions: the ordered set P is called a **retract** of the ordered set Q iff there are order-preserving maps $r : Q \to P$ and $c : P \to Q$ such that $r \circ c = \mathrm{id}_P$. r is called the **retraction** and c is called the **coretraction**. Prove that $c[P]$ is a retract of Q in the sense of our definition and that $c[P]$ is isomorphic to P.

    This means that while the above definition is more general than the one we will commonly use, we can use either description of retractions (after the proper exchanges if necessary).

2. Prove that if r is a retraction of P onto $P \setminus \{a\}$, then a is retractable to $r(a)$.

3. Let P be a finite ordered set. Prove that a four-crown-tower that contains a maximal chain is a retract of P.

4. Give an example of an ordered set with more than one good subset.

5. Prove Lemma 4.5.6.

6. Find the isomorphism types of all retracts of fences and crowns.

7. Prove that every ordered set of height 1 with at least one comparability $x < y$ has at least $2^{\frac{n}{2}-1}$ retractions and at least $2^{\frac{n}{2}}$ order-preserving self-maps.

8. Prove that if r is a retraction that removes a retractable point a and $r(x) \sim x$ for all $x \in P$, then a is in fact an irreducible point.

9. Let P be an ordered set of finite width w.

    (a) Prove that for each $p \in P$ we have $f^{2w!}(p) \sim f^{w!}(p)$.
    (b) Conclude that if P is chain-complete and of finite width w, then for every order-preserving self map f there is a retraction r_f such that for any two order-preserving self maps f and g on P, if $f(p) \le g(p)$ for all $p \in P$, then $r_f(p) \le r_g(p)$ for all $p \in P$.

10. Find an example of an ordered set P, a retraction $r : P \to P$ and a sub-set $B \subseteq P$ such that B has a supremum in P, but $r[B]$ does not have a supremum in $r[P]$.

11. Finish the proof of Lemma 4.2.10 by proving that r is a retraction.

12. Finish the proof of Theorem 4.2.11 by proving that F is order-preserving and that $\mathcal{H}$ is inductively ordered by $\supseteq$.

13. Let P be a finite ordered set such that there is a non-maximal $p \in P$ that has only one maximal upper bound. Prove that P has an irreducible point.

14. Prove that an ordered set P is C-dismantlable to Q iff P is $(\mathcal{U} \cup \mathcal{D})$-dismantlable to Q.

15. Finish the proof of the uniqueness of $\mathcal{U}$-cores (cf. Theorem 4.3.8) by show-ing that $P \setminus F_P$ is a $\mathcal{U}$-core.

16. Let P be an ordered set and let $f : P \to P$ be an order-preserving map such that there is a $k \in \mathbb{N}$ such that

$$|\{p \in P : f(p) \ne p\}| \le k.$$

Prove that $f^{k!}$ is a retraction and $|\{p \in P : f^{k!}(p) \ne p\}| \le k$.

17. Prove that $\mathcal{R}_k$-cores are unique.

18. (Exploring Exercise 16.)

   (a) Prove that if P is an ordered set and $f : P \to P$ is an order-preserving mapping with $|\{p \in P : f(p) \neq p\}| = 1$, then f is a retraction.

   (b) Find an ordered set P and an order-preserving map $f : P \to P$ such that $|P \setminus f[P]| = 1$ and f is not a retraction.

   (c) Find an ordered set P and an order-preserving map $f : P \to P$ such that $|\{p \in P : f(p) \neq p\}| = 2$, and f is not a retraction.

19. (An alternative proof for Theorem 4.3.8, cf. [79].) Let P be a finite ordered set. Let Q be the set of isomorphism classes of retracts of P that were obtained by repeated removal of irreducible points.

   (a) Prove that $\leq$ defined as follows is an order for Q. $[A] \leq [B]$ iff some representative of $[A]$ is obtained from some representative of $[B]$ by repeated removal of irreducible points.

   (b) Use Theorem 3.2.3 to prove that Q has a smallest element.

   (c) Conclude that C-cores are unique for finite ordered sets.

   (d) Use the same technique to prove that $\mathcal{R}_1$-cores are unique for finite ordered sets.

   (e) Try to use the same technique to prove that $\mathcal{R}_k$-cores are unique for finite ordered sets.

20. Call a fence in an ordered set **spanning** iff it contains only extremal elements. Call a fence $F \subseteq P$ **isometric** iff for any two points $a, b \in F$ we have $\text{dist}_P(a, b) = \text{dist}_F(a, b)$. Prove that an isometric spanning (finite or infinite) fence in an ordered set is a retract.

21. Prove that any ordered set that contains no crowns, infinite chains or infinite fences must have an irreducible point.

22. Prove that for finite ordered sets $\mathcal{R}_1 \cap C = \mathcal{I}$.

23. Prove directly that every retract of a finite $\mathcal{I}$-dismantlable set is again $\mathcal{I}$-dismantlable. (There also is a proof by combining two results in this chapter.)

24. (Folklore) Let λ be an ordinal and let P be a set. Let $\{f_\alpha : P_\alpha \to P_{\alpha+1}\}_{\alpha < \lambda}$ be a family of surjective functions such that $P_{\alpha+1} \subseteq P_\alpha \subseteq P$, $P_0 = P$ and such that for each limit ordinal $\gamma \leq \lambda$ we have $P_\gamma = \bigcap_{\alpha < \gamma} P_\alpha$. Inductively define $F_\alpha : P \to P_\alpha$ by

$$F_0 := id_P,$$
$$F_{\alpha+1} := f_\alpha \circ F_\alpha,$$

$$F_\gamma(p) \quad := \quad F_\beta(p), \text{ if } \gamma \text{ is a limit ordinal and}$$
$$\beta \text{ is such that } F_\alpha(p) = F_\beta(p) \text{ for}$$
$$\text{all } \beta \leq \alpha < \gamma.$$

Stop when F_γ is not totally defined for some limit ordinal γ or after having defined F_λ for all $p \in P$. If F_λ can be totally defined we will call $\{f_\alpha\}_{\alpha<\lambda}$ **infinitely composable** and F_λ is the **infinite composition** of $\{f_\alpha\}_{\alpha<\lambda}$.

Let $\mathcal{R}$ be a class of retractions and let P be a chain-complete ordered set. Then P is called $\mathcal{R}$-**infinite-dismantlable to** $Q \subseteq P$ iff there is an infinitely composable family $\{r_\alpha\}_{\alpha<\lambda}$ of retractions in $\mathcal{R}$ such that $R_\lambda[P] = Q$.

(a) Prove that if P is $\mathcal{C}$-infinite-dismantlable to a finite ordered set with the fixed point property, then P has the fixed point property.

(b) Prove that if P has no infinite chains and is $\mathcal{C}$-infinite-dismantlable to a singleton, then P has the relational fixed point property.

(c) Conclude that the "spider" in Figure 1.1 part f) has the relational fixed point property.

(d) Find an ordered set for which nontrivial $\mathcal{C}$-retractions can be iterated indefinitely, but which is not $\mathcal{C}$-infinite-dismantlable to any $\mathcal{C}$-core.

(e) Prove for an ordered set P that any two $\mathcal{C}$-cores C_1, C_2 to which P is $\mathcal{C}$-infinite-dismantlable must be isomorphic. (Hint: Push the proof in Theorem 4.3.8 past the limit ordinal.)

(f) Prove that an ordered set of height 1 has the fixed point property iff it is $\mathcal{C}$-infinite-dismantlable to a singleton, which is the case iff it does not contain any infinite fences or crowns.

25. This exercise investigates how different notions of dismantlability are related in infinite ordered sets.

(a) Prove that P is $\mathcal{C}$-dismantlable to Q iff P is $(\mathcal{U} \cup \mathcal{D})$-dismantlable to Q (note that P or Q or both could be infinite).

(b) Construct an $\mathcal{I}$-core that has a nontrivial good subset, that is, a nontrivial $\mathcal{G}$-retraction.

(c) Construct a $\mathcal{G}$-core that has a nontrivial $\mathcal{C}$-retraction.

26. Prove that a connectedly collapsible ordered set of height 1 is $\mathcal{C}$-dismantlable.

27. Prove that a stack of two six-crowns (as in Proposition 4.4.7) is minimal automorphic.

28. Call a set $A \subseteq P$ **retractable** to $b \in P \setminus A$ iff for every $p \in P \setminus A$ with an $a \in A$ such that $p > a$ (resp. $p < a$) we have $p \geq b$ (resp. $p \leq b$).

(a) Prove that $P \setminus A$ is a retract of P.

(b) Suppose P is chain-complete and $P \setminus A$ has the fixed point property. Prove that if for every $a \in A$, every map from $(\uparrow a) \setminus A$ to $\uparrow a$ has a point that is comparable to its image, then P has the fixed point property.

29. Let α be a cardinal and let β be an ordinal number with $\beta < \alpha$. Prove that $\alpha \setminus \beta$ is isomorphic to α.

30. Prove that every infinite chain has a cofinal well-ordered subchain that is isomorphic to a cardinal.

31. Give a direct proof that if P is a $\mathcal{G}$-core that is chain-complete with no infinite antichains, then P must be finite.

32. Prove that every infinite ordered set of finite width has a comparative retraction.

    Hints. First prove the following. Let P be an ordered set and let $C := \{c_\alpha : \alpha < \xi\}$ be an increasing well-ordered chain in P. Assume there is a $y \in P$ such that there is an $\alpha_0 < \xi$ with $y < c_{\alpha_0}$, $y \notin C$ and such that for all $d \geq y$ we have
    $$d \geq C \text{ or } \exists c \in C : c \geq d.$$
    Then there is a retraction $r : P \to P$ such that $r \neq \mathrm{id}_P$ and for all $p \in P$ we have $r(p) \geq p$.

    Then show that if the only comparative retraction is the identity, then the set must be finite.

33. (A partial converse to Lemma 4.6.7.) Let P be a finite ordered set such that $\Phi(p) := \{x \in P : p \not> x\}$ is an isotone relation.

    (a) Prove that P contains no points a, b, c such that b is the unique upper cover of a, $c < b$ and $c \not\leq a$.

    (b) Give an example that shows that P could contain points a and b, such that b is the unique upper cover of a and a is the unique lower cover of b.

34. Prove part 2 of Proposition 4.6.4.

35. Prove Lemma 4.6.5.

36. Prove directly that if a finite ordered set of width 2 has a fixed point free isotone relation, then it also has a fixed point free order-preserving map. Use this fact and Theorem 4.6.8 to give an alternative proof of Theorem 4.4.3.

37. Let P be a finite ordered set. Then an isotone relation $\phi : P \to \mathcal{P}(P) \setminus \{\emptyset\}$ is called a **connected isotone relation** iff for all $p \in P$ the set $\phi(p)$ is connected. P is said to have the **connected relational fixed point property** iff for each connected isotone relation $\phi : P \to \mathcal{P}(P) \setminus \{\emptyset\}$ there is a fixed point $p \in P$, that is, a $p \in P$ such that $p \in \phi(p)$.

   (a) Prove that if $a \in P$ is retractable to $b \in P$, then P has the connected relational fixed point property iff
      i. $P \setminus \{a\}$ has the connected relational fixed point property, and
      ii. One of $(\uparrow a) \setminus \{a\}$ or $(\downarrow a) \setminus \{a\}$ has the connected relational fixed point property.

   (b) Conclude that connectedly collapsible ordered sets have the connected relational fixed point property.

   (c) Take two finite ordered sets P and Q without any irreducible points. Choose a minimal element $m_P \in P$ and a minimal element $m_Q \in Q$ and form the set R by identifying m_P and m_Q. Show that $\Phi(p) := \{x \in P : x \not> p\}$ is an isotone relation on P but not a connected isotone relation.

   (This shows that Φ cannot be used as the canonical candidate for a fixed point free connected isotone relation.)

38. (Cf. [235].) Prove that all ordered sets with the fixed point property and at most nine elements are connectedly collapsible.

Remarks and Open Problems

Most of the open problems in this section are motivated by my own strong interest in the fixed point property and retractions.

1. For more on retractions cf. [217].

2. While irreducible points are presented here as special types of retractable points, they are actually the more widespread notion. We will see more on irreducible points when we discuss their role in lattices in Section 5.4 and subsequently the correspondence between ordered sets and distributive lattices in Section 5.5.

3. The idea of minimal automorphic sets has been present since the characterization of the fixed point property for ordered sets of height 1. Yet it is not widely used. One interesting use related to the product problem is presented in [43]. (Also cf. Remark 13 in Chapter 10.)

4. It is tempting to hope that ordered sets that are defined recursively via removals of single points, such as ordered sets that are dismantlable to a singleton and collapsible ordered sets, are easily proved to be reconstructible.

Unfortunately, this hope so far has not produced any fruit, leading to our first open problems.

Are ordered sets that are C-dismantlable to a singleton reconstructible? What about collapsible ordered sets?

5. Let P be a chain-complete ordered set with no infinite antichains. What is the smallest number of retractions onto good subsets that will lead to a finite ordered set?

   In [81], Figure 5, Farley shows a chain-complete set with no infinite antichains and an infinite good subset. However the good subset of the good subset already is finite. In [237] Rutkowski shows that perfect sequences in general can be arbitrarily long, but his examples contain infinite antichains. A related question would be how many C-retractions it would take to reach a finite ordered set (in Farley's example this is possible in the first step).

6. Is infinite dismantlability equivalent to the relational fixed point property for ordered sets with no infinite chains?

7. Is there a polynomial algorithm that decides if a finite ordered set all of whose subsets are $\mathcal{R}_1$-dismantlable to a singleton has the fixed point property?

8. For a given finite ordered set P find the smallest number k such that for all order-preserving maps $f : P \rightarrow P$ the map f^k is a retraction.

9. Let Q be a fixed, given ordered set. How hard is it to determine if Q is isomorphic to a retract of an arbitrary given ordered set? (The answer will depend on Q.)

10. Characterize those ordered sets for which for each order-preserving map $f : P \rightarrow P$ the set $\mathrm{Fix}(f)$ is connected. Is this "connected fixed point property" preserved by products (cf. Section 10.2.1)?

11. For infinite ordered sets is it true that the $\mathcal{R}_k$-core is unique or doesn't exist? Is it true that the $\mathcal{F}$-core is unique or doesn't exist?

   Uniqueness of cores is subtle in infinite ordered sets. In [246] it is shown that in chain-complete ordered sets the C-core is unique if it exists, while the $\mathcal{I}$-core need not be unique.

12. Characterize the connected relational fixed point property.

   I do not know how hard this problem is. Exercise 37c shows that there is no analogue of Lemma 4.6.7. Indeed, the connected relational fixed point property is not equivalent to connected collapsibility as Exercise 4 in Chapter 9 shows.

13. Find a nontrivial characterization of all minimal automorphic sets of width 3.

14. Is there a way to prove the Li–Milner theorem without using the Dushnik–Miller–Erdös theorem?

15. Are cores in the sense of Li and Milner (cf. [159]) unique?

5
Lattices

Lattices are (after chains) the most common ordered structures in mathematics. The reason probably is that the union and intersection of sets are the lattice operations "supremum" and "infimum" in the power set ordered by inclusion (cf. Example 3.3.4 part 5) and that many function spaces can be viewed as lattices (cf. Example 3.3.4 part 6 and Appendix B). Lattice theory is a much more developed branch of mathematics than the theory of ordered sets in general. Since there are many excellent texts on lattice theory (cf., e.g., [17, 47, 49, 90, 100]) we will concentrate here only on some core topics and on the aspects that relate to unsolved problems and work presented in this text.

5.1 Definition and Examples

What sets lattices apart from other ordered sets is an abundance of suprema and infima.

Definition 5.1.1 *Let L be an ordered set. Then L is called a* **lattice** *iff any two elements of L have a supremum and an infimum. L is called a* **complete lattice** *iff any subset of L has a supremum and an infimum.*

Clearly, complete lattices are stronger structures than (incomplete) lattices. In this chapter we will first mainly focus on issues related to completeness. General lattice ideas come to the forefront starting in Section 5.4.

Example 5.1.2

1. Every chain is a lattice, but not every chain is a complete lattice (consider $\mathbb{N}$).

2. If X is a set, then the power set $\mathcal{P}(X)$ ordered by set inclusion is a complete lattice (cf. Example 3.3.4, part 5).

3. The space $C([0, 2], \mathbb{R})$ is a lattice that is not complete (cf. Example 3.3.4, parts 6 and 7).

4. Every set of sets that is closed under unions and intersections is a lattice.

5. The closed subspaces of a Hilbert space form a lattice when ordered by inclusion. The supremum of two subspaces X and Y is their direct sum $X \oplus Y$ and their infimum is (of course) $X \cap Y$.

The existence of suprema of doubleton sets easily implies the existence of suprema of finite nonempty sets, thus yielding the fact that there are no incomplete finite lattices.

Proposition 5.1.3 *Let L be a finite lattice. Then L is complete.*

Proof. Let L be a finite lattice. We shall prove that every nonempty subset of L has a supremum. If $A \neq \emptyset$ is a subset of L with $|A| \in \{1, 2\}$ there is nothing to prove. Proceeding by induction we assume that every subset A of L with $|A| < n$ has a supremum. Let $B \subseteq L$ with $|B| = n$. Let $b \in B$ and let $B' := B \setminus \{b\}$. Then B' has $n - 1$ elements and thus it has a supremum $\bigvee B'$. But then by Proposition 3.3.3 $b \vee \bigvee B'$ is the supremum of B.

By duality we obtain that every nonempty subset of L has an infimum. The supremum and infimum of the empty set are now obtained via $\bigvee \emptyset = \bigwedge L$ and $\bigwedge \emptyset = \bigvee L$. ∎

Proposition 5.1.4 *Every complete lattice L has a largest and a smallest element. The largest element will be denoted $\mathbf{1}$ and the smallest element will be denoted $\mathbf{0}$.*

Proof. The largest element is $\mathbf{1} := \bigvee L = \bigwedge \emptyset$ and the smallest element is $\mathbf{0} := \bigwedge L = \bigvee \emptyset$. ∎

Corollary 5.1.5 *Every finite lattice is reconstructible.*

Proof. Easy consequence of the above and Proposition 1.5.11. ∎

The existence of infima and suprema in the definition of complete lattices is somewhat redundant and can be replaced solely by the existence of suprema. This fact is helpful when proving that an ordered set is a complete lattice, as it allows us to focus on suprema only.

Proposition 5.1.6 *Let P be an ordered set. If every subset of P has a supremum, then P is a complete lattice.*

Proof. We need to prove that every $A \subseteq L$ has an infimum. First note that L has a smallest element $\bigvee \emptyset$. Thus for all $A \subseteq L$ we have that $\downarrow A \neq \emptyset$. However then $\bigvee \downarrow A = \bigwedge A$ and we are done. ∎

The proof of Proposition 5.1.3 shows that in any lattice any nonempty finite set of elements has a supremum and an infimum. The question arises what types of infinite sets must have suprema and infima in order to make a lattice complete. The answer is provided in the following.

Proposition 5.1.7 *Let L be a lattice. Then the following are equivalent.*

1. *L is complete.*

2. *L is chain-complete.*

3. *Every maximal chain of L is a complete lattice.*

Proof. The implication "1⇒2" is obvious and "1⇒3" follows from Exercise 4 and part 6 of Example 4.1.3.

To prove "3⇒2", let L be a lattice such that every maximal chain in L is a complete lattice. Let $C \subseteq L$ be a nonempty chain. Let K be a maximal chain that contains C and let $c := \bigvee_K C$. Then c is an upper bound of C in L. If $d \geq C$ was an upper bound with $d \ngeq c$, then we would have $C \leq d \wedge c < c$, which would imply that K was not maximal, a contradiction. Thus $c = \bigvee_L C$.

To prove "2⇒1", let L be a chain-complete lattice and let $A \subseteq L$ be a nonempty subset. Construct a chain that has the same upper bounds as A as follows. Well-order A to obtain an indexing $A = \{a_\alpha : \alpha < \xi\}$ for some ordinal number ξ. Let $c_0 := a_0$ and for $0 < \alpha < \xi$ define

$$c_\alpha := \begin{cases} c_{\alpha^-} \vee a_\alpha; & \text{if } \alpha \text{ has an immediate predecessor } \alpha^-; \\ \bigvee_{\beta < \alpha} c_\beta; & \text{if } \alpha \text{ is a limit ordinal.} \end{cases}$$

Since L was assumed to be chain-complete, c_α is defined for all $\alpha < \xi$. The set $C := \{c_\alpha : \alpha < \xi\}$ is a chain. Moreover C and A have the same upper bounds. Indeed, $p \geq C$ trivially implies $p \geq A$, and if $q \geq A$, we can prove $q \geq C$ inductively. Thus since L is chain-complete, C has a supremum and hence A has a supremum.

Infima of nonempty subsets are constructed dually. Since all nonempty subsets of L (in particular L itself) have a supremum and an infimum, L must be a complete lattice. ∎

5.2   Fixed Point Results/The Tarski–Davis Theorem

The fixed point problem (cf. Open Question 1.4.3) has a complete solution when we restrict our scope to the class of lattices. This solution is due to A. Tarski and A. Davis. While the proof presented here is deceptively short, the reader should note that it heavily relies on strong results that we built earlier on.

Theorem 5.2.1 (The Tarski–Davis theorem, cf. [51, 260].) *Let L be a lattice. Then L has the fixed point property iff L is a complete lattice. In this situation, for each order-preserving map $f : L \to L$ the set $Fix(f)$ is a complete lattice.*

Proof. To show the direction "⇒", let L be a lattice with the fixed point property and assume L is not complete. Then by Proposition 5.1.7 there is a maximal chain $B \subseteq L$ that is not a complete lattice.

Thus there is a chain $C' \subseteq B$ that does not have a supremum in B. Let $K := \uparrow_B C'$ and let $C := \downarrow_B K$. Then K has no infimum in B, C has no supremum in B and $B = C \cup K$. Thus the (C, K)-core (see Definition 4.2.9) must be empty, since otherwise B would not be maximal. Hence the (C, K)-core does not have the fixed point property. By Theorem 4.2.11 this means that L does not have the fixed point property, a contradiction. Thus L must be complete.

The direction "⇐" and the fact that $Fix(f)$ has a smallest element is an easy consequence of the Abian–Brown theorem applied to L and the smallest element of L.

To show that $Fix(f)$ is a complete lattice, we now have to show that every nonempty set $A \subseteq Fix(f)$ has a supremum in $Fix(f)$. Let $p := \bigvee A$ be the supremum of A in L. Then every fixed point of f that is above A is also above p. Moreover

$$f(p) = f\left[\bigvee A\right] \geq \bigvee f[A] = \bigvee A = p.$$

Again by the Abian–Brown theorem there is a smallest fixed point a above p and a must be the supremum of A in $Fix(f)$.   ∎

We shall now first show a nice set-theoretical application of the Tarski–Davis theorem and then also investigate the structure of sets of fixed points in a more general setting. The reader who wishes to focus on lattices exclusively may wish to go on to Section 5.3.

5.2.1   *Preorders/The Bernstein–Cantor–Schröder Theorem*

A natural order relation for sets (besides containment) is size. For finite sets one simply counts the elements. For infinite sets we cannot count elements, but injective functions will allow us to make size comparisons. If there is an injective function $f : A \to B$, then we say A is of smaller cardinality than B and denote this by $A \unlhd B$. This relation defines what is called a preorder on any set of sets (and also on classes of sets, if we want to extend our definitions in that direction).

Definition 5.2.2 *Let P be a set and let $\trianglelefteq$ be a reflexive and transitive relation on P. Then $(P, \trianglelefteq)$ is called a* **preordered set** *and $\trianglelefteq$ is called a* **preorder**.

Preorders are a weaker notion than orders. One could say there is a "glitch" when it comes to antisymmetry. We have seen one example of such a glitch earlier.

Example 5.2.3 The relation defined in Proposition 1.1.3 is a preorder.

One can translate from preorders to orders simply by forcing antisymmetry. This is done by identifying any two points a and b such that $a \trianglelefteq b$ and $b \trianglelefteq a$. The next result shows that this simple idea is formally sound.

Proposition 5.2.4 *Let $(P, \trianglelefteq)$ be a preordered set. Then the relation $a \equiv b$ iff $a \trianglelefteq b$ and $a \trianglerighteq b$ is an equivalence relation on P. Let $Q := P/\equiv$ be the set of equivalence classes in P with respect to $\equiv$. The relation $\leq$ on Q defined by $[a] \leq [b]$ iff $a \trianglelefteq b$ is an order relation on Q.*

Proof. It is clear that $\equiv$ is reflexive, symmetric and transitive. To see that $\leq$ is well-defined, let $a, a' \in [a]$ and $b, b' \in [b]$. Then $a \trianglelefteq b$ together with $a' \trianglelefteq a$ and $b \trianglelefteq b'$ implies that $a' \trianglelefteq b'$ and the converse is proved similarly. Thus the definition of $\leq$ is independent of the choice of representatives.

Reflexivity and transitivity of $\leq$ follow from reflexivity and transitivity of $\trianglelefteq$. Finally if $[a] \leq [b]$ and $[b] \leq [a]$, then $a \trianglelefteq b$ and $b \trianglelefteq a$, which means $a \equiv b$, that is $[a] = [b]$. ∎

We can therefore conclude that our preorder relation $\trianglelefteq$ which measures size on any set or class of sets can be turned into an order relation by identifying any sets for which $A \trianglelefteq B$ and $B \trianglelefteq A$. While this is formally correct, it would be very unsatisfying if two sets such that A is at most as large as B and B is at most as large as A are not of the same size. Equal size for sets of course means that there is a bijective function between them. Thus it is natural to look for a proof that any two sets that are equivalent as above are in fact of equal size.

The idea for proving this result is quite simple, and the implementation is a beautiful application of fixed point theory for ordered sets. If we have injective maps $f : A \to B$ and $g : B \to A$, then what we need to be concerned with is surjectivity. (Unless of course one of f and g already is surjective, in which case there is nothing to prove.) The elements in $B \setminus f[A]$ are the elements that f "misses". However, these are all inverse images of elements of A under g^{-1}. Thus we should be able to use g^{-1} to "fill in" the holes that f leaves in B. To do this we need to partition A into two subsets X and $A \setminus X$. X will be mapped into B with f, while $A \setminus X$ will be mapped with g^{-1}. The trick is how to choose X.

If g^{-1} has to fill up $B \setminus f[X]$, then we must have that $A \setminus X = g[B \setminus f[X]]$, or $X = A \setminus g[B \setminus f[X]]$. Finding X is a now the task of finding a fixed point in a complete lattice. Subsequently we have to make sure that the map we had in mind truly is the desired bijection. As it turns out, all the needed pieces fall into place.

Theorem 5.2.5 (The Bernstein–Cantor–Schröder[1] theorem.) *Let A and B be sets such that there is an injective map $f : A \to B$ and an injective map $f : B \to A$. Then there is a bijective map $h : A \to B$.*

Proof. (Guided by Exercise 1J in [275], which is in turn inspired by [145].) Define $F : \mathcal{P}(A) \to \mathcal{P}(A)$ by $F(C) := A \setminus g[B \setminus f[C]]$. To see that F is order-preserving, let $C \subseteq D$. Then $f[C] \subseteq f[D]$, $B \setminus f[C] \supseteq B \setminus f[D]$, $g[B \setminus f[C]] \supseteq g[B \setminus f[D]]$, and finally

$$F(C) = A \setminus g[B \setminus f[C]] \subseteq A \setminus g[B \setminus f[D]] = F(D),$$

which means F is order-preserving.

Now, since $\mathcal{P}(A)$ is a complete lattice, by the Tarski–Davis theorem, F has a fixed point

$$E = F(E) = A \setminus g[B \setminus f[E]].$$

Define $h : A \to B$ by

$$h(x) := \begin{cases} f(x); & \text{if } x \in E, \\ g^{-1}(x); & \text{if } x \in A \setminus E. \end{cases}$$

First of all, h is overall defined, since $A \setminus E = g[B \setminus f[E]]$. To see that h is injective, let x and y be two distinct elements of A. Since f and g^{-1} are both injective, the only case we need to consider is $x \in E$ and $y \in A \setminus E$. $y \in A \setminus E = g[B \setminus f[E]]$ means there is a $z \in B \setminus f[E]$ such that $y = g(z)$. This in turn implies $h(y) = g^{-1}(y) = z \notin f[E]$ and since $h(x) = f(x) \in f[E]$, we have $h(x) \neq h(y)$.

To show that h is surjective, let $b \in B$. If $b \in f[E]$, there is nothing to prove. If $b \in B \setminus f[E]$, then $g(b) \in g[B \setminus f[E]] = A \setminus E$. But then $b = g^{-1}(g(b)) = h(g(b))$. ∎

5.2.2 Other Results on the Structure of Fixed Point Sets

Establishing the fixed point property for an ordered set means showing that for all order-preserving maps $f : P \to P$ we have that the set Fix(f) of fixed points is not empty. It would now also be nice to know more about Fix(f). As seen in Theorem 5.2.1, a sufficient condition for the fixed point property may be strong enough to allow us to determine more properties of Fix(f).

We start here by first weakening the hypotheses of the Tarski–Davis theorem. Then we consider other properties such as dismantlability and connected collapsibility. While the latter two results do not connect with lattices, it is natural to present them here, as the investigation of the structure of the fixed point sets clearly is motivated by the Tarski–Davis theorem.

---

[1] Not a relative of the author.

Theorem 5.2.6 *Let P be an ordered set with the fixed point property. Then for every order-preserving map $f : P \to P$ we have that every maximal chain in $\text{Fix}(f)$ is a complete lattice.*

Proof. Let $f : P \to P$ be an order-preserving function and let $C \subseteq \text{Fix}(f)$ be a maximal chain. Suppose that $B \subseteq C$ has no supremum in C. Then $U := \uparrow_C B$ has no infimum in C and $L := \downarrow_C U$ has no supremum in C. Clearly $C = L \cup U$ and since C is maximal in $\text{Fix}(f)$, the restriction of f to $\{x \in P : L \leq x \leq U\} = (L, U)$-core does not have any fixed points. However if P has the fixed point property, then by Theorem 4.2.11 the (L, U)-core also has the fixed point property, which is a contradiction. Thus every maximal chain in $\text{Fix}(P)$ must be a complete lattice. ∎

A natural weakening of the definition of a complete lattice is to take away the top and bottom elements. We will discuss these structures and the related truncated lattices in more detail in Chapter 6.

Definition 5.2.7 *An ordered set L is called **conditionally complete** iff any subset $S \subseteq L$ that has an upper bound and is not empty has a supremum.*

Theorem 5.2.8 *Let P be a conditionally complete ordered set with the fixed point property. Then for each order-preserving map $f : P \to P$ the set $\text{Fix}(P)$ is conditionally complete.*

Proof. Let $f : P \to P$ be order-preserving and let $A \subseteq \text{Fix}(f)$ be such that there is a $p \in P$ with $p \geq A$. Then f maps $\uparrow \left(\bigvee A \right)$ to itself. Since P is conditionally complete and has the fixed point property, P must be chain-complete. Thus $\uparrow \left(\bigvee A \right)$ is chain-complete. Therefore the Abian–Brown theorem implies that f has a smallest fixed point in $\uparrow \left(\bigvee A \right)$. We conclude that A has a supremum in $\text{Fix}(f)$. ∎

Theorem 5.2.9 (Cf. [65], Theorem 3.) *Let P be a finite ordered set that is $\mathcal{I}$-dismantlable to a singleton. Then for each order-preserving $f : P \to P$ the set $\text{Fix}(f)$ is $\mathcal{I}$-dismantlable to a singleton.*

Proof. The proof is an induction on $n := |P|$. There is nothing to prove for $|P| = 1$. For the induction step, let P be a finite ordered set, $\mathcal{I}$-dismantlable to a singleton, and assume the result has already been proved for ordered sets with fewer elements than P. Let $f : P \to P$ be order-preserving. Then by Proposition 4.1.4 the map $f^{n!}$ is a retraction on P. Note that $f|_{f^{n!}[P]}$ is order-preserving and that by Proposition 4.6.4 and Theorem 4.6.8 (or by Exercise 23 in Chapter 4) $f^{n!}[P]$ is $\mathcal{I}$-dismantlable to a singleton. Thus since $\text{Fix}(f) = \text{Fix}(f|_{f^{n!}[P]})$ we are done by induction hypothesis unless $f^{n!}[P] = P$.

In this case we have that $f : P \to P$ is an automorphism. First suppose that there is an irreducible point $p \in P$ that is not in $\text{Fix}(f)$. Then for all $i \in \mathbb{N}$ we have $f^i(p) \notin \text{Fix}(f)$. Since f is an automorphism, $\{f^i(p) : i \in \mathbb{N}\}$ is an antichain and all $f^i(p)$ are irreducible. Thus $P' := P \setminus \{f^i(p) : i \in \mathbb{N}\}$ is $\mathcal{I}$-dismantlable to a singleton and f maps P' to itself. Since we have $\mathrm{Fix}(f) = \mathrm{Fix}(f|_{P'})$ we are done by induction hypothesis.

This leaves the case in which f is an automorphism and all irreducible points of P are in $\mathrm{Fix}(f)$. Without loss of generality, let $p \in P$ be a point with a unique lower cover l. Since f is an automorphism, we must have that $f(l)$ also is a lower cover of p. Thus $f(l) = l$. This means that $f|_{P \setminus \{p\}}$ is an order-preserving map such that $\mathrm{Fix}(f|_{P \setminus \{p\}})$ is $\mathcal{I}$-dismantlable and $l \in \mathrm{Fix}(f|_{P \setminus \{p\}})$. Since $\mathrm{Fix}(f) = \mathrm{Fix}(f|_{P \setminus \{p\}}) \cup \{p\}$ we conclude that $\mathrm{Fix}(f)$ is $\mathcal{I}$-dismantlable to a singleton. ∎

For the weaker condition of connected collapsibility we can at least prove that fixed point sets will be connected.

Theorem 5.2.10 *Let P be a finite connectedly collapsible ordered set. Then for each order-preserving map $f : P \to P$ the set $\mathrm{Fix}(f)$ is nonempty and connected.*

Proof. Nonemptyness of $\mathrm{Fix}(F)$ is proved in Theorem 4.3.13. The proof of connectedness is an induction on $n := |P|$. For $n = 1$ there is nothing to prove. For the step $\{1, \ldots, n\} \to (n + 1)$ let P be an $(n + 1)$-element connectedly collapsible ordered set and let $x \in P$ be as in the definition of connected collapsibility. Let $r : P \to P \setminus \{x\}$ be a retraction and let $b := r(x)$. By definition $P \setminus \{x\}$ and $(\uparrow x \cup \downarrow x) \setminus \{x\}$ are connectedly collapsible and clearly both sets have $\leq n$ elements. Thus $H := \mathrm{Fix}(r \circ f|_{P \setminus \{x\}})$ is connected. Clearly $H \setminus \{b\} \subseteq \mathrm{Fix}(f)$, and thus $\mathrm{Fix}(f)$ must be one of the following four sets: $H, H \setminus \{b\}, (H \cup \{x\}) \setminus \{b\}$, $H \cup \{x\}$. The case $\mathrm{Fix}(F) = H$ is trivial. In all the other cases we have to show that any two elements of $\mathrm{Fix}(f)$ are joined by a fence.

In case $\mathrm{Fix}(f) = H \setminus \{b\}$ we are trivially done if $b \notin H$, so we will assume $b \in H$. Since $r(f(b)) = b$ and $f(b) \neq b$ we infer $f(b) = x$. Moreover our assumption on $\mathrm{Fix}(f)$ means $f(x) \neq x$. Let $p, q \in H \setminus \{b\}$ and let $F := \{p = y_0, \ldots, y_k = q\}$ be a fence in H. If $b \notin F$, then $F \subseteq H \setminus \{b\}$ and we are done. If $b \in F$, we can assume without loss of generality that there is exactly one index m such that $b = y_m$ and we can assume that $b < y_{m-1}, y_{m+1}$. By $f(b) = x$ we infer that $x \leq y_{m-1}, y_{m+1}$. If $f(x)$ is not related to x, then (since $f(b) = x$) f maps $(\uparrow x \cup \downarrow x) \setminus \{x\}$ to itself and thus since $(\uparrow x \cup \downarrow x) \setminus \{x\}$ is connectedly collapsible with $\leq n$ elements, $\mathrm{Fix}(f) \cap [(\uparrow x \cup \downarrow x) \setminus \{x\}]$ is connected. Hence there is a fence from y_{m-1} to y_{m+1} that lies entirely in $\mathrm{Fix}(f)$ and thus there is a fence from p to q in $\mathrm{Fix}(f)$. If $f(x)$ is related to x, there is a smallest fixed point of f that is above x or a largest fixed point of f that is below x. Call this fixed point c. Then $y_{m-1} > c < y_{m+1}$ and p and q are joined by a fence in $\mathrm{Fix}(f)$.

In case $\mathrm{Fix}(f) = (H \cup \{x\}) \setminus \{b\}$ again we will assume that $b \in H$ (the case $b \notin H$ is treated when $Fix(f) = H \cup \{x\}$). Again we infer $f(b) = x$. If $H = \{b\}$, then we must have $\mathrm{Fix}(f) = \{x\}$ and we are done. Otherwise there is a fixed point $d \in H$ of f that is related to b and not equal to b or x. Since $f(b) = x$, we have that d is related to x also. Let $p, q \in \mathrm{Fix}(f)$. Since d is related to x we can assume that $p, q \neq x$. Let $F := \{p = y_0, \ldots, y_k = q\}$ be a fence in H. Again we are done unless $b = y_m$ for exactly one m and (without loss of generality) $b < y_{m-1}, y_{m+1}$. Since $f(b) = x$, we have $x \leq y_{m-1}, y_{m+1}$ and p and q are joined by a fence in Fix(f).

Finally in case Fix(F) = $H \cup \{x\}$ let us first assume $b \in H$. Then f maps $(\uparrow x \cup \downarrow x) \setminus \{x\}$ to itself, so x and b have a common upper bound in Fix(f) and thus Fix(f) is connected. In case $b \notin H$, we have that $f(b) \neq x$. If $f(b)$ is related to x (and thus to b), then there is a point in H that is above x or below x and thus $H \cup \{x\}$ is connected. If $f(b)$ is not related to x, then f maps $(\uparrow x \cup \downarrow x) \setminus \{x\}$ to itself. Hence again there is a point in H that is above x or below x and thus $H \cup \{x\}$ is connected. ∎

5.3 Embeddings/The Dedekind–MacNeille Completion

Since complete lattices have many nice properties it can be helpful to embed a given ordered set in a complete lattice.

Definition 5.3.1 (Cf. [49], Definition 2.29.) *Let P be an ordered set and let L be a complete lattice. If there is an embedding $\phi : P \to L$, then L is called a* **completion** *of P.*

In Proposition 3.3.7 we have seen that completions are possible. Every ordered set can be embedded not just into a complete lattice, but into a nice complete lattice, namely a power set. However in some ways such an embedding is quite unsatisfactory. The power set of an ordered set is a lot bigger than the set itself. Thus even sets that are "close" to being complete lattices are embedded into a much bigger structure. For example, consider the natural numbers $\mathbb{N}$ with their canonical order. The easiest way to complete $\mathbb{N}$ is by attaching a top element ∞ to it. Completing via Proposition 3.3.7 leads to an embedding of $\mathbb{N}$ into $\mathcal{P}(\mathbb{N})$. The power set of $\mathbb{N}$ does not have the same cardinality as $\mathbb{N}$ and it is also not a chain.

The question whether there is a more "economical" way to embed an ordered set into a complete lattice is thus natural to ask. As we will see there is a "smallest" completion, which is the Dedekind–MacNeille completion.

Definition 5.3.2 *Let P be an ordered set. We define the* **Dedekind–MacNeille completion** *of P to be*

$$DM(P) := \{A \subseteq P : A = \downarrow\uparrow A\}$$

ordered by inclusion. It is also referred to as the **MacNeille completion** *or the* **completion by cuts**.

Example 5.3.3

1. The Dedekind–MacNeille completion of $\mathbb{N}$ is isomorphic to $\mathbb{N} \cup \{\infty\}$.

2. The Dedekind–MacNeille completion of $\mathbb{Q}$ is $\mathbb{R} \cup \{\pm\infty\}$.

Proof. For part 1 note that for all $A \subseteq \mathbb{N}$ we have $\downarrow\uparrow A = \left\{1, \ldots, \bigvee A\right\}$ if A has a largest element and $\downarrow\uparrow A = \mathbb{N}$ if A is nonempty and unbounded. Since $\downarrow\uparrow \emptyset = \{1\}$, we have that $DM(\mathbb{N}) = \{\{1, \ldots, n\} : n \in \mathbb{N}\} \cup \{\mathbb{N}\}$, which is isomorphic to $\mathbb{N} \cup \{\infty\}$.

For part 2 suffice it to say that one way to construct $\mathbb{R}$ from $\mathbb{Q}$ in set theory is to perform a completion by cuts construction that introduces all points except the largest and the smallest element. ∎

Theorem 5.3.4 (Cf. [49], Theorem 2.33.) *Let P be an ordered set. Then $DM(P)$ is a complete lattice. Moreover, the map $\phi_{DM} : P \to DM(P)$, which is defined by $\phi_{DM}(p) := \downarrow p$, is an embedding that preserves all suprema and infima that exist in P.*

Proof. The smallest element of $DM(P)$ is $\downarrow\uparrow \emptyset$. Thus by Proposition 5.1.6 we only need to show that all nonempty subsets of $DM(P)$ have a supremum to prove $DM(P)$ is complete. Let $\{A_i\}_{i \in I}$ be a nonempty family of subsets of P with $A_i = \downarrow\uparrow A_i$. Then $\downarrow\uparrow \bigcup_{i \in I} A_i$ is an upper bound of $\{A_i\}_{i \in I}$ in $DM(P)$. (It is an element of $DM(P)$ by Exercise 17 in Chapter 3.) Now suppose $B = \downarrow\uparrow B \in DM(P)$ is another upper bound of $\{A_i\}_{i \in I}$ in $DM(P)$. Then $\bigcup_{i \in I} A_i \subseteq B$, which means $\uparrow \bigcup_{i \in I} A_i \supseteq \uparrow B$ and then $\downarrow\uparrow \bigcup_{i \in I} A_i \subseteq \downarrow\uparrow B$. This proves that $DM(P)$ is complete.

It is easy to see that ϕ_{DM} is an embedding. Now let $A \subseteq P$ be a set that has a supremum $\bigvee A$ in P. Then by part 3 of Lemma 3.3.6 we have $\downarrow\uparrow A = \downarrow \bigvee A$. This implies that

$$\bigvee_{DM(P)} \phi_{DM}[A] = \downarrow\uparrow \bigcup_{a \in A} \downarrow a = \downarrow\uparrow A = \downarrow \bigvee A = \phi_{DM}\left(\bigvee A\right).$$

Thus ϕ_{DM} preserves suprema and dually it preserves infima. ∎

The above immediately shows that the Dedekind–MacNeille completion is very economical when we start with a complete lattice. The Dedekind–MacNeille completion leaves complete lattices essentially unchanged.

Proposition 5.3.5 *Let L be a complete lattice. Then $DM(L)$ is isomorphic to L via ϕ.*

Proof. If L is a complete lattice, then by part 3 of Lemma 3.3.6 for every $A \subseteq L$ we have that $\downarrow\uparrow A = \downarrow \bigvee A$. Thus $\phi_{DM} : L \to DM(L)$ is surjective and hence it is an isomorphism. ∎

Continuing along the idea of a smallest completion, all elements of $L \setminus \phi[P]$ should be "packed tightly" around $\phi[P]$. A notion that formalizes this idea is the following.

Definition 5.3.6 *Let L be an ordered set and let $P \subseteq L$. Then P is called* **join-dense** *in L iff for all $l \in L$ there is a set $A \subseteq P$ such that $l = \bigvee_L A$. The dual notion is called* **meet-dense**.

Proposition 5.3.7 *Let P be an ordered set. Then $\phi_{DM}[P]$ is join-dense and meet-dense in $DM(P)$.*

Proof. Every element of $DM(P) \setminus \{0\}$ is of the form $\downarrow\uparrow A$ for some nonempty $A \subseteq P$. Since $\bigvee_{DM(P)} \phi_{DM}[A] = \downarrow\uparrow \bigcup_{a \in A} \downarrow a = \downarrow\uparrow A$ this means that $\phi_{DM}[P]$ is join-dense in $DM(P)$. (0 is of course the supremum of the empty set.)

It would save work if we could infer meet-density by duality. However this is not possible, since we have not proved that the definition of the MacNeille completion is self-dual. Thus we continue as follows.

First we find a representation of infima in the Dedekind–MacNeille completion. Let $\{A_i\}_{i \in I}$ be a nonempty family of subsets of P with $A_i = \downarrow\uparrow A_i$. Then $\downarrow\uparrow \bigcap_{i \in I} A_i$ is a lower bound of $\{A_i\}_{i \in I}$ in $DM(P)$. Now suppose $B = \downarrow\uparrow B \in DM(P)$ is another lower bound of $\{A_i\}_{i \in I}$ in $DM(P)$. Then $\bigcap_{i \in I} A_i \supseteq B$, which means $\uparrow \bigcap_{i \in I} A_i \subseteq \uparrow B$ and then $\downarrow\uparrow \bigcap_{i \in I} A_i \supseteq \downarrow\uparrow B$. Therefore

$$\downarrow\uparrow \bigcap_{i \in I} A_i = \bigwedge_{DM(P)} \{A_i : i \in I\}.$$

Now let $A = \downarrow\uparrow A$ be an element of $DM(P)$. Then

$$A = \downarrow\uparrow (\downarrow\uparrow A) = \downarrow\uparrow \bigcap_{p \in \uparrow A} \downarrow p = \bigwedge_{DM(P)} \{\downarrow p : p \in \uparrow A\}.$$

Thus $\phi_{DM}[P]$ is also meet-dense in $DM(P)$.   ∎

In what sense is the Dedekind–MacNeille completion the smallest completion? The following theorem shows that every completion of an ordered set will contain a copy of its Dedekind–MacNeille completion. This is why we say the Dedekind–MacNeille completion is the smallest possible completion of an ordered set.

Theorem 5.3.8 *Let P be an ordered set and let L be a completion of P. Let $\Phi : P \to L$ be an embedding. Then there is an embedding $\Phi' : DM(P) \to L$ such that $\Phi = \Phi' \circ \phi_{DM}$.*

Proof. For $A = \downarrow_P \uparrow_P A$, define $\Phi'(A) := \bigvee_L \downarrow_L \uparrow_L \Phi[A]$. Clearly Φ' is order-preserving from $DM(P)$ to L. To show that Φ' is an embedding, let $\Phi'(A) \leq_L \Phi'(B)$.

Suppose $A \not\subseteq B$. Then there is an $a \in A \setminus B$. In particular, there is an upper bound u of $B = \downarrow_P \uparrow_P B$ such that $a \not\leq u$. However, by our assumption we have

$$\Phi(a) \leq \bigvee_L \downarrow_L \uparrow_L \Phi[A] = \Phi'(A) \leq \Phi'(B) = \bigvee_L \downarrow_L \uparrow_L \Phi[B] \leq \Phi(u),$$

a contradiction. Thus $A \subseteq B$ and Φ' is an embedding.   ■

5.4   Irreducible Points in Lattices

With suprema and infima abounding in lattices, it is natural to ask which points are absolutely needed to know as much as possible about the lattice. One possible viewpoint is that points that can be reconstructed as joins or meets of other elements could be considered less essential than those that cannot.

Definition 5.4.1 *Let P be an ordered set. If $x \in P$ is such that for all sets $X \subseteq P$ the equality $\bigvee X = x$ implies $x \in X$, then x will be called* **join-irreducible**. *If $x \in P$ is such that for all sets $X \subseteq P$ the equality $\bigwedge X = x$ implies $x \in X$, then x will be called* **meet-irreducible**. *$J(P)$ and $M(P)$ will denote the sets of join- and meet-irreducible elements respectively. If $x \in P$ is either join- or meet-irreducible, we will call x* **irreducible**.

To show consistency between this definition and our earlier notion of irreducibility, we note the following.

Proposition 5.4.2 *Let L be a finite lattice and let $x \in L$. Then x is join-irreducible iff x has a unique lower cover.*

Proof. To prove "$\Leftarrow$", let l be the unique lower cover of x and let $X \subseteq L$ be a set such that $\bigvee X = x$. Then $X \subseteq_\downarrow x$. Suppose $x \notin X$. Then $X \subseteq_\downarrow l$ and $\bigvee X \leq l$, a contradiction. Thus we must have $x \in X$ and x is join-irreducible.

For "$\Rightarrow$" we will prove the contrapositive, that is that every element that has no or more than one lower cover is not join-irreducible. Every element with more than one lower cover is the supremum of its lower covers and $\mathbf{0}$ (the only element without lower covers) is the supremum of the empty set.   ■

The above explains the choice of language for irreducible points in the first place. Since lattice theory has attracted attention earlier than order theory in general some notation was adopted from lattice theory. However, the original motivation behind the notation can be mysterious when the subject is not approached through lattice theory. Irreducible points are a prime example.

As it turns out, not only can irreducible elements not be broken down, they can also be used to represent all elements of a finite lattice.

Proposition 5.4.3 *Let L be a finite lattice. Then for every element x of $L \setminus \{0\}$ we have that $x = \bigvee[(\downarrow x) \cap J(L)]$.*

Proof. Assume the contrary. Then there is a point $x \in L \setminus \{0\}$ such that $x \neq \bigvee[(\downarrow x) \cap J(L)]$ and such that for all $y < x$ that are not equal to $\mathbf{0}$ we have $y = \bigvee[(\downarrow y) \cap J(L)]$. In particular, $x \notin J(L)$, so x has more than one lower cover. Let $l_1, \ldots, l_k$ be the lower covers of x. Then $x = \bigvee_{i=1}^{k} l_i$ and for each l_i we have $l_i = \bigvee[(\downarrow l_i) \cap J(L)]$. This implies

$$x = \bigvee_{i=1}^{k} l_i = \bigvee_{i=1}^{k} \bigvee[(\downarrow l_i) \cap J(L)] = \bigvee[(\downarrow x) \cap J(L)],$$

a contradiction.   ∎

A first consequence of the join-density of the join-irreducible elements is that on finite lattices automorphisms are completely determined by their restriction to $J(L)$.

Proposition 5.4.4 *Let L be a finite lattice. If f and g are automorphisms of L and $f|_{J(L)} = g|_{J(L)}$, then $f = g$.*

Proof. First note that for any lattice automorphism Φ and all subsets $A \subseteq L$ we have $\Phi\left(\bigvee A\right) = \bigvee \Phi[A]$ and $\Phi(\mathbf{0}) = \mathbf{0}$. Thus $f(\mathbf{0}) = \mathbf{0}$ and for all $x \in L \setminus \{0\}$ we have

$$f(x) = f\left(\bigvee[(\downarrow x) \cap J(L)]\right) = \bigvee f[(\downarrow x) \cap J(L)]$$
$$= \bigvee g[(\downarrow x) \cap J(L)] = g\left(\bigvee[(\downarrow x) \cap J(L)]\right) = g(x).$$

∎

5.5   Finite Ordered Sets vs. Distributive Lattices

A nice property of union and intersection in set systems is that they distribute over each other. Since the lattice operations $\vee$ and $\wedge$ are modeled after union and intersection, it is natural to give special consideration to lattices for which $\vee$ and $\wedge$ satisfy a distributive law. We will only discuss some basic ideas about distributive lattices here. For further information on this topic the reader is referred to [17, 47, 49, 90, 100]. (Lattices as algebraic objects is a vast topic.)

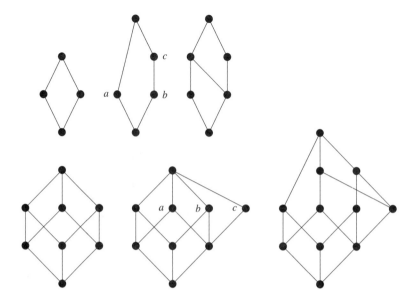

Figure 5.1: Some examples of distributive and non-distributive lattices.

Definition 5.5.1 *A lattice L is called* **distributive** *iff for all* $x, y, z \in L$ *we have*

- $x \vee (y \wedge z) = (x \vee y) \wedge (x \vee z)$, *and*
- $x \wedge (y \vee z) = (x \wedge y) \vee (x \wedge z)$.

Example 5.5.2

1. Every power set is a distributive lattice.

2. Figure 5.1 shows some examples of distributive and non-distributive lattices. In the non-distributive lattices three points for which distributivity fails are marked. Note that in each row we take a distributive lattice and through adding of points we first obtain a non-distributive lattice and then a distributive lattice again.

Join-irreducible elements have another characterization in distributive lattices. This characterization will be important for us when we prove Theorem 5.5.6, the characterization theorem of finite distributive lattices.

Lemma 5.5.3 *Let L be a finite distributive lattice. Then* $x \in L$ *is join-irreducible iff for all* $a, b \in L$ *we have that* $x \leq a \vee b$ *implies* $x \leq a$ *or* $x \leq b$.

Proof. To prove "$\Leftarrow$" let $x \in L$ be such that for all $a, b \in L$ we have that $x \leq a \vee b$ implies $x \leq a$ or $x \leq b$. Then by induction for all $Y \subseteq L$ we have that

$x \leq \bigvee Y$ implies there is a $y \in Y$ with $x \leq y$. However then for $x = \bigvee X$ there must be a $y \in X$ such that $x \leq y$. Since x cannot be strictly less than y we must have $x = y \in X$ and x is join-irreducible.

To prove "$\Rightarrow$" let $x \in L$ be join-irreducible and let $a, b \in L$ with $x \leq a \vee b$. Then

$$x = x \wedge (a \vee b) = (x \wedge a) \vee (x \wedge b).$$

Since x is join-irreducible we must have $x \in \{x \wedge a, x \wedge b\}$, say, $x = x \wedge a$. Then $x \leq a$ and we are done.   ∎

Distributive lattices arise naturally when considering subsets of ordered sets. Before we can state the next theorem, we need to generalize our notion of a down-set.

Definition 5.5.4 *Let P be an ordered set. Then $A \subseteq P$ is called a **down-set** iff for all $a \in A$ we have $\downarrow a \subseteq A$. An **up-set** is defined dually.*

Proposition 5.5.5 *Let P be a finite ordered set and let $D(P)$ be the set of down-sets of P ordered by inclusion. Then*

1. *$D(P)$ is a distributive lattice.*

2. *The join-irreducible elements of $D(P)$ are exactly the ideals $\downarrow p$ with $p \in P$.*

3. *The map $\phi : p \mapsto \downarrow p$ is an isomorphism between P and $J(D(P))$.*

Proof. Since the intersection and the union of two down-sets is again a down-set (cf. Exercise 16), $D(P)$ is a lattice. Union and intersection are the join and meet operations on $D(P)$. Since union and intersection are distributive, $D(P)$ is a distributive lattice. This proves 1.

To prove 2, let $p \in P$ and consider $\downarrow p$. The set $(\downarrow p) \setminus \{p\}$ is the unique lower cover of $\downarrow p$ in $D(P)$. Conversely, let $D \in D(P)$ be join-irreducible. If D had two maximal elements m_1 and m_2, then $D \setminus \{m_1\}$ and $D \setminus \{m_2\}$ would both be lower covers of D in $D(P)$, which is not possible. Thus D has exactly one maximal element m, which means $D = \downarrow m$.

Finally to prove 3 note that ϕ trivially is injective and order-preserving both ways and by 2 ϕ is surjective.   ∎

Distributive lattices and lattices that arise as above are closely related indeed. The next theorem shows that every finite distributive lattice arises as in Proposition 5.5.5.

Theorem 5.5.6 (Birkhoff's characterization theorem of finite distributive lattices.) *Let L be a finite distributive lattice. Then L and $D(J(L))$ are isomorphic via the map $\eta : x \to (\downarrow x) \cap J(L)$. Here $D(\cdot)$ is as in Proposition 5.5.5 and $J(L)$ is the ordered subset of join-irreducible elements.*

Proof. By Proposition 5.4.3, η is injective (note that $\eta(\mathbf{0}) = \emptyset$). Moreover since $x \leq y$ implies $\downarrow x \subseteq \downarrow y$, we have that η is order-preserving. To show η is an embedding, note that $\eta(a) \subseteq \eta(b)$ implies $a = \bigvee \eta(a) \leq \bigvee \eta(b) = b$. (Note that we did not use the distributivity of L yet. This means η is an embedding for all finite lattices.)

To prove surjectivity of η we proceed as follows. Let $A \in D(J(L))$ and let $a := \bigvee A$. We want to show that $\eta(a) = A$. To do so, first note that we have $A \subseteq (\downarrow a) \cap J(L) = \eta(a)$. For the reverse inclusion, let $x \in \eta(a)$. Then $x \leq \bigvee A$ is join-irreducible. Thus by Lemma 5.5.3 (extended inductively to sets of arbitrary size) there is a $y \in A$ with $x \leq y$. Since A is a down-set this means that $x \in A$. Thus $A \subseteq (\downarrow a) \cap J(L) = \eta(a)$ and η is surjective. ∎

The above means that there is a one-to-one correspondence between isomorphism classes of finite ordered sets and the isomorphism classes of finite distributive lattices. Proposition 5.5.5 shows that $[P] \mapsto [D(P)]$ is injective and Theorem 5.5.6 shows that it is surjective with inverse $[L] \mapsto [J(L)]$.

This means that any problem for ordered sets can also be formulated as a problem for distributive lattices and vice versa. Some of these translations will appear awkward and may not have much impact. (For example, try to translate the statement "P has the fixed point property" into its analogue for $D(P)$.) Other translations lead to nice problems and results for distributive lattices and for ordered sets. For example, cf. Exercise 30 and Remark 6.

5.6   More on Distributive Lattices

Distributivity and similar algebraic properties have received much attention in lattice theory. In this section we will give a pictorial characterization of distributive lattices. This characterization is useful in determining from the diagram if a given lattice is distributive or not. Rather than doing algebraic computations one only needs to search for certain substructures. We start by proving some equations that are true in arbitrary lattices. This will show more clearly what distinguishes distributive lattices from other lattices. We also take this opportunity to state some other (obvious) algebraic properties of lattices.

Lemma 5.6.1 *Let L be a lattice and let $a, b, c \in L$. Then*

1. $a \wedge (b \vee c) \geq (a \wedge b) \vee (a \wedge c)$,

2. $a \vee (b \wedge c) \leq (a \vee b) \wedge (a \vee c)$,

3. $a \vee b = b \vee a$ and $a \wedge b = b \wedge a$ (commutativity of $\vee$ and $\wedge$),

4. $(a \vee b) \vee c = a \vee (b \vee c)$ and $(a \wedge b) \wedge c = a \wedge (b \wedge c)$ (associativity of $\vee$ and $\wedge$),

5. $a = a \vee b$ iff $a \geq b$ and $a = a \wedge b$ iff $a \leq b$.

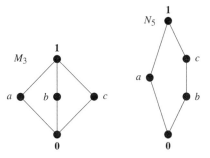

Figure 5.2: The sublattices that are forbidden in distributive lattices.

Proof. Left as Exercise 22.   ■

This means that demanding distributivity amounts to demanding that two general inequalities actually become equalities. Moreover, the equalities in the definition of distributivity actually are equivalent.

Proposition 5.6.2 *Let L be a lattice. The following are equivalent.*

1. *For all $a, b, c \in L$ we have $a \wedge (b \vee c) = (a \wedge b) \vee (a \wedge c)$.*

2. *For all $a, b, c \in L$ we have $a \vee (b \wedge c) = (a \vee b) \wedge (a \vee c)$.*

Proof. We will only need to prove the direction "1⇒2", since the other direction follows by duality. Assume that for all $a, b, c \in L$ we have

$$a \wedge (b \vee c) = (a \wedge b) \vee (a \wedge c).$$

Then

$$
\begin{aligned}
(a \vee b) \wedge (a \vee c) &= [(a \vee b) \wedge a] \vee [(a \vee b) \wedge c] \\
&= a \vee [(a \vee b) \wedge c] = a \vee [c \wedge (a \vee b)] \\
&= a \vee [(c \wedge a) \vee (c \wedge b)] = [a \vee (c \wedge a)] \vee (c \wedge b) \\
&= a \vee (b \wedge c).
\end{aligned}
$$

■

This means to prove distributivity we only need to prove that one of the two inequalities in Lemma 5.6.1 parts 1 and 2 actually is an equality. Next we need the notion of a sublattice.

Definition 5.6.3 *Let L be a lattice and let $S \subseteq L$. Then S is called a **sublattice** iff for all $x, y \in S$ we have that $x \vee y \in S$ and $x \wedge y \in S$.*

Clearly every sublattice is an ordered subset of L and a lattice. Note however that an ordered subset S that is a lattice need not be a sublattice. This is because the supremum of two elements in S could be different from the supremum of the same two elements in L. The reader is asked in Exercise 23 to find an example.

Using sublattices we can now give a pictorial characterization of distributive lattices in terms of sublattices. This is one of the nice features of order and lattice theory. Algebraic concepts often can be translated into natural pictures.

Theorem 5.6.4 *Let L be a lattice. Then L is distributive iff L does not have any sublattices isomorphic to either of the two forbidden lattices M_3 and N_5 in Figure 5.2.*

Proof. To prove "$\Rightarrow$" note that every sublattice of a distributive lattice must again be distributive. Yet we have

In the left lattice (M_3) $\quad a \vee (b \wedge c) = a \neq \mathbf{1} = (a \vee b) \wedge (a \vee c),$

In the right lattice (N_5) $\quad b \vee (a \wedge c) = b \neq c = (b \vee a) \wedge (b \vee c).$

For the direction "$\Leftarrow$" we prove the contrapositive. (This proof is essentially the proof of [49], 6.10.) Let L be a non-distributive lattice. Then by Lemma 5.6.1 part 1 and by Proposition 5.6.2 there must be $a, b, c \in L$ such that

$$a \wedge (b \vee c) > (a \wedge b) \vee (a \wedge c).$$

Equality of any two of a, b, c, or comparability of b and c, or $a \leq b$, or $a \leq c$, or $a \geq b, c$ imply equality above. Thus we must have that

1. The three points a, b, c are distinct,

2. The points b and c are not comparable,

3. a is comparable to at most one of b and c and if it is comparable to one of them, then a is larger than that point.

First we show that if $a > c$, then L contains a sublattice isomorphic to N_5. By the above we have that $b \not> a, c$. Thus $a \wedge b < b < b \vee c$. Moreover

$$b \vee c > a \wedge (b \vee c) > (a \wedge b) \vee (a \wedge c) = (a \wedge b) \vee c > a \wedge b,$$

where the middle inequality is by assumption and the outer inequalities are via the incomparability of b with a, c. Moreover the incomparability of b with a, c implies b is not comparable to $a \wedge (b \vee c)$ or to $(a \wedge b) \vee c$. Therefore

$$B := \{a \wedge b, b, b \vee c, a \wedge (b \vee c), (a \wedge b) \vee c\}$$

is isomorphic to N_5. To prove that B is a sublattice, by Exercise 26 we need to show that $b \vee [(a \wedge b) \vee c] = b \vee c$ and $b \wedge [a \wedge (b \vee c)] = a \wedge b$. By duality we only need to show the first equality. However this equality is easily seen to be

$$b \vee [(a \wedge b) \vee c] = [b \vee (a \wedge b)] \vee c = b \vee c.$$

This leaves the case in which none of a, b, c are comparable to each other. Moreover we can assume that whenever any two of x, y, z are comparable, then

$$x \wedge (y \vee z) = (x \wedge y) \vee (x \wedge z).$$

Note that our situation is not symmetric in a, b and c. Thus it is unlikely that we will find a copy of M_3, which is a very symmetric structure, as directly as we found a copy of N_5. We first generate a situation that is symmetric in a, b and c. Consider the points

$$
\begin{aligned}
s &:= (a \wedge b) \vee (a \wedge c) \vee (b \wedge c), \\
l &:= (a \vee b) \wedge (a \vee c) \wedge (b \vee c).
\end{aligned}
$$

It is easy to see that $l \geq s$. Now note that $a \wedge l = a \wedge (b \vee c)$, while

$$
\begin{aligned}
a \wedge s &= a \wedge [[(a \wedge b) \vee (a \wedge c)] \vee (b \wedge c)] \\
&= [a \wedge [(a \wedge b) \vee (a \wedge c)]] \vee [a \wedge (b \wedge c)] \\
&= [(a \wedge b) \vee (a \wedge c)] \vee [a \wedge b \wedge c] \\
&= (a \wedge b) \vee (a \wedge c).
\end{aligned}
$$

This means that $s = l$ is not possible. We now consider the three symmetrically defined points

$$(a \wedge l) \vee s, \qquad (b \wedge l) \vee s, \qquad (c \wedge l) \vee s.$$

Clearly all three are between s and l. Now note that

$$
\begin{aligned}
[(a \wedge l) &\vee s] \wedge [(b \wedge l) \vee s] \\
&= [[(a \wedge l) \vee s] \wedge (b \wedge l)] \vee [[(a \wedge l) \vee s] \wedge s] \\
&= [[(l \wedge a) \vee (l \wedge s)] \wedge (b \wedge l)] \vee s \\
&= [l \wedge (a \vee s) \wedge (b \wedge l)] \vee s \\
&= [l \wedge (b \wedge l) \wedge (a \vee s)] \vee s \\
&= [(b \wedge l) \wedge (a \vee s)] \vee s \\
&= [b \wedge (a \vee b) \wedge (b \vee c) \wedge (a \vee c) \wedge (a \vee (a \wedge b) \vee (a \wedge c) \vee (b \wedge c))] \vee s \\
&= [b \wedge ((a \vee c) \wedge (a \vee (b \wedge c)))] \vee s \\
&= [b \wedge [((a \vee c) \wedge a) \vee ((a \vee c) \wedge (b \wedge c))]] \vee s \\
&= [b \wedge [a \vee (b \wedge c)]] \vee s \\
&= [(b \wedge a) \vee (b \wedge (b \wedge c))] \vee s \\
&= s.
\end{aligned}
$$

Moreover, by duality we can assume that whenever any two of x, y, z are comparable, then $x \vee (y \wedge z) = (x \vee y) \wedge (x \vee z)$. (Indeed, otherwise L contains a sublattice isomorphic to N_5 and we are done.) This implies

$$[(a \wedge l) \vee s] \vee [(b \wedge l) \vee s]$$
$$= s \vee (a \wedge l) \vee (b \wedge l)$$
$$= (a \wedge b) \vee (a \wedge c) \vee (b \wedge c) \vee (a \wedge (b \vee c)) \vee (b \wedge (a \vee c))$$
$$= [(a \wedge b) \vee (a \wedge c) \vee (a \wedge (b \vee c))] \vee [(b \wedge c) \vee (b \wedge (a \vee c))]$$
$$= (a \wedge (b \vee c)) \vee (b \wedge (a \vee c))$$
$$= (b \wedge (a \vee c)) \vee (a \wedge (b \vee c))$$
$$= [(b \wedge (a \vee c)) \vee a] \wedge [(b \wedge (a \vee c)) \vee (b \vee c)]$$
$$= [a \vee (b \wedge (a \vee c))] \wedge (b \vee c)$$
$$= [(a \vee b) \wedge (a \vee (a \vee c))] \wedge (b \vee c)$$
$$= l$$

Symmetric arguments show that the other infima are equal to s and that the other suprema are equal to l. Now suppose any two of $(a \wedge l) \vee s$, $(b \wedge l) \vee s$, and $(c \wedge l) \vee s$ are equal. Then their supremum and infimum would be equal. Since $s \neq l$ this cannot be and our three points must be distinct. Finally if any one of them was equal to s or l, then the other two would have to be equal to l or s respectively, which cannot be. Thus $\{l, s, (a \wedge l) \vee s, (b \wedge l) \vee s, (c \wedge l) \vee s\}$ is a sublattice of L that is isomorphic to M_3.   ■

Exercises

1. Let P be a three-dimensional polyhedron. The **faces** of P are P, the empty set, the vertices, the edges, the sides and the whole polyhedron. The faces are ordered by inclusion.

   (a) Prove that the faces form a lattice.

   (b) Draw the face lattices for the Platonian solids (tetrahedron, octahedron, icosahedron, cube, dodecahedron).

2. A **topology** is a set τ of subsets of a set X, such that

   - $\emptyset, X \in \tau$.
   - If $A \subseteq \tau$, then $\bigcup A \in \tau$.
   - If $A_1, \ldots, A_n \in \tau$, then $\bigcap_{i=1}^{n} A_i \in \tau$.

   Clearly every topology is a lattice.

   (a) Give an example of a topology that is not a complete lattice.

   (b) Prove that the set of topologies on a set X forms a complete lattice.

3. Prove that if $L \subseteq P$ is a complete lattice, then L is a retract of P.

4. Prove that every retract of a lattice is a lattice and that every retract of a complete lattice is a complete lattice.

5. (Lattices as algebraic objects.) Let L be a set with the binary operations $\vee : L \times L \to L$ and $\wedge : L \times L \to L$ so that for all $a, b, c \in L$ we have

   - $a \vee a = a, a \wedge a = a$ (idempotency),
   - $a \vee b = b \vee a, a \wedge b = b \wedge a$ (commutativity),
   - $a \vee (b \vee c) = (a \vee b) \vee c, a \wedge (b \wedge c) = (a \wedge b) \wedge c$ (associativity),
   - $a \vee (b \wedge a) = a, a \wedge (b \vee a) = a$ (absorption laws).

   (a) Prove that in a lattice P as in Definition 5.1.1, the above holds for $\vee$ denoting the supremum and $\wedge$ denoting the infimum.

   (b) Prove that in a structure as defined above $a \vee b = b$ iff $a \wedge b = a$.

   (c) Prove that if for a structure as defined above we define $a \leq b$ iff $a \vee b = b$, then L equipped with this order is a lattice as in Definition 5.1.1 and $a \vee b$ is the supremum of a and b and $a \wedge b$ is the infimum of a and b.

6. Let P be a finite ordered set such that P has a largest element, a smallest element and such that P does not contain any covering four crowns. That is, there are no subsets $\{l_1, l_2, u_1, u_2\}$ in P such that the u_i $(i = 1, 2)$ are upper covers of the l_j $(j = 1, 2)$. Give an example that shows that P need not be a lattice.

   Then show that if for each four crown $\{l_1, l_2, u_1, u_2\}$ there is an element m so that $l_1, l_2 \leq m \leq u_1, u_2$, then P must be a lattice.

7. Let L and M be lattices. Then $f : L \to M$ is called a **lattice homomorphism** iff for all $x, y \in L$ we have $f(x \vee y) = f(x) \vee f(y)$ and $f(x \wedge y) = f(x) \wedge f(y)$.

   (a) Show that every lattice homomorphism is order-preserving.

   (b) Show that not every order-preserving map between lattices is a lattice homomorphism.

   (c) Show that lattice homomorphisms need not preserve infinite suprema and infima.

   (d) Show that a bijective lattice homomorphism is an order isomorphism.

8. (Join and meet irreducibility in infinite lattices.) In an infinite complete lattice an element x is called **join-irreducible** iff for all subsets X with $x = \bigvee X$ we have $x \in X$.

(a) Prove that there are infinite complete lattices that do not have any join-irreducible elements. In particular this means that there is no analogue of Propositions 5.4.3 and 5.4.4 in general infinite lattices.

(b) Let L be a lattice. Then $a \in L$ is called an **atom** iff a is an upper cover of the smallest element **0**. Prove that in a finite lattice every element except **0** is above an atom. Give an example that not every complete lattice has an atom.

(c) A lattice L is called **atomic** iff every element of L is a supremum of atoms. Prove that a finite lattice is atomic iff no element of L that is not an atom is join-irreducible.

(d) Let S be an infinite set.

  i. Prove that the set of infinite-coinfinite subsets of S, that is, the set of infinite subsets $A \subset S$ such that $S \setminus A$ is also infinite, is a complete lattice.

  ii. Prove that the set of finite or cofinite subsets of S, that is, the set of subsets $A \subset S$ such that A or $S \setminus A$ is finite, is an atomic lattice.

(e) Recall that a complete lattice satisfies the **descending chain condition** iff it contains no copies of the dual of $\mathbb{N}$. Prove that if $a \not\leq b$ in a complete lattice with the descending chain condition, then there is a join-irreducible element below a that is not below b.

(f) Prove analogues of Propositions 5.4.3 and 5.4.4 in complete lattices with the descending chain condition.

9. Let P be a finite ordered set and let $\Phi : P \to P$ and $\Psi : P \to P$ be two automorphisms.

  (a) Let $a \in P$ be minimal in the set $\{p \in P : \Phi(p) \neq p\}$. Prove that there is a point $b \neq a$ such that a and b have the same lower covers.

  (b) Conclude that $\Phi = \Psi$ iff $\Phi(p) = \Psi(p)$ for all $p \in P$ such that there is a $q \in P$ that has the same lower covers as p.

10. (Alternative proof for one direction of the Tarski–Davis theorem.) Use Exercise 26 in Chapter 3, Example 4.1.3, part 6 and Proposition 5.1.7 to prove that an incomplete lattice does not have the fixed point property.

11. Let us analyze the Bernstein–Cantor–Schröder theorem in the context of order theory, where the natural mappings should always be order-preserving.

  (a) Show that for finite ordered sets there is a Bernstein–Cantor–Schröder theorem. That is, if A and B are finite ordered sets, such that there are injective order-preserving functions $f : A \to B$ and $g : B \to A$, then A and B are order-isomorphic.

    (b) Show that there is no Bernstein–Cantor–Schröder theorem for ordered sets in general. That is, if A and B are (necessarily infinite) ordered sets, such that there are injective order-preserving functions $f : A \to B$ and $g : B \to A$, then A and B need not necessarily be order-isomorphic.

12. Find the Dedekind–MacNeille completion of

    (a) A three-element fence,

    (b) A four crown,

    (c) An antichain,

    (d) The ordered set in part e) of Figure 1.4.

13. For every $n \in \mathbb{N}$ find an ordered set P of width n for which the width of $DM(P)$ is larger than $2n$.

14.   (a) For every $n \in \mathbb{N}$ find a finite ordered set P_n of height 1 such that $DM(P_n)$ has height $\geq n$.

    (b) Find an ordered set of height 1 whose Dedekind–MacNeille completion has infinite height.

15. Let P and Q be ordered sets and let $f : P \to Q$ be order-preserving.

    (a) Prove there is an order-preserving map $f_{DM} : DM(P) \to DM(Q)$ such that $f_{DM} \circ \phi_{DM(P)} = \phi_{DM(Q)} \circ f$.

    (b) Give an example that shows that f_{DM} is not unique.

16. Let P be an ordered set and let $\{D_i\}_{i \in I}$ be a family of down-sets in P. Prove that $\bigcup_{i \in I} D_i$ and $\bigcap_{i \in I} D_i$ are down-sets.

17. Show that Lemma 5.5.3 does not hold in arbitrary lattices.

18. Let P be a finite ordered set. Prove that $h(D(P)) = |P|$.

19. Let P be a finite ordered set and let $A \subseteq P$ be an antichain. Prove that
$$w(D(P)) \geq \binom{|A|}{\left\lfloor \frac{|A|}{2} \right\rfloor}.$$

20. Let L be an arbitrary lattice. Prove that the following are equivalent.

    (a) For all $a, b \in L$ we have that $x \leq a \vee b$ implies $x \leq a$ or $x \leq b$.

    (b) For all finite sets $Y \subseteq L$ we have that $x \leq \bigvee Y$ implies there is a $y \in Y$ with $x \leq y$.

21. Permutation lattices. Let S_n be the set of permutations σ of $\{1, \ldots, n\}$. The number of descents in σ is the number of elements k such that $\sigma(k) > \sigma(k+1)$. Define $\sigma \prec \mu$ iff there is a transposition τ of two adjacent elements such that $\sigma \circ \tau = \mu$ and such that μ has more descents than σ.

    (a) Prove that the transitive closure of $\prec$ is an order relation on S_n with $\prec$ being its lower cover relation.

    (b) Prove that the thus obtained ordered set is a lattice.

22. Prove Lemma 5.6.1.

23. Give an example of a lattice L and an ordered subset $S \subseteq L$ such that S is a lattice, but not a sublattice of L.

24. Let L be the set of infinite subsets S of $\mathbb{N}$ such that the complement $\mathbb{N} \setminus S$ also is infinite. Prove that $L \cup \{\emptyset, \mathbb{N}\}$ is a non-distributive lattice.

25. Let L be a distributive lattice. Prove that if $a \geq c$, then $a \wedge (b \vee c) = (a \wedge b) \vee c$.

26. Let L be a lattice and let $a, b, c \in L$. Prove that if $b \leq c$ and $a \vee b \geq c$, then $a \vee b = a \vee c$.

27. The implication $[a \geq c] \Rightarrow [a \wedge (b \vee c) = (a \wedge b) \vee c]$ in Exercise 25 is called the **modular law** and lattices satisfying this law are called **modular lattices**.

    (a) Prove that a lattice is modular iff it does not contain a sublattice isomorphic to the lattice on the right in Figure 5.2.

    (b) Prove that if L is a modular lattice and $a, b \in L$ are such that $a \wedge b$ is a lower cover of a and b, then $a \vee b$ is an upper cover of a and b.

    (c) A lattice L is called **upper semi-modular** iff for all $a, b \in L$ we have that if $a \wedge b$ is a lower cover of a and b, then $a \vee b$ is an upper cover of a and b. **Lower semi-modular** lattices are defined dually. Prove that a lattice is modular iff it is upper semi-modular and lower semi-modular.

    (d) A lattice that is lower semi-modular or upper semi-modular is called **semi-modular**. Give an example of a semi-modular lattice that is not modular.

    (e) Show that the closed subspaces of a normed vector space ordered by inclusion form a modular lattice that is not distributive. Use the intersection as the infimum and the direct sum as the supremum of two spaces.

28. Prove that in a finite semi-modular lattice all maximal chains have the same length.

   This is called the **Jordan–Dedekind chain condition**. Lattices that satisfy this condition are also called **graded**, which in general is a weaker concept. The connection between being graded and satisfying the Jordan–Dedekind chain condition is examined further in Chapter 11, Exercise 3.

29. A lattice with largest element **1** and smallest element **0** is called **complemented** iff there is a map $x \mapsto x'$ such that $x \leq y$ implies $x' \geq y'$ and such that $x \vee x' = \mathbf{1}$. A **Boolean lattice** is a complemented distributive lattice. (When considered as algebraic objects they are also called **Boolean algebras**.)

   (a) Prove that in a Boolean lattice every element has exactly one complement.

   (b) Prove that if B is a Boolean lattice and $x \in B$ is join-irreducible, then $\mathbf{0} \prec x$.

   (c) Prove that every finite Boolean lattice is isomorphic to a power set of a finite set.

   (d) Give an example that shows that complete Boolean lattices need not be isomorphic to power sets.

   (e) Give an example that shows that atomic Boolean lattices need not be isomorphic to power sets.

   (f) Prove that every complete atomic Boolean lattice is isomorphic to a power set.

30. In [24] an efficient method for checking if two distributive lattices were isomorphic was suggested. We investigate some of the details here and discuss consequences in Remark 6 below.

   (a) Let D be a distributive lattice and let $m \in D$ be minimal. Prove that $a \in D$ is join-irreducible in D iff $m \vee a$ is join-irreducible in $\uparrow m$.

   (b) Let P be a finite ordered set and let $m \in P$ be a minimal element. Prove that $D(P \setminus \{m\})$ is isomorphic to $\uparrow_{D(P)} \{m\}$.

Remarks and Open Problems

1. For more on lattices cf., for example, [17, 47, 49, 90, 100].

2. A variation on the lattice theme is the idea of a directed ordered set. Here we only look at upper bounds and we do not demand existence of a least upper bound, but only the existence of some upper bound. For more on directed sets cf. [49].

3. Is it true that for P finite and connectedly collapsible and $f : P \to P$ order-preserving we must have that Fix(f) is connectedly collapsible?

Theorem 5.2.9 shows that the stronger condition of dismantlability allows for a similar result there. Theorem 5.2.10 shows that connected collapsibility allows some further inferences about the set of fixed points. If we could prove that every retract of a connectedly collapsible ordered set is again connectedly collapsible, then we should be able to mimic the proof of Theorem 5.2.9.

4. Is there an analogue of Theorem 5.2.9 for infinite dismantlability?

5. Let P be an ordered set without a largest or a smallest element. If P has the fixed point property, must $DM(P) \setminus \{0, 1\}$ have the fixed point property?

The other direction is false as the reader will show in Exercise 5a in Chapter 6.

6. Consider the results of Exercise 30. Recall that any finite distributive lattice D is the lattice $D(P)$ for some finite ordered set P and that $D(P)$ is isomorphic to $D(Q)$ iff P is isomorphic to Q. Exercise 30 shows that the decks of minimal ideals of $D(P)$ and $D(Q)$ are isomorphic iff the minimal decks of P and Q are isomorphic. Thus, if ordered sets were reconstructible from their minimal decks, isomorphism of finite distributive lattices could be checked through isomorphism of their minimal ideal decks. In [24] this (false) conjecture is translated into a fast (wrong) algorithm to check isomorphism of finite distributive lattices.

While the results of [249, 250] apparently show that the approach in [24] is not salvageable, even with stronger hypotheses, it still is remarkable that reconstruction results can be translated into results and algorithms regarding isomorphism. This possibility should merit future investigation.

6
Truncated Lattices

How much does a (finite) lattice change when we remove the trivially always present elements **0** and **1**? Pictorially there is very little change, since only the top and the bottom are gone. However in terms of order-theoretical properties there is a significant change. Note that both the proof of reconstructibility of finite lattices as well as the characterization of the fixed point property for lattices heavily relied on the existence of the smallest (or the largest) element. (We will again see the importance of the smallest element in Theorem 11.5.5, which settles the automorphism conjecture for finite lattices.) Thus in terms of two of our main open questions (and with respect to at least one future open question) the loss of **0** and **1** is significant. The question arises what "intrinsic" parts of the lattice structure can be used to tackle problems such as reconstruction or the fixed point property. To this end, in this chapter we investigate lattices from which top and bottom element have been removed.

6.1 Definition and Examples

The basics of truncated lattices and the related idea of conditional completeness (cf. Definition 5.2.7) are unsurprisingly quite uncomplicated. We give a few facts that are easily translated from lattices in this section.

Definition 6.1.1 *An ordered set T is called a **truncated lattice** iff any two elements of T that have a common upper bound have a supremum and any two elements of T that have a common lower bound have an infimum.*

Scholium 6.1.2 *Let P be a finite ordered set. Then P is a truncated lattice iff each nonempty subset $A \subseteq P$ that has an upper bound has a lowest upper bound $\bigvee A$.*

Proof. Easy induction on $n = |A|$ as in Proposition 5.1.3. ∎

The corresponding proof of the concept of conditional completeness is similar.

Scholium 6.1.3 *The ordered set P is conditionally complete, iff any nonempty subset $S \subseteq P$ that has a lower bound has an infimum.*

Proof. Similar to the proof of Proposition 5.1.6. ∎

Example 6.1.4

1. Any lattice is also a truncated lattice and any complete lattice is also conditionally complete.

2. For any finite lattice the set $T = L \setminus \{\mathbf{0}, \mathbf{1}\}$ is a truncated lattice.

3. A finite ordered set P is a truncated lattice iff it is conditionally complete.

4. The set of continuous real-valued functions on $[0, 1]$ is a truncated lattice, but it is not conditionally complete. Indeed, the pointwise supremum of two continuous functions is again continuous. However the family $\{f_n\}_{n \geq 2}$ with

$$
f_n(x) := \begin{cases} 0; & \text{for } x \in \left[0, \frac{1}{2}\right], \\ 1; & \text{for } x \in \left[\frac{1}{2} + \frac{1}{n}, 1\right], \\ n\left(x - \frac{1}{2}\right); & \text{for } x \in \left[\frac{1}{2}, \frac{1}{2} + \frac{1}{n}\right], \end{cases}
$$

has an upper bound, but no supremum in $C([0, 1], \mathbb{R})$.

5. The possibly most important example of a conditionally complete ordered set is given in Theorem B.1.4.

6.2 Recognizability and More

Considering the proof of the reconstructibility of finite lattices (cf. Corollary 5.1.5), one realizes that for reconstruction of finite truncated lattices we will need to go a completely different route. Fortunately there is enough structure in finite truncated lattices to allow us to recognize them and to almost reconstruct them. We start by proving some general facts about the deck of a truncated lattice. These center around finding out how the property of being a truncated lattice is preserved or lost when going to a one-point-deleted subset.

Lemma 6.2.1 *Let T be a finite truncated lattice, let $x \in T$ and let $C := T \setminus \{x\}$.*

1. *If $A \subseteq C$ and $\bigvee_T A \neq x$, then $\bigvee_C A = \bigvee_T A$.*

2. *If x has a unique upper cover y, then C is a truncated lattice.*

3. *If $A \subseteq C$ is such that $\bigvee_T A = x$ and $y := \bigvee_C A$ exists, then y is the unique upper cover of x and C is a truncated lattice.*

4. *If $\bigvee_T (\downarrow x) \setminus \{x\} \neq x$, then x has a unique lower cover and C is a truncated lattice.*

5. *If x is maximal, then C is a truncated lattice.*

6. *If x is not maximal, minimal or irreducible, then C is not a truncated lattice.*

Proof. To prove 1, note that $\bigvee_T A \in C$ is an upper bound of A in C. Now each upper bound b of A in C is also in T and hence $b \geq \bigvee_T A$.

For 2 assume that $\emptyset \neq A \subseteq C$ has an upper bound in C. Then A has an upper bound in T and $\bigvee_T A$ exists. If $\bigvee_T A \neq x$, then by 1 A has a supremum in C. If $\bigvee_T A = x$, then $y \in C$ is an upper bound of A and if $b \in C$ is an upper bound of A, then $b > x$ and hence $b \geq y$. Hence in this case $y = \bigvee_C A$. This means C is a truncated lattice.

To prove 3, let $A \subseteq C$ be such that $\bigvee_T A = x$ and $y := \bigvee_C A \neq x$ exists. Since $y \in T$, clearly $y > x$. Now let $b > x$ be an upper bound of x in T. Then $b \in C$ and b is an upper bound of A. Hence $b \geq \bigvee_C A = y$ and thus y is the unique upper cover of x. Now by 2, C is a truncated lattice.

For 4 note that the condition means that x is join-irreducible. By Proposition 5.4.2, x has a unique lower cover. By the dual of 2 we conclude C is a truncated lattice.

To prove 5 let $x \in T$ be a maximal element and let $\emptyset \neq A \subseteq C$. If A has an upper bound b in C, then $b \neq x$. Hence $\bigvee_T A \neq x$ and thus by 1, A has a supremum in C. Therefore C is a truncated lattice.

Finally to prove 6 let $x \in T$ be non-maximal, non-minimal and non-irreducible. Let U be the set of upper covers of x and let D be the set of lower covers of x. Then $|U|, |D| \geq 2$, U and D are antichains, all elements of U are upper bounds of D and there is no point in C that is an upper bound of D and a lower bound of U. Thus C is not a truncated lattice as D has upper bounds and no supremum in C. ∎

In particular we obtain the following characterization of when the property of being a finite truncated lattice is hereditary.

Proposition 6.2.2 *Let T be a finite truncated lattice and let $x \in T$. Then the card $C = T \setminus \{x\}$ is a truncated lattice iff x is maximal, minimal or irreducible.* ∎

Proposition 6.2.3 *Let P be a finite ordered set that is not a truncated lattice. Then P has at most four cards that are truncated lattices.*

Proof. Since P is not a truncated lattice, there are two elements $a, b \in P$ that have an upper bound but not a supremum. Let $c, d \in P$ be two minimal upper bounds of a and b. Then $\{a, b, c, d\}$ is a four crown, which is contained in all cards $C_{x\notin\{a,b,c,d\}} := P \setminus \{x\}$ with $x \notin \{a, b, c, d\}$. There is no element in P that is between $\{a, b\}$ and $\{c, d\}$. Thus on all the cards $C_{x\notin\{a,b,c,d\}}$, a and b have upper bounds, but not a supremum. Therefore none of the cards $C_{x\notin\{a,b,c,d\}}$ is a truncated lattice. That leaves at most four cards, $P \setminus \{a\}$, $P \setminus \{b\}$, $P \setminus \{c\}$, and $P \setminus \{d\}$ that could be truncated lattices. ∎

Theorem 6.2.4 *Finite truncated lattices with* > 3 *elements are recognizable.*

Proof. Ordered sets with a largest or smallest element and disconnected ordered sets are reconstructible. Thus we shall prove recognizability of connected truncated lattices with at least two maximal and at least two minimal elements. Any such truncated lattice has at least four cards that are truncated lattices also. By Proposition 6.2.3 any deck with ≥ 5 cards that are truncated lattices is the deck of a truncated lattice.

Thus the only truncated lattices that are not trivially recognizable from their decks would have to have two minimal elements, two maximal elements and no non-extremal irreducible elements. Let T be such a truncated lattice. Let $a, b \in T$ be the minimal elements of T. If T has no non-extremal elements, then, since T is connected, T must be the fence with four elements, which is also called "N".

In case T has non-extremal elements, recall that T has no irreducible non-extremal elements. Thus any non-extremal element of T must be above both a and b. (Indeed, otherwise by the dual of Exercise 13 in Chapter 4, T would have an irreducible non-extremal point.) This means a and b have a non-extremal supremum $a \vee b$ and all elements of $T \setminus \{a, b\}$ are above $a \vee b$. The upper covers of $a \vee b$ thus have only one lower cover. This means all upper covers of $a \vee b$ must be maximal, which in turn means that T has exactly five elements $a, b \leq a \vee b \leq c, d$.

Thus for a finite truncated lattice we have the following cases.

- It can be disconnected, in which case it is reconstructible.

- It can have a unique largest or a unique smallest element, in which case it is reconstructible.

- It can have at least five cards that are truncated lattices, in which case its deck is recognizable as the deck of a truncated lattice.

- It is equal to "N" or the five element set constructed above, each of which are recognizable (even reconstructible) from their decks.

Thus truncated lattices are recognizable. ∎

Theorem 6.2.5 *Truncated lattices with* > 3 *elements and a non-extremal, non-irreducible element are reconstructible.*

Proof. Truncated lattices are recognizable by Theorem 6.2.4. Truncated lattices with a non-extremal, non-irreducible element are recognizable as exactly those truncated lattices that have a card that is not a truncated lattice.

Let T be a truncated lattice with a non-extremal, non-irreducible element. Find a card $C = T \setminus \{x\}$ that is not a truncated lattice. Then by Lemma 6.2.1, part 4 we have that $\bigvee_T (\downarrow x) \setminus \{x\} = x$. Since by Lemma 6.2.1 part 5, x cannot be maximal, $(\downarrow x) \setminus \{x\}$ has an upper bound in C. Since x is not irreducible, by Lemma 6.2.1, part 3, $(\downarrow x) \setminus \{x\}$ does not have a supremum in C. Now let $A \subseteq C$ be a subset with an upper bound in C and no supremum. If there was an $a \in A$ with $a \not< x$, then $\bigvee_T A \neq x$ and thus by Lemma 6.2.1, part 1, A would have a supremum in C contrary to the assumption. Thus every $A \subseteq C$ that has an upper bound and no supremum is contained in $(\downarrow x) \setminus \{x\}$. However this means that

$$(\downarrow x) \setminus \{x\} = \bigcup \{A \subseteq C : A \text{ has an upper bound and no supremum in } C\},$$

and the right-hand side can be reconstructed from C. Dually we obtain $(\uparrow x) \setminus \{x\}$. Hence T is reconstructible.   ∎

For an alternative proof of Theorem 6.2.5, cf. Exercise 6.

6.3   The Fixed Clique Property

To start the discussion on the fixed point property for truncated lattices we will first consider an analogous property for graphs and for simplicial complexes. The natural connection between these properties and truncated lattices is described in Theorem 6.3.16. The idea is simple. Any lattice can be represented in various ways as a set system. Thus the same is true for truncated lattices. The trick is to find the right "representation". The set system we will ultimately choose is not necessarily isomorphic to the truncated lattice we start with, but it will have the fixed point property if and only if the original truncated lattice does.

As we have seen in Section 4.4.4 the notion of a fixed point property for graphs is not very interesting. Corollary 4.4.10 showed that if we consider fixed vertices and fixed edges, then the theory becomes a little more interesting. What is the next step beyond fixing vertices or edges? Vertices and edges are the smallest representatives of the class of complete graphs. We define

Definition 6.3.1 *A* **clique** *in a graph* $G = (V, E)$ *is a nonempty set of vertices* K *such that for all distinct* $x, y \in K$ *we have* $\{x, y\} \in E$.

Definition 6.3.2 (Cf. [15], p.10; for earlier results in this direction cf. [41], Section 1, [191, 196, 206], for recent contributions cf. [195].) *Let* $G = (V, E)$ *be a graph without infinite cliques. We will say that* G *has the* **fixed clique property** *iff for each graph endomorphism* $f : V \to V$ *there is a clique* K *of* G *with* $f[K] = K$.

We will say that an arbitrary graph G *has the* **invariant clique property** *iff for each endomorphism* $f : V \to V$ *there is a clique* C *such that* $f[C] \subseteq C$.

The closest analogue to a clique we have in order theory is a chain. By the Abian–Brown theorem whenever an order-preserving map maps a complete chain to itself, then the map must have a fixed point. This is the reason why in order theory there is no "fixed chain property" that has drawn much attention. In most interesting cases this property would be equivalent to the fixed point property, which is easier to formulate. For graphs, as we have seen, there is quite a difference

Example 6.3.3

1. Any finite complete graph has the fixed clique property.

2. Any complete graph has the invariant clique property, but infinite complete graphs do not have the fixed clique property.

3. By Corollary 4.4.10 any finite connected graph that has no cliques of size 3 and no cycles has the fixed clique property.

The fixed clique property also is interesting for certain graphs related to ordered sets.

Definition 6.3.4 *Let P be an ordered set. Let $G = (V, P)$ be the graph with vertices $V = P$ and edges $E := \{\{p, q\} : p \neq q, p \sim q\}$. Then G is called the* **comparability graph** $G_C(P)$ *of P.*

We will investigate comparability graphs further in Section 9.3. The question which graphs are comparability graphs is answered in [93]. This characterization is beyond the scope of this text.

Note that it is quite obvious that the fixed clique property for comparability graphs is a stronger property than the fixed point property for the ordered set. That is, as long as we have the Abian–Brown theorem available.

Proposition 6.3.5 *Let P be a chain-complete ordered set. If $G_C(P)$ has the invariant clique property, then P has the fixed point property.*

Proof. Let $f : P \to P$ be an order-preserving map. Then f is also a graph endomorphism for the comparability graph of P. Therefore there is a clique C in $G_C(P)$ such that $f[C] \subseteq C$. In particular this means that there is a $p \in P$ such that $f(p) \sim p$. By the Abian–Brown theorem this means that f must have a fixed point. ∎

The question beckons if the fixed clique property for the comparability graph is equivalent to the fixed point property. Unfortunately this is not the case.

Example 6.3.6 *Even in finite ordered sets the fixed clique property for the comparability graph is not equivalent to the fixed point property for the ordered set.* Consider the ordered set in part b) of Figure 1.1. The map f such that $a \mapsto i \mapsto b \mapsto k \mapsto a$, $c \leftrightarrow h$, and $d \mapsto g \mapsto e \mapsto f \mapsto d$ is an endomorphism of the comparability graph that does not fix any cliques.

If this map seems "strange" to the reader, consider the following. Visual work with the fixed clique property is made challenging by the fact that there is no notion of "up" or "down" in a graph. Thus, for one who is used to working with fixed points in ordered sets, all the usual references are gone. Especially when going from an ordered set to its comparability graph it is tempting to forget the comparabilities induced by transitivity, which are omitted in the Hasse diagram. It is also tempting to try to maintain the up-down direction of the ordered set, though it has no meaning in the comparability graph.

6.3.1 Simplicial Complexes

The set of cliques in a graph is a special example of a more general structure, called a simplicial complex.

Definition 6.3.7 *A **simplicial complex** is a pair $K = (V, S)$ of a set V of vertices and a set S of nonempty subsets of V called **simplexes** such that:*

1. *Every singleton subset of V is a simplex.*

2. *Every nonempty subset of a simplex is a simplex.*

*We will call sets S of simplexes ordered by inclusion **truncated face lattices** also denoted $TFL(K)$. A **simplicial map** on a simplicial complex is a map $f : V \rightarrow V$ such that for all $S \in S$ we have $f[S] \in S$.*

Simplicial complexes originate in geometry, which motivates the following choice of language.

Definition 6.3.8 *A q-**simplex** is a simplex with exactly $q + 1$ elements. We also say that the **dimension** of the simplex S is q iff S is a q-simplex. If the (q-dimensional) simplex S is contained in the simplex T, then we also say that S is a (q-**dimensional**) **face** of T.*

Geometrically, the 0-simplex is a point, the 1-simplex is a line segment between two points, the 2-simplex is a triangle, the 3-simplex is a tetrahedron, etc. Simplicial complexes are then geometric figures made up of simplexes. They are triangulated by their simplexes and can thus be analyzed easier than smooth figures. We will first work with simplicial complexes from a purely algebraic point-of-view. For the topological realization of a simplicial complex cf. Section 6.4 and also Figure 6.1 which shows a triangulation of the 2-dimensional unit sphere S^2. The truncated face lattice of the complex is shown on the bottom of the figure.

Any truncated face lattice is conditionally complete with the supremum being the union and the infimum being the intersection. We still speak of a truncated (face) lattice, since our main focus will be on finite simplicial complexes. Also note that the minimal elements in a truncated face lattice are the singleton sets and that every element is the supremum of the minimal elements below it. Coming from the fixed clique property our most natural example of a simplicial complex is the following.

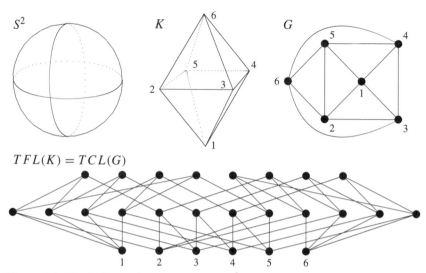

Figure 6.1: The 2-dimensional unit sphere S^2, a simplicial complex K that triangulates S^2 (the simplexes are the points, the edges and all triangular faces), the unique graph G whose truncated clique lattice equals the truncated face lattice of K (such a graph need not always exist) and the truncated clique/face lattice $TFL(K) = TCL(G)$.

Example 6.3.9 Every graph $G = (V, E)$ induces a simplicial complex with vertices V and simplexes being the cliques of G. This complex is called the **clique complex** of G. Every graph endomorphism f is a simplicial map on the clique complex. Moreover G has the fixed clique property iff its clique complex has the fixed simplex property. The truncated face lattice of the clique complex will also be called the **truncated clique lattice** $TCL(G)$ of G.

Note that this also means that every ordered set P induces a simplicial complex, namely the clique complex of its comparability graph. Described directly, the vertices are the set P and the simplexes are the chains of P. This complex is called the P-**chain complex**[1] of P and the truncated face lattice of the P-chain complex is called the **truncated chain lattice** $TCL(P)$ of P. There is no conflict of notation, since the argument of $TCL(\cdot)$ reveals whether we are looking at a truncated chain lattice or a truncated clique lattice.

The simplicial complex in Figure 6.1 is indeed the clique complex of the graph that is shown. However, not every simplicial complex is a clique complex. (Just consider the complex with vertices $\{a, b, c\}$ and simplexes $\{a\}$, $\{b\}$, $\{c\}$, $\{a, b\}$,

---

[1] We have to be careful here. This natural naming convention is distinct from the idea of a chain complex in algebraic topology as introduced in Definition A.1.1.

$\{a, c\}, \{b, c\}$.) So which simplicial complexes arise from graphs and which do not? The following result answers this question.

Definition 6.3.10 *Let T be a truncated lattice and let M be the set of minimal elements of T. T is said to satisfy the* **clique condition** *iff for all $C \subseteq M$ we have that if any two elements of C have a supremum in T, then C has a supremum in T.*

Proposition 6.3.11 (Cf. [96], Proposition 4.4, or originally [255].) *Let S be a truncated face lattice. Then there is a graph G such that S is the truncated clique lattice of G iff S satisfies the clique condition.*

Proof. First suppose that S is the truncated clique lattice of a graph $G = (V, E)$. Let A be a set of minimal elements of S (that is, singletons). If any two elements of A have a supremum (=union) in S, then for any $\{v\}, \{w\} \in A$ we have $v \sim w$ in G. Thus $\bigcup A$ is a clique in G and thus an element of S. The set $\bigcup A$ is the supremum of A in S and hence S satisfies the clique condition.

For the other direction suppose S satisfies the clique condition. Let $V := \{x : \{x\} \in S\}$ and let $E := \{\{x, y\} : \{x, y\} \in S\}$. Then $G = (V, E)$ is a graph. Clearly every $C \in S$ is a clique in G. Moreover by the clique condition every clique in G is in S and thus S is the truncated clique lattice of G. ∎

The property that is analogous to the fixed clique property on simplicial complexes is naturally defined as the following.

Definition 6.3.12 *A simplicial complex has the* **fixed simplex property** *iff for each simplicial map $f : V \to V$ there is a simplex S with $f[S] = S$. A simplicial complex has the* **invariant simplex property** *iff for each simplicial map $f : V \to V$ there is a simplex S with $f[S] \subseteq S$.*

Every order-preserving self-map is also a graph endomorphism of the comparability graph. Every graph endomorphism is a simplicial map of the clique complex. We have thus so far presented two properties that are sufficient for the fixed point property of an ordered set. We can also see that the fixed simplex property is very close to a property for truncated lattices. After all, it focuses on (simplicial) maps, which are essentially maps on truncated lattices. This is the connection we will investigate now.

As we have seen in Proposition 3.1.1 it can be helpful to modify a given map slightly when analyzing the existence of fixed points. The more properties a map has, the easier it should be to analyze. The property that is natural to consider in the context of simplicial complexes is the preservation of unions.

Definition 6.3.13 *If $Q \subseteq P(X)$ is ordered by inclusion, we will call a map $f : Q \to Q$* **union-preserving** *iff for all families $\{A_i\}_{i \in I}$ of sets with $\bigcup_{i \in I} A_i \in Q$ we* *have that*

$$f\left(\bigcup_{i\in I} A_i\right) = \bigcup_{i\in I} f(A_i) \in Q.$$

Lemma 6.3.14 *Let $K = (V, S)$ be a simplicial complex.*

1. *If $f : V \to V$ is a simplicial map, then $f^*(S) := f[S]$ defines an order-preserving, union-preserving map on S that maps minimal elements to minimal elements.*

2. *If $g : S \to S$ is an order-preserving, union-preserving map on S that maps minimal elements to minimal elements, then defining $g_*(v)$ to be the unique element of $g(\{v\})$ defines a simplicial map for K such that $(g_*)^* = g$.*

Proof. To prove part 1 let $f : V \to V$ be a simplicial map of K. Then $f^* : S \to S$ certainly is well-defined and maps minimal elements to minimal elements. Moreover if for $i \in I$ the sets $A_i \in S$ are simplexes of K with $\bigcup_{i\in I} A_i \in S$, then $f\left[\bigcup_{i\in I} A_i\right] \in S$. Hence

$$f^*\left(\bigcup_{i\in I} A_i\right) = f\left[\bigcup_{i\in I} A_i\right] = \bigcup_{i\in I} f[A_i] = \bigcup_{i\in I} f^*(A_i).$$

Thus f^* is union-preserving and a union-preserving map is order-preserving.

For part 2 let $g : S \to S$ be an order-preserving, union-preserving map on S that maps minimal elements to minimal elements, and let $g_*(v)$ be the element of $g(\{v\})$. Let $S \in S$. To show that g_* is a simplicial map, we need to prove that $g_*[S] \in S$. However this is trivial since

$$g_*[S] = \bigcup_{s\in S}\{g_*(s)\} = \bigcup_{s\in S} g(s) = g(S) \in S.$$

Thus g_* is a simplicial map.

Moreover clearly $(g_*)^*(\{v\}) = g(\{v\})$ for all $v \in V$. Since $(g_*)^*$ and g are both union-preserving and every element of S is the union of a unique set of minimal elements of S, this means that $g = (g_*)^*$.   ∎

Proposition 6.3.15 *Suppose $K = (V, S)$ is a simplicial complex without infinite simplexes. Then K has the fixed simplex property iff S has the fixed point property.*

Proof. First note that if K has no infinite simplexes, then S has no infinite chains. Thus if $g : S \to S$ is order-preserving, then existence of an element that is comparable to its image implies existence of a fixed point.

Assume that K has the fixed simplex property. We claim that for each simplicial map $f : V \to V$ the order-preserving map $f^* : S \to S$ has a fixed point. Indeed, if $S \in S$ is a simplex with $f[S] = S$, then $f^*(S) = S$. Thus by part

2 of Lemma 6.3.14 each order-preserving, union-preserving map on S that maps minimal elements to minimal elements has a fixed point. Now let $g : S \to S$ be an order-preserving map. For each $v \in V$ choose $h(\{v\})$ to be a singleton contained in $g(\{v\})$. For each non-singleton $S \in S$ let $h(S) := \bigcup\{h(\{v\}) : v \in S\}$. Let $U \subseteq S$ have a union in S. Then

$$h\left(\bigcup_{S \in U} S\right) = \bigcup\left\{h(\{v\}) : v \in \bigcup_{S \in U} S\right\}$$
$$= \bigcup_{S \in U}\bigcup \{h(\{v\}) : v \in S\} = \bigcup_{S \in U} h(S).$$

Thus h is order- and union-preserving and hence $h = (h_*)^*$ has a fixed point. Thus there is an $S \in S$ such that $g(S) \supseteq h(S) = S$. But this implies that g has a fixed point. Since g was arbitrary, S has the fixed point property.

Conversely suppose that S has the fixed point property. Let $f : V \to V$ be a simplicial map. Then f^* has a fixed point $S \in S$, which is a simplex of K such that $f[S] = f^*(S) = S$. Thus K has the fixed simplex property.   ∎

6.3.2 The Fixed Point Property for Truncated Lattices

Proposition 6.3.15 shows how to encode the fixed simplex property into the fixed point property for a truncated lattice. The interesting fact now is that the fixed point property for a truncated lattice can also be encoded as the fixed simplex property for the appropriate simplicial complex.

Theorem 6.3.16 *Let T be a truncated lattice such that:*

1. *T has no elements with unique lower covers.*

2. *T has no infinite chains.*

3. *No element of T is above infinitely many minimal elements.*

Let $M \subseteq T$ be the set of minimal elements of T. We define B_T to be the set of all nonempty subsets H of M that have an upper bound. Order B_T by set inclusion. Then B_T is a truncated face lattice and the following are equivalent.

1. *T has the fixed point property.*

2. *B_T has the fixed point property.*

3. *The simplicial complex (M, B_T) has the fixed simplex property.*

Moreover if T satisfies the clique condition, then there is a graph G_T that has the fixed clique property iff T has the fixed point property.

Proof. Clearly B_T is a truncated face lattice. Our first step towards proving the equivalence is to turn the original truncated lattice T into a system of sets. For each $x \in T$ let $M_x := (\downarrow x) \cap M$. Order

$$A_T := \{M_x : x \in T\}$$

by inclusion. Then

$$\Phi : T \to A_T; \qquad x \mapsto M_x$$

is order-preserving and surjective. To see that Φ is an isomorphism note that since all non-minimal points of T have at least two lower covers we have $x = \bigvee[(\downarrow x) \cap M]$ for all $x \in T$. Now $M_x \subseteq M_y$ implies that

$$x = \bigvee (\downarrow x) \cap M = \bigvee M_x \leq \bigvee M_y = \bigvee (\downarrow y) \cap M = y.$$

Thus Φ is an isomorphism and in particular T has the fixed point property iff A_T does.

For $Y \in B_T$ let $r(Y)$ be the supremum of $\{\{y\} : y \in Y\}$ in A_T. The function r is well-defined, since every $Y \in B_T$ has an upper bound in A_T, no element of A_T is above infinitely many minimal elements and since A_T is a truncated lattice. Clearly r is a retraction and $r(Y) \supseteq Y$ for all Y. Since no element of T is above infinitely many minimal elements, B_T has no infinite chains. Hence A_T has the fixed point property iff B_T does. This is by Proposition 6.3.15 the case iff (M, B_T) has the fixed simplex property. We have thus proved the claimed equivalences.

Finally if T satisfies the clique condition, then by Proposition 6.3.11, the simplicial complex (M, B_T) is induced by a graph G_T. This proves the "Moreover"-part.   ∎

Remark 6.3.17 Proposition 6.3.11 distinguishes truncated clique lattices from general truncated face lattices. It seems difficult to try to work without the clique condition and obtain fixed simplex theorems. For the fixed simplex property not even a weak analogue of the Abian–Brown theorem like Exercise 11b is available. To see this consider the simplicial complex with vertices a, b, c and simplexes $\{a\}$, $\{b\}, \{c\}, \{a, b\}, \{a, c\}, \{b, c\}$. We have $N(a) = N(b) = N(c) = \{a, b, c\}$ and this simplicial complex does not have the fixed simplex property (its truncated face lattice is the six crown).

The connections between the properties investigated in this section are summarized in Figure 6.2.

6.4   Triangulations of S^n

As was mentioned earlier, simplicial complexes are originally geometrical objects. Indeed, using a sufficiently fine triangulation, geometrical objects in any dimension can be approximated through a union of line segments, triangles, tetrahedra and higher-dimensional simplexes. Coming to this subject from an algebraic

Topological FPP (Topological Spaces)	$\Rightarrow$	Fixed Simplex Property (Simplicial Complexes)	$\Rightarrow$	Fixed Clique Property (Graphs)
		$\Updownarrow$		$\Downarrow$
		Fixed Point Property (Truncated Lattices)		Fixed Point Property (Ordered Sets)

Figure 6.2: Relation between the properties discussed in Sections 6.3 and 6.4. Any of the shown implications is valid whenever the representative of the more general structure was induced by a representative of the more restrictive structure.

or combinatorial direction, we will first need to define what the topological realization of a simplicial complex is. (In this section we have no choice but to assume some background in topology.)

Definition 6.4.1 *Let* $K = (V, S)$ *be a finite simplicial complex. We define the* **topological realization** $|K|$ *of* K *to be the metric space consisting of the set of functions* $\alpha : V \to [0, 1]$ *such that for each* α,

1. $\{v \in V : \alpha(v) \neq 0\} \in S$,

2. $\sum_{v \in V} \alpha(v) = 1$,

with the metric

$$d(\alpha, \beta) := \sqrt{\sum_{v \in V} (\alpha(v) - \beta(v))^2}.$$

Clearly the topological realization of a finite simplicial complex can be embedded as a subspace in $\mathbb{R}^{|V|}$ with the usual topology. Note that there may be lower dimensional spaces into which the complex embeds. This is the case for the simplicial complex shown in Figure 6.1, which has six vertices, but which can be realized in $\mathbb{R}^3$.

Simplicial maps between simplicial complexes can be extended to affine maps by affine interpolation. A function $f : S \to T$ from a subset $S \subseteq \mathbb{R}^n$ to a subset $T \subseteq \mathbb{R}^m$ is called an **affine map** iff for all $x_1, \ldots, x_k \in S$ and $\alpha_1, \ldots, \alpha_k \in \mathbb{R}$ such that $\sum_{i=1}^{k} \alpha_i = 1$ and $\sum_{i=1}^{k} \alpha_i x_i \in S$ we have that $f\left(\sum_{i=1}^{k} \alpha_i x_i\right) = \sum_{i=1}^{k} \alpha_i f(x_i)$. Affine interpolation now is the following process. A simplicial map g is defined on the "corners" x_i of the simplexes. We can extend g to an affine map g_a by defining $g_a\left(\sum_{i=1}^{k} \alpha_i x_i\right) := \sum_{i=1}^{k} \alpha_i g(x_i)$, where we took the liberty of identifying the elements of the simplicial complex with their topological realizations. The fact that g is a simplicial map together with $\sum_{i=1}^{k} \alpha_i = 1$ assures that the new map g_a does not map to any points outside the topological realization. Moreover, any affine map on the topological realization can be constructed in this fashion. It is easy to see that a simplicial complex has the fixed simplex property iff every affine map as above fixes a topological realization of a simplex.

Since every affine map is continuous, we can now connect to the fixed point property from topology. A topological space has the **fixed point property** iff each continuous self map has a fixed point. (Topological) Retractions (idempotent *continuous* self maps in topology) play a similar role in topology as in ordered sets and the topological fixed point property is for example preserved by retractions.

If the topological realization of a simplicial complex has the topological fixed point property, then the simplicial complex will have the fixed simplex property. Indeed, first note that every affine map is continuous. If every continuous map on the topological realization has a fixed point, then every affine map has a fixed point. An affine map induced by a simplicial map can only have a fixed point if it fixes some simplex.

The converse of the above however is false. If a simplicial complex has the fixed simplex property, then its topological realization can still fail to have the fixed point property. This may not be too surprising as not every continuous function on a given topological realization of a simplicial complex can be approximated closely by affine maps.

For the examples we wish to present in this section, we need to consider the n-dimensional unit sphere.

Definition 6.4.2 *Let S^n denote the n-dimensional unit sphere, that is,*

$$S^n := \{(x_1, \ldots, x_{n+1}) : x_1^2 + \cdots + x_{n+1}^2 = 1\}.$$

We will always consider S^n with the standard topology inherited from $\mathbb{R}^{n+1}$.

The unit sphere has a property analogous to being minimal automorphic in ordered sets. Itself, it does not have the topological fixed point property (consider the map that maps every point to its antipode), yet every one of its nontrivial retracts does have the fixed point property (cf. Lemma 6.4.3). This should allow us to construct a wealth of examples of minimal automorphic ordered sets. Indeed, since the topological fixed point property is stronger than the combinatorial fixed point/clique/simplex properties, we will be able to obtain simplicial complexes for which every retract has the fixed simplex property. If we then structure the simplicial complex so we can capture a fixed point free continuous map on it, we are done.

Lemma 6.4.3 *Let $n \geq 2$. Every retract of S^n that is not equal to S^n has the topological fixed point property.*

Proof. Let $r : S^n \to S^n$ be a (topological) retraction with $r[S^n] \neq S^n$. Then $r[S^n]$ is isomorphic to a retract of the n-dimensional unit ball which has the topological fixed point property by Brouwer's fixed point theorem.   ■

Definition 6.4.4 *Let K be a finite simplicial complex. Then K is called a* **triangulation** *of S^n iff the topological realization $|K|$ of K is homeomorphic to S^n.*

An example of a triangulation of S^2 is given in Figure 6.1.

Lemma 6.4.5 *Every nontrivial retract of the truncated face lattice T_K of a triangulation of S^n with $n \geq 2$ has the (order-theoretical) fixed point property.*

Proof. Let T_K be the truncated face lattice of the triangulation K of S^n. Let $r : T_K \to T_K$ be a nontrivial retraction with $r[T_K] \neq T_K$.

We claim there is a minimal element m of T_K that is not in $r[T_K]$. Suppose otherwise. Then r fixes all minimal elements of T_K, which implies $r(x) \geq x$ for all $x \in T_K$. This implies there is a $y \in T_K$ such that $r(y) > y$ and $r(p) = p$ for all $p \geq y$. Then y has exactly one upper cover. Therefore there is a maximal element $M \in T_K$ such that the simplex $M \setminus y$ has exactly one point (otherwise y has more than one upper cover). We have established that M is the only upper bound of y.

Now we need to switch our viewpoint to topology. Recall that for any x and $\varepsilon > 0$, $B_\varepsilon(x)$ denotes the solid ball of radius ε with center x. The topological realization $|y|$ is contained in the boundary of exactly one n-dimensional simplex $|M|$. Thus every point x that is in the interior of the topological realization $|y| \subseteq |K|$ of y is such that for all $\varepsilon > 0$ small enough we have that $B_\varepsilon(x) \cap |K|$ is homeomorphic to the upper half space $\{(x_1, \ldots, x_n) : x_j \in \mathbb{R}, x_n \geq 0\}$. Since S^n is an n-dimensional manifold without boundary, this is a contradiction to $|K|$ being homeomorphic to S^n. Thus there must be a minimal element $m \in T_K \setminus r[T_K]$, which proves the claim.

Now let $f : r[T_K] \to r[T_K]$ be an order-preserving map. Let $F := f \circ r$. For each minimal element $p \in T_K$ choose a minimal element $G(p) \in r[T_K]$ such that $G(p) \leq F(p)$. For $q \in T_k$ not minimal let

$$G(q) := \bigvee \{G(p) : p \in T_K, \ p \leq q \text{ is minimal }\}.$$

Then G is order-preserving on T_K. By Proposition 4.1.4 $G^{|T_K|!}$ is a retraction. Above we have shown above that the retract does not contain all minimal elements. Thus $G^{|T_K|!}$ induces a simplicial map on K that is a *nontrivial* retraction. This map can be extended to a continuous retraction $R : |K| \to |K|$. Now by Lemma 6.4.3, $R[|K|] = |G^{|T_K|!}[K]|$ has the topological fixed point property. Thus the continuous map G_c on $R[|K|]$ induced by G has a fixed point p. Let S be the smallest simplex in $G^{|T_K|!}[K]$ such that $p \in G_c[|S|] \subseteq R[|K|]$. Then G maps S to a sub-simplex of S, that is, $G(S) \supseteq S$ and thus G has a fixed point. Thus $F \geq G$ has a fixed point, which must be a fixed point of f.   ∎

Theorem 6.4.6 *Let K be a triangulation of S^n that has a realization $|K|$ such that the antipodal map $x \mapsto -x$ maps k-simplexes of K to k-simplexes of K. Then the truncated face lattice T_K of K is minimal automorphic.*

Proof. For each simplex S of K let $A(S) := -S$. By hypothesis this is well-defined. Since no simplex is equal to its antipode, $A : T_K \to T_K$ is a fixed point free order-preserving automorphism of T_K. By Lemma 6.4.5 all nontrivial retracts of T_K have the (order-theoretical) fixed point property.   ∎

For an example of a triangulation and truncated face lattice as in Theorem 6.4.6, consider Figure 6.1.

Remark 6.4.7 It is clear that the above construction can be carried out for other kinds of polyhedral complexes such as Bucky balls etc. Also it is clear that this is a result about the fixed clique and the fixed simplex properties as well as the fixed point property.

■

6.5 Cutsets

We are now going to prove some results that depend on the use of algebraic topology, homology to be precise. The reader not versed in these fields has the option to read Appendix A to gain the necessary preliminaries or to simply use Theorems 6.5.1, 6.5.3, and 6.5.4 as axioms. Especially on first reading the second approach may be more efficient.

The main idea from algebraic topology that we will use is the notion of acyclicity. Acyclicity is defined through the homology complex (cf. Definition A.3.8; in particular, though the word may suggest it, it has nothing to do with the absence of cycles in the graph itself). Acyclicity implies the fixed clique and fixed simplex properties.

Theorem 6.5.1 (This is exactly Lemma A.3.9.) *Let G be a finite acyclic graph. Then G has the fixed clique property.*

The key combinatorial notion will be Constantin and Fournier's notion of a (weakly) escamotable point.

Definition 6.5.2 *Let $G = (V, E)$ be a finite graph. For $v \in V$, let the* **neighborhood** *of v be $N(v) := \{x \in V : \{v, x\} \in E\}$. Call $v \in V$* **weakly escamotable**[2] *iff $G[N(v) \setminus \{v\}]$ is acyclic.*

Escamotable points in graphs are such that their adding or removal does not affect whether the graph is acyclic or not. This is an approach we know well from Section 4.2.1.

Theorem 6.5.3 (This is exactly Corollary A.4.3.) *Let $G = (V, E)$ be a finite connected graph and let $v \in V$ be weakly escamotable. Then for all $q \in \mathbb{N}$ we have $H_q(G) = H_q(G[V \setminus \{v\}])$ (these are the homology groups defined in Definition A.3.2). In particular, $G[V \setminus \{v\}]$ is acyclic iff G is acyclic.*

The last result whose proof we relegate to Appendix A is that (unsurprisingly) sufficiently nice graphs are acyclic.

---

[2]The original notion is that of an *escamotable* point, cf. [41] Définition 3.1. A point is escamotable iff its pointed neighborhood is contractible. We use the weaker notion here, as it allows us to avoid the use of topology in our proofs. We also obtain slightly stronger fixed point results. The word "escamotable" is French and means "retractable". To avoid confusion we use the original French term.

Theorem 6.5.4 (Easy consequence of Lemma A.4.4.) *A finite graph that has a vertex that is adjacent to all other vertices is acyclic.*

The main theorem that applies algebraic topology to the fixed point theory for ordered sets is now a removal theorem that reduces the analysis of a graph to the analysis of a certain subgraph. It appears to be a standard technique in algebraic topology to remove excess points from complexes under investigation to ease the analysis.

Definition 6.5.5 *Let $K = (V, S)$ be a simplicial complex. Then the simplicial complex $K' := K[V'] := (V', S')$ is called a* **full subcomplex** *of K iff $V' \subseteq V$ and for all simplexes $S \in S$ with $S \subseteq V'$ we have $S \in S'$.*

Theorem 6.5.6 (Compare [41], Théorème 4.1.) *Let $K = (V, S)$ be the clique complex of a finite graph. Let $B \subseteq V$ be such that for every nonempty simplex $S \subseteq V \setminus B$ the simplicial complex $K[\{v \in B : S \cup \{v\} \in S\}]$ is acyclic. Then every $x \in V \setminus B$ is weakly escamotable. Moreover $H(K)$ is isomorphic to $H(K[B])$ and K is acyclic iff $K[B]$ is acyclic.*

Proof. This is an induction on $n := |V \setminus B|$. For $n = 1$ this is Theorem 6.5.3. So now let $n > 1$ and suppose the result holds for all $k < n$. Let $x \in V \setminus B$ and let $V' := \{v \in V : \{v, x\} \in S\} \setminus \{x\}$, $K' = K[V']$ and let $B' := V' \cap B \neq \emptyset$. We claim K' and B' satisfy the hypothesis of the theorem. In fact if $S \subseteq V' \setminus B'$ is a simplex in K', then by the clique condition $S \cup \{x\}$ is a simplex in K and again via the clique condition

$$K[\{v \in B' : S \cup \{v\} \in S\}] = K[\{v \in B : (S \cup \{x\}) \cup \{v\} \in S\}],$$

where the latter simplicial complex is acyclic. Thus K' and B' satisfy the hypothesis of the theorem and since $|V' \setminus B'| < n$ the induction hypothesis can be applied.

Now note that $K[B'] = K[\{v \in B : \{v, x\} \in S\}]$ is acyclic by assumption. Thus by induction hypothesis K' is acyclic and thus x is weakly escamotable.

To prove the "moreover" part we need to prove that K is "dismantlable via weakly escamotable points" to $K[B]$. To see this it is good enough to show that $K[V \setminus \{x\}]$ and B satisfy the assumption of the theorem. Let $S \subseteq V \setminus (\{x\} \cup B)$ be a simplex. Then $K[\{v \in B : S \cup \{v\} \in S\}]$ is acyclic by assumption. This finishes the induction. ∎

Clearly a set B as above must intersect every maximal simplex, which is exactly the notion of a cutset.

Definition 6.5.7 *Let $K = (V, S)$ be a finite simplicial complex. Then $A \subseteq V$ is called*

1. *A* **cutset** *iff A intersects every maximal simplex,*

2. *An* **astral subset** *iff there is a $k \in V$ such that for all $a \in A$ we have $\{a, k\} \in S$,*

3. *An* **astral subset with a center** *c iff*

  (a) *A is astral,*

  (b) *For all a* $\in$ *A we have* $\{a, c\} \in S$,

  (c) *If k* $\in$ *V is such that for all a* $\in$ *A we have* $\{a, k\} \in S$, *then* $\{k, c\} \in S$,

4. *A* **coherent cutset** *iff A is a cutset and every astral subset of A has a center,*

5. **Semi-bounded by** *B iff A* $\subseteq$ *B, every element of B is a center of a subset of A and for every astral subset S of A the set B contains one center of S.*

The name "cutset" is well motivated, as a cutset essentially "cuts into" all parts of a simplicial complex. Astral subsets *A* can be visualized like stars with rays emanating from *k* to the elements of *A*. However note that the element *k* need not be in *A*. In ordered sets the points *k* often are upper or lower bounds, though they need not be (cf. Exercise 16). The center of an astral subset *A* is essentially the most natural choice for the heart of the "star" made up by an astral subset. In ordered sets the centers will normally be a supremum or an infimum, though, again, they need not be. Semi-boundedness assures that everything that should have a center actually has one. This is similar to the idea of conditional completeness (everything that needs a supremum has one). Semi-boundedness allows us to choose a smaller number of all the centers that might be available (normally one for each astral set), which makes combinatorial work easier.

The above leads to the first result, which says that an ordered set with a sufficiently nice cutset has the fixed point property. While Theorem 6.5.6 provides a general framework in case one has a good handle on the property of acyclicity, Theorem 6.5.8 puts us back on more solid combinatorial footing. All hypotheses are stated in combinatorial terms, which makes the visualization of the result slightly easier. For a special case that may clarify a few things, cf. Exercise 17. The proof relies once again on algebraic topology however.

Theorem 6.5.8 (Cf. [12], Corollary 2.6; [41], Thèoréme 4.6; [73], main theorem on p. 117; [233], Theorem 3; [242], Theorem 5.3.) *Let K* = (*V, S*) *be the clique complex of a finite graph. Suppose C is a coherent cutset of K that is semi-bounded by B. Then every x* $\in$ *V* \ *B is weakly escamotable in K. In particular K is acyclic iff K[B] is and we also have that H(K) is isomorphic to H(K[B]).*

Proof. We will prove that the simplicial complex satisfies the assumption of Theorem 6.5.6. Let *S* $\subseteq$ *V* \ *B* be a simplex. Then the set $\{v \in B : S \cup \{v\} \in S\}$ has a center, which is contained in the set (recall that we are considering a graph, so the clique condition holds). Thus by Theorem 6.5.4 it is acyclic. ∎

6.6 Truncated Noncomplemented Lattices

We conclude our presentation of truncated lattices with one of the most intriguing results in the fixed point theory of ordered sets. It is Baclawski and Björner's result that every finite truncated noncomplemented (cf. Definition 6.6.1) lattice has the fixed point property. The original proof (cf. [12], Corollary 3.2) is quite complex (it ultimately goes back to the intricate topological arguments in [18], Theorem 3.2, and to the arguments in [10], Corollary 6.3, respectively) and seems to be not algorithmic. We use the arguments of Section 6.5 to give another proof of this result. Analysis of Section 6.5 and Appendix A reveals that we have a completely algebraic proof of the result. The problem of finding a combinatorial proof was claimed to be solved by Baclawski in [11]. Even though Baclawski and Björner's result is topological and should thus have analogues for the fixed simplex and the fixed clique property, we need to assume that we are working with ordered sets for our proof.

Definition 6.6.1 *Two points x, y in a lattice L with largest element $\mathbf{1}$ and smallest element $\mathbf{0}$ are called* **complements** *iff*

$$x \vee y = \mathbf{1} \qquad \text{and} \qquad x \wedge y = \mathbf{0}.$$

A lattice L with largest element $\mathbf{1}$ and smallest element $\mathbf{0}$ is called **noncomplemented** *iff there is an $x \in L$ such that for all $y \in L$ we have $x \vee y < \mathbf{1}$ or $x \wedge y > \mathbf{0}$.*

The next notion is a somewhat natural generalization of noncomplementedness.

Definition 6.6.2 (Cf. [41], Définition 5.1 and preceeding remarks.) *Let P be a finite ordered set. We define the* **lower distance** $d(x, y)$ *from $x \in P$ to $y \in P$ to be the length n of the shortest fence $x = x_0 \geq x_1 \leq x_2 \geq \cdots x_n = y$ from x to y. We set*

$$x \bigvee y := \{x_{n-1} : x = x_0 \geq x_1 \leq x_2 \geq \cdots x_{n-1} \sim x_n = y \text{ and } n \text{ minimal}\}.$$

$r_x := \max\{n \in \mathbb{N} : d(x, y) = n \text{ for some } y \in P\}$ *will be called the* **lower x-radius** *of P. P will be called* **weakly noncomplemented from below** *("mal complémenté par le bas" in the original) iff there is an $x \in P$ such that for all $y \in P \setminus \{x\}$ the set $x \bigvee y$ is acyclic.*

Baclawski and Björner's classical result now easily follows.

Theorem 6.6.3 (Cf. [41], Théorème 5.2.) *Every finite ordered set that is weakly noncomplemented from below is acyclic and thus has the fixed point property.*

Proof. The proof is an induction on r_x, the lower x-radius of P, with x as in the definition of weak noncomplementedness from below. For $r_x \in \{0, 1\}$ there is nothing to prove as P has a largest element. Now assume that $r_x = n + 1$ and the theorem holds for all weakly noncomplemented from below sets P with $r_x \leq n$. Then $B := \{y \in P : d(x, y) \leq n\}$ is weakly noncomplemented from below and thus acyclic. It is thus sufficient to verify that the conditions of Theorem 6.5.6 are satisfied. (This is again a nice exercise in the translations between ordered sets, graphs and simplicial complexes.) We can assume without loss of generality that n is even.

Let C be a chain in $P \setminus B$ and let y be the largest element of C. Then there is a fence $x = x_0 \geq x_1 < x_2 > \cdots < x_n > x_{n+1} = y$ from x to y. We will now show that $\{v \in B : C \cup \{v\} \text{ is a chain }\} = x \bigvee y$, which is acyclic, thus concluding the proof. Let $v \in B$ be such that $C \cup \{v\}$ is a chain. Then there is a fence $x = v_0 \geq v_1 < v_2 > \cdots \sim v_{k-1} \sim v_k = v$ with k minimal. Moreover since v is comparable to y we infer $k = n$ and that $v_{n-2} > v_{n-1} < v_n = v > y$ (otherwise $d(x, y) \leq n$). Thus $v \in x \bigvee y$. Conversely if $v \in x \bigvee y$, then $d(x, v) \leq n$, so $v \in B$. Moreover we must have $v > y$ (otherwise $d(x, y) \leq n$), which means that $C \cup \{v\}$ is a chain. This proves that $\{v \in B : C \cup \{v\} \text{ is a chain }\} = x \bigvee y$ and we are done.   ∎

Theorem 6.6.4 *Every finite truncated noncomplemented lattice T is weakly noncomplemented from below. Thus every finite truncated noncomplemented lattice has the fixed point property.*

Proof. Let $x \in T$ be an element that has a supremum or an infimum with any other element of T. For all y with $x \sim y$ we have that $x \bigvee y$ has a largest element, which means $x \bigvee y$ is acyclic. If $x \not\sim y$, then x and y have a supremum or an infimum. If $x \wedge y$ exists, it is the largest element of $x \bigvee y$, which is then acyclic. Otherwise $x \vee y$ exists. Then every element of $x \bigvee y$ has an infimum with $x \vee y$, which means $x \bigvee y$ is C-dismantlable to the singleton $\{x \vee y\}$. Again $x \bigvee y$ is acyclic. This proves that T is weakly noncomplemented from below and thus T has the fixed point property.   ∎

There is an interesting fact to be noted here. By Theorem 6.3.16 there is a direct connection between the fixed point property for finite truncated lattices and the fixed simplex property for simplicial complexes. Yet to prove that finite truncated noncomplemented lattices have the fixed point property we are going "the long route" through the P-chain complex, rather than the direct route through Theorem 6.3.16. Are there stronger direct applications of algebraic topology through Theorem 6.3.16?

Exercises

1. Prove that every retract of a truncated lattice is again a truncated lattice. Prove the same result for conditionally complete ordered sets.

2. Prove that a finite ordered set L is a truncated lattice iff for any two antichains $\{a, b\}$ and $\{c, d\}$ with $\{a, b\} \leq \{c, d\}$ there is an element $m \in L$ that is above $\{a, b\}$ and below $\{c, d\}$.

3. Let T be a finite truncated lattice and let $f, g : T \to T$ be automorphisms. Prove that if $f|_{J(T) \cup \mathrm{Min}(T)} = g|_{J(T) \cup \mathrm{Min}(T)}$, then $f = g$.

4. Let P be a conditionally complete ordered set that has no largest and no smallest element. Prove that the Dedekind–MacNeille completion of P is the set P with a largest element $\mathbf{1}$ and a smallest element $\mathbf{0}$ attached.

5. Truncated Dedekind–MacNeille completions and the fixed point property.

   (a) Find a finite ordered set P that does not have the fixed point property such that $DM(P) \setminus \{\mathbf{0}, \mathbf{1}\}$ has the fixed point property.

   (b) Find a finite ordered set P that has the fixed point property such that $DM(P) \setminus \{\mathbf{0}, \mathbf{1}\}$ does not have the fixed point property.

6. Let C be the card used in the proof of Theorem 6.2.5. Prove that if T has at least two maximal and two minimal elements, then $T = DM(C) \setminus \{\mathbf{0}, \mathbf{1}\}$. (Hint: Use Theorem 5.3.8.)

7. Let L be a finite lattice and let $x \in L$. Prove that $L \setminus \{x\}$ is a lattice iff x is irreducible.

8. Let P, Q be ordered sets and let $f : P \to Q$ be order-preserving. Prove that f is a graph homomorphism from $G_C(P)$ to $G_C(Q)$.

9. Show that the graph in Figure 6.1 is not a comparability graph.

10. Show that the truncated clique/face lattice in Figure 6.1 is minimal automorphic. (Warning: The author believes this is very tedious. Hence computer searches should be permissible. For more on algorithmic approaches, cf. Section 12.5.)

11. (Abian–Brown type theorems for the fixed clique property.) Let $G = (V, E)$ be a graph and let $f : V \to V$ be a graph endomorphism.

    (a) Prove that if there is a finite clique C with $f[C] \subseteq C$, then there is a clique $K \subseteq C$ with $f[K] = K$.

    (b) Define $N(v) := \{w \in V : w \sim v\}$. Let $v \in V$ be such that $f[N(v)] \subseteq N(v)$. Prove that there is a clique $K \subseteq V$ such that $f[K] \subseteq K$.

12. Prove that the truncated clique lattice of a $2n$-crown is a $4n$-crown.

13. Find the truncated clique lattice of

    (a) An n-fence,

    (b) An n-chain,

    (c) An n-antichain.

14. Fixed cliques vs. invariant cliques.

    (a) Show that if the graph G has no infinite cliques and the invariant clique property, then G has the fixed clique property.

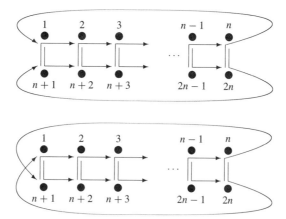

Figure 6.3: Two nonisomorphic ternary relations with the same decks. Every ternary edge (a, b, c) as indicated contains the points nearest its ends and the point nearest its corner. The point in the corner is b. The rightmost point in the figure is c, except for the edges that wrap around. For these edges the leftmost point in the figure is c.

(b) Show that no graph with an infinite clique can have the fixed clique property.

15. Prove that if T is a finite truncated lattice, $x \in T$, $f : T \to T$ is order-preserving, $f^n(x) = x$ and $\{x, f(x), \ldots, f^n(x) = x\}$ has an upper bound, then f has a fixed point.

16. Find an ordered set that has an astral subset A with a center such that the center is not part of A and is not an upper or a lower bound of A.

17. Let P be a finite ordered set that has a cutset C such that all nonempty subsets of C have a supremum or an infimum. Prove that P has the fixed point property. Can you find a proof that does not rely on algebraic topology (cf. [73])?

Remarks and Open Problems

1. It has been proved in [128] that if two ordered sets have the same comparability graph and the same decks, then they must be isomorphic. Thus graph reconstruction implies order reconstruction. It is not known if order reconstruction is equivalent to graph reconstruction. The following problem is equivalent to graph reconstruction. Is every truncated clique lattice T of a graph reconstructible from its deck of subsets $T \setminus (\uparrow m)$ where m is minimal

in T? It is now natural to ask if ordered sets in general are reconstructible from these maximal filter deleted subsets.

Unfortunately for ordered sets in general the answer is negative. In [197] it is shown that ternary and higher order relations are not reconstructible from their decks. A counterexample for ternary relations is given in Figure 6.3. The key idea is that a Möbius band untwists when it is cut and that for higher order relations one can model a cut with the removal of a point.

From this example we can construct nonisomorphic ordered sets whose maximal filter deleted decks are isomorphic. The points of the respective ordered sets are the singletons $\{1\}, \ldots, \{2n\}$, for each ternary edge (a, b, c) the triples $(\{a, b, c\}, 1, 1), (\{a, b, c\}, 2, 1), (\{a, b, c\}, 2, 2), (\{a, b, c\}, 3, 1),$ $(\{a, b, c\}, 3, 2), (\{a, b, c\}, 3, 3),$ and the set $\{a, b, c\}$. The comparabilities are that each $(\{a, b, c\}, i, j)$ is below $(\{a, b, c\}, i, j')$ iff $j \leq j'$ and each $(\{a, b, c\}, i, j)$ is below $\{a, b, c\}$; each $\{x\}$ is below $(\{x, y, z\}, i, 1)$ iff there is a ternary edge with elements x, y, z such that x is the ith element. These, plus the comparabilities dictated by transitivity are all comparabilities. The two possible sets constructed as above from the example in Figure 6.3 are not isomorphic, yet their decks of maximal filter deleted subsets are.

Note that this is an example involving general ordered sets, not truncated lattices. The question if truncated lattices or truncated clique lattices are reconstructible from their maximal filter deleted subsets remains thus (of course) open. Are there any lattice theoretical insights that can help to advance knowledge on graph reconstruction through this formulation of the problem? The reader who is interested in counterexamples in reconstruction should also consider the proof in [256] that tournaments are not reconstructible.

2. For more on algebraic topology cf. [253] and Appendix A.

3. For more on fixed clique properties in graphs also cf. [13, 129]. There is also a connection to the Helly property, cf. [206, 207, 208].

4. It would be nice to have a combinatorial proof of Theorem 6.5.8, which might provide us with stronger fixed point theorems that are also applicable in the infinite setting. Combinatorial proofs of the special case of Theorem 6.5.8 in which the only maximal astral subset of C is C itself are given in [73], [233], Theorem 3 and [242], Proposition 5.2.

   In [11] a combinatorial proof of related results is announced.

5. There is no known good characterization of which finite truncated lattices have the fixed point property. The problem is conjectured to be "intractable", but it has not (yet?) been proved to be co-NP-complete.

6. It is notable that while the results in Section 6.5 rely on algebraic topology, the fact that the complexes in question are induced by graphs is indispens-

able in the proofs. Is there a way to prove the results of Section 6.5 for simplicial complexes in general?

7. Underneath our algebraic/homological investigation of the fixed point property lurks the topological fixed point property and a sufficient condition called contractibility. While the author decided to keep the presentation discrete (to keep the involved quantities computable), the topological approach can yield powerful results.

Consider an ordered set P such that the topological realization $|P|$ of the P-chain complex of P has the topological fixed point property. Then the topological realization $|TCL(P)|$ of the chain-complex of the truncated chain lattice of P also has the topological fixed point property. This follows from the fact that the $TCL(P)$-chain complex of $TCL(P)$ is a barycentric refinement of the P-chain complex of P. This means that $|P|$ and $|TCL(P)|$ are isomorphic, which implies the result.

This means that iteration of the $TCL(\cdot)$ operator to the right starting set leads to a sequence of ordered sets that rapidly grows in size and such that every member has the fixed point property. It can be shown that if we start with a finite ordered set that is $\mathcal{I}$-dismantlable to a singleton, then all sets in this sequence are also $\mathcal{I}$-dismantlable to a singleton. However if we start with a finite ordered set that is not $\mathcal{I}$-dismantlable to a singleton, then the size of the $\mathcal{I}$-cores of these sets tends to infinity.

8. Part 7 shows that iterating the $TCL(\cdot)$ operator yields examples of large nontrivial truncated lattices with the fixed point property. There is another operator whose iteration leads to a sufficient condition for the fixed point property.

The **clique graph** of a graph G has as vertices the maximal (with respect to inclusion) cliques of G and two vertices are connected by an edge iff their intersection is not empty. It is shown in [200] that for every $\mathcal{I}$-dismantlable finite ordered set there is an n such that the nth iterated clique graph of the comparability graph is a singleton. In [111] it is shown that for finite ordered sets, if the nth iterated clique graph of the comparability graph is a singleton, then the ordered set has the fixed point property. Moreover in [111] examples are given which show that this "collapsing of the iterated clique graphs" is not equivalent to dismantlability.

What is the relation between the "collapsing of the iterated clique graphs" and the fixed point property? The properties appear to be not equivalent, but there is no proof that the author is aware of. What is the relation between the "collapsing of the iterated clique graphs" and the other properties discussed in this section?

7

The Dimension of Ordered Sets

Dimension theory is a prominent area in ordered sets. So far, we have looked at specific ordered structures and investigated their properties. In dimension theory one represents orders using the orders that occur most frequently, namely total orders.

As for Chapter 5 on lattices it must be said that this chapter can only provide brief exposure to the basics of dimension theory. For a thorough presentation of this subject consider [263].

7.1   (Linear) Extensions of Orders

In many situations a given order must be turned into a linear order. For example, one may have several tasks to be processed on a machine (say the whole job is to produce a booklet on a sophisticated copier[1]). Some of these tasks depend on the output of other tasks (booklets cannot be stapled before all pages are produced and sorted), but there are also pairs of tasks that could be processed in either order (if some pages are to have a different color, either color could be run first). The relation "task y needs the output of task x" defines an order relation on the set of tasks. Tasks that are independent of each other are not comparable. Still, to perform the job on one machine, all tasks have to be put in one linear order.

Similarly, in evaluation of employees to assign raises, bonuses, etc. an employer may have comparable and incomparable performances. For example, the

---

[1] If this is not interesting enough, consider scheduling computations on a processor.

quality of papers of two mathematicians working in the same research area might be comparable. Yet the same department may also have two members that have published outstanding works in entirely different areas. Say, one on nonlinear partial differential equations and the other on algebraic topology. How would one rank these accomplishments? To be fair, it has to be said that a valuation of one subject over the other often depends on personal preferences. In an order which puts greater accomplishments higher than lesser ones, the two works would be incomparable. And yet, these incomparabilities have to be resolved when raises are assigned since any two raises (being numbers) are comparable. Thus here also an order has to be (or is inadvertently) turned into a linear order.[2] Ties are possible, but for our work we will assume ties are broken somehow.

It will be our first task to investigate how an order can be turned into a linear order. The first step is to show that to any order we can add another comparability of previously incomparable elements. For the first time in this text we need to consider the order relation separate from the underlying set. This is because we will need to keep track of several different order relations on the same underlying set.

Lemma 7.1.1 *Let $(P, \leq)$ be an ordered set and let $p, q \in P$ be such that $p \not\sim q$. Then there is an order $\leq'$ on P such that $\leq$ is contained in $\leq'$ and $p \leq' q$.*

Proof. Let $\leq'$ be the transitive closure of $\leq \cup \{(p, q)\}$. Clearly $\leq'$ is reflexive and transitive. Antisymmetry remains to be proved. Let $a \leq' b$ and $b \leq' a$. We need to prove that $a = b$ and we will do so by showing $a \leq b$ and $b \leq a$.

First assume that $a \not\leq b$. Then $a \leq' b$ implies $a \leq p$ and $q \leq b$. Now, since $p \not\sim q$, we must have $b \not\leq a$. This, in turn, since $b \leq' a$, implies $b \leq p$ and $q \leq a$. We can now conclude $q \leq b \leq p$, a contradiction. Hence we must have $a \leq b$.

The symmetric argument shows that $b \leq a$ also must hold. Now $a \leq b$ and $b \leq a$ naturally imply $a = b$.

Thus $\leq'$ is an order relation that contains $\leq$ and also the new comparability $p <' q$. ∎

Lemma 7.1.1 guarantees that for any order relation that is not a total order, there are larger order relations that contain it. We shall now investigate the structure of the set of orders that contain a given order $\leq$. The last part of the following theorem is also known as Szpilrajn's Theorem (cf. [259], every order $\leq$ is contained in a linear order).

Theorem 7.1.2 *Let $(P, \leq)$ be an ordered set and let*

$$\mathcal{E}_\leq := \{\sqsubseteq : \sqsubseteq \text{ is an order on } P \text{ and } \sqsubseteq \text{ contains } \leq\}.$$

---

[2] And, depending on the climate at the institution, the resulting order may be the subject of heated discussion.

Then $\mathcal{E}_\le$ is a conditionally complete ordered set when ordered by inclusion. The maximal elements of $\mathcal{E}_\le$ are linear orders. Finally (Szpilrajn's Theorem) every element $\sqsubseteq \in \mathcal{E}_\le$ is below a maximal element.

Proof. To prove $\mathcal{E}_\le$ is conditionally complete, let $\mathcal{A} \subseteq \mathcal{E}_\le$ be a subset of $\mathcal{E}_\le$ with upper bound $\sqsubseteq$. Let $\sqsubseteq_A$ be the transitive closure of $\bigcup \mathcal{A}$. Then $\le$ is a subset of $\sqsubseteq_A$, $\sqsubseteq$ is a superset of $\sqsubseteq_A$, and $\sqsubseteq_A$ contains all relations in $\mathcal{A}$. $\sqsubseteq_A$ is transitive by definition, reflexivity follows because $\sqsubseteq_A$ contains all orders in $\mathcal{A}$ and antisymmetry follows because $\sqsubseteq_A$ is contained in $\sqsubseteq$. Since every order that contains all orders in $\mathcal{A}$ must contain the transitive closure of the union of $\mathcal{A}$, $\sqsubseteq_A$ is the supremum of $\mathcal{A}$ in $\mathcal{E}_\le$.

Since $\mathcal{E}_\le$ has $\le$ as its smallest element, the above already shows $\mathcal{E}_\le$ is conditionally complete. To shed another light on the structure of $\mathcal{E}_\le$ it is worth mentioning that infima are obtained via intersections. This is because the intersection of any family of orders is again an order (cf. Chapter 1, Exercise 2).

To show that the maximal elements of $\mathcal{E}_\le$ are all linear orders, note that if $\sqsubseteq$ is not a linear order, we can use Lemma 7.1.1 to find a superset $\sqsubseteq'$ of $\sqsubseteq$ that is in $\mathcal{E}_\le$.

Finally to show that every element $\sqsubseteq \in \mathcal{E}_\le$ is below a maximal element, note that $\uparrow_{\mathcal{E}_\le} \sqsubseteq$ is inductively ordered. Indeed for every chain $\mathcal{C}$ of orders that contain $\sqsubseteq$, the union $\sqsubseteq_u := \bigcup \mathcal{C}$ is an upper bound of $\mathcal{C}$ in $\mathcal{E}_\le$. Thus by Zorn's lemma $\uparrow_{\mathcal{E}_\le} \sqsubseteq$ has maximal elements. ∎

Knowing that every order relation can be extended to a linear order, the following definition is sensible.

Definition 7.1.3 *Let $(P, \le)$ be an ordered set. Then $\sqsubseteq$ is called a* **linear extension** *of $\le$ iff $\sqsubseteq$ is a total order and it contains $\le$.*

We conclude with a note on the number of linear extensions of a finite ordered set. The proof also gives an idea how to compute all linear extensions of an order.

Proposition 7.1.4 *For a finite ordered set P, let $L(P)$ be the number of linear extensions of P. Then*

$$L(P) = \sum_{m \text{ minimal}} L(P \setminus \{m\}).$$

Proof. For an ordered set Q let $\mathcal{L}(Q)$ be the set of all linear extensions of Q. The smallest element in a linear extension of $\le$ must be a minimal element of P. Thus

$$\mathcal{L}(P) = \bigcup_{m \text{ minimal}} \mathcal{L}(P \setminus \{m\}).$$

Since the sets on the right side are pairwise disjoint, the result follows. ∎

7.2   Balancing Pairs

Clearly if $a \leq b$, then a will be below b in any linear extension of $\leq$. What can be said for incomparable elements? Is there a bound for how many times one element can occur above another in a linear extension? Would such a result have applications in the ordering of tasks?

To address this issue, (before we continue on to dimension theory) let us consider the subject of sorting from an order-theoretical point-of-view. Essentially, whenever a set of objects has to be sorted, the task is to find a linear order $\leq_l$ for the objects. This order $\leq_l$ already exists and can be evaluated for any pair of objects. However, the whole order, in the form of a list of the objects in that order, is not yet known to us. Moreover the decision if $a <_l b$ for given elements a, b may require some effort, computational or otherwise. (Consider sorting the words in a dictionary or listing computer files in order of increasing dates.)

The goal of an efficient sorting algorithm is then to find a list of the objects in the order $\leq_l$ by using as few evaluations of $\leq_l$ as possible (that is, as little unnecessary effort as possible). At intermediate stages in the sorting, the prior evaluations of $\leq_l$ induce a (partial) order $\leq$ on the elements.

Clearly $\leq_l$ is a linear extension of $\leq$. Thus sorting a list can also be seen as the task of finding a particular linear extension of a given order relation. Every evaluation of $\leq_l$, say for a, b with $a \leq_l b$, rules out all the linear extensions $\leq'$ in which $b \leq' a$. The sorting is most efficient if at each stage the evaluation of $\leq_l$ rules out many linear extensions. It is inefficient if an evaluation of $\leq_l$ rules out few linear extensions. Structurally a sorting algorithm that functions efficiently produces a very short chain in $\mathcal{E}_\leq$ from $\leq$ to $\leq_l$.

How fast could we guarantee such a search algorithm to be in principle? To tackle this problem we define for any two incomparable elements the proportion of linear extensions of $\leq$ in which a is less than b.

Definition 7.2.1 *Let P be a finite ordered set and let $a \not\sim b$ be elements of P. We define* **the proportion of linear extensions of $\leq$ in which a is less than b** *as*

$$P(a < b) := \frac{\text{number of all linear extensions } \leq_e \text{ of } \leq \text{ with } a <_e b}{\text{number of all linear extensions of } \leq}.$$

What one would want for efficient sorting is to find lots of pairs a, b such that $a <_l b$ and $P(a < b)$ is small. Yet this cannot be guaranteed. Moreover, if $P(a < b)$ is small and $a >_l b$, then the comparison was essentially wasted. Thus, to obtain a performance guarantee for sorting one needs the existence of pairs a, b such that $P(a < b)$ is near $\frac{1}{2}$. Independent of whether a is on the "right" or the "wrong" side of b, the comparison will reduce the remaining effort by a factor near 2. Pairs with $P(a < b)$ near $\frac{1}{2}$ are called **balancing pairs**.

Example 7.2.2 Regarding the proportion $P(a < b)$ we can record the following.

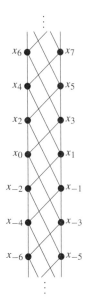

Figure 7.1: An ordered set Q that shows that existence of points a, b such that $\dfrac{5 - \sqrt{5}}{10} \le P(a < b) \le \dfrac{5 + \sqrt{5}}{10}$ is best possible when using a certain infinitary generalization of the proportion $P(a < b)$.

1. $P(a < b)$ is always a rational number. The following shows that every rational number $\dfrac{n}{d} \in (0, 1)$ occurs as a proportion $P(a < b)$ in some ordered set. Let P_d be an ordered set that consists of a $(d - 1)$-chain $1, \ldots, d - 1$ with another element e added that is incomparable with any element except itself. Then $P(e > d - n) = \dfrac{n}{d}$.

2. (Cf. [27].) This example is an infinite ordered set for which we will compute the proportion $P(a < b)$ in a sequence of subintervals. In a way this example shows that the currently best result regarding the $\dfrac{1}{3} - \dfrac{2}{3}$ problem (cf. Open Question 7.2.3 below) is optimal if the scope of the problem is widened slightly. The ordered set Q has a countable ground set $\{x_n : n \in \mathbb{Z}\}$ ordered such that $x_i < x_j$ iff $i + 1 < j$. (For an illustration, cf. Figure 7.1.) Let Q_n be the ordered subset of Q with ground set $\{x_{-2n}, x_{-(2n-1)}, \ldots, x_{(2n-1)}, x_{2n}\}$. We shall prove that

$$\lim_{n \to \infty} P_{Q_n}(x_0 > x_1) = \frac{5 - \sqrt{5}}{10}.$$

Let a_n be the number of linear extensions of the subset of Q_n with ground set $\{1, \ldots, 2n\}$ and let b_n be the number of linear extensions of the subset of

Q_n with ground set $\{2, \ldots, 2n\}$. The point x_0 has three possible positions in a linear extension of Q_n: above x_1, below x_{-1} or between x_{-1} and x_1. Let the numbers of such linear extensions be $L_{x_0 > x_1}$, $L_{x_0 < x_{-1}}$, and $L_{x_{-1} < x_0 < x_1}$, respectively. Then, using symmetry, duality and the fact that adding or removing a smallest element does not change the number of linear extensions we obtain

$$L_{x_0 > x_1} = a_n b_n, \quad L_{x_0 < x_{-1}} = a_n b_n, \quad L_{x_{-1} < x_0 < x_1} = a_n^2.$$

Consequently we have

$$P_{Q_n}(x_0 > x_1) = \frac{L_{x_0 > x_1}}{L_{x_0 > x_1} + L_{x_0 < x_{-1}} + L_{x_{-1} < x_0 < x_1}}$$

$$= \frac{a_n b_n}{a_n b_n + a_n b_n + a_n^2} = \frac{1}{\frac{a_n}{b_n} + 2}.$$

Moreover, a_n and b_n satisfy the recursion

$$a_n = b_n + a_{n-1} \quad \text{and} \quad b_n = a_{n-1} + b_{n-1}.$$

This implies

$$\frac{a_{n-1}}{b_n} = \frac{1}{1 + \frac{b_{n-1}}{a_{n-1}}} \quad \text{and then} \quad \frac{a_n}{b_n} = \frac{1}{1 + \frac{b_{n-1}}{a_{n-1}}} + 1,$$

which implies

$$P_{Q_n}(x_0 > x_1) = \frac{1}{\frac{a_n}{b_n} + 2} = \frac{1}{\frac{1}{1 + \frac{b_{n-1}}{a_{n-1}}} + 3}.$$

Moreover, we have

$$\frac{b_n}{a_n} = \frac{1}{1 + \frac{1}{1 + \frac{b_{n-1}}{a_{n-1}}}}.$$

From these equations $P_{Q_n}(x_0 > x_1)$ can be computed. More importantly for what follows, we have $\lim\limits_{n \to \infty} \dfrac{b_n}{a_n} = \dfrac{-1 + \sqrt{5}}{2}$, which finally implies that

$$\lim_{n \to \infty} P_{Q_n}(x_0 > x_1) = \frac{5 - \sqrt{5}}{10}. \qquad \blacksquare$$

Now assume that there is a $c > 0$ such that for each ordered set there exist a, b with $c \leq P(a < b) \leq 1 - c$. Then a bound on the efficiency of a sorting algorithm can be given. If, at every step, we were able to find the points a, b with $c \leq P(a < b) \leq 1 - c$, then the number of remaining linear extensions is reduced by at least a factor $(1 - c)$ in every step of the search. The search stops when the resulting order is total. If n is the number of steps, then we must have that $(1 - c)^n L(P, \leq) \leq 1$ or $n \leq \log_{1-c} \left(\dfrac{1}{L(P, \leq)} \right) = -\dfrac{\ln(L(P, \leq))}{\ln(1 - c)}$. The closer $1 - c$ is to $\dfrac{1}{2}$, the smaller this quantity is. It has been proved that such numbers c exist, though none of the proofs fits our scope. The question that remains is how large a c is possible. The following problem addresses this issue.

Open Question 7.2.3 The $\dfrac{1}{3}$-$\dfrac{2}{3}$ problem (Cf. [139, 239].) *If P is a finite ordered set that is not a chain, is it true that there is a pair of incomparable elements $a, b \in P$ such that*

$$\frac{1}{3} \leq P(a < b) \leq \frac{2}{3}?$$

If the above question could be answered in the affirmative (which is the $\dfrac{1}{3}$-$\dfrac{2}{3}$ conjecture, originally in [139]), the bounds would be sharp. Indeed the bounds are realized in the three element ordered set $\{a, b, c\}$ with $a < b$ and no further comparabilities except reflexivity. In this set, all the proportions are either $\dfrac{1}{3}$ or $\dfrac{2}{3}$. For the latest on open question 7.2.3, cf. [30], where Brightwell, Felsner and Trotter prove that in any finite ordered set there are a, b such that

$$\frac{5 - \sqrt{5}}{10} \leq P(a < b) \leq \frac{5 + \sqrt{5}}{10}.$$

The paper [30] also contains a good overview on the history and the motivation for the study of balancing pairs.

The $\dfrac{5 - \sqrt{5}}{10} \leq P(a < b) \leq \dfrac{5 + \sqrt{5}}{10}$ bounds can be seen as optimal in a certain, slightly generalized sense. An infinite ordered set is called **thin** iff there is a $k > 0$ such that each element is incomparable with at most k others. The ordered set in Example 7.2.2, part 2 is thin. In [27] it is shown that for thin ordered sets the limit of the $P_n(a < b)$, computed for increasing sequences of subintervals $(\uparrow x_n) \cap (\downarrow y_n)$ such that eventually each point is in an interval, exists and is independent of the sequence of subintervals. For infinite thin ordered sets $P(a < b)$ is defined as that limit.

In the ordered set in Example 7.2.2, part 2 for any pair of incomparable elements a, b we have $P(a < b) \in \left\{ \dfrac{5 - \sqrt{5}}{10}, \dfrac{5 + \sqrt{5}}{10} \right\}$. Thus, in a way the above bounds are the best possible.

Structurally we can conclude with the following.

Proposition 7.2.4 *Let* $(P, \leq)$ *be a finite ordered set. Then for any linear extension* $\leq_l$ *of* $\leq$ *there is a chain in* $\mathcal{E}_\leq$ *from* $\leq$ *to* $\leq_l$ *such that the chain has at most*
$$-\frac{\ln(L(P, \leq))}{\ln\left(\frac{5+\sqrt{5}}{10}\right)} + 1 \text{ elements.}$$

7.3 Defining the Dimension

By Lemma 7.1.1 it is a trivial observation that every order is the intersection of all orders that contain it. In dimension theory one now wants to use "nice" (that is, in this case, linear) orders in the intersection, and one wants to use as few orders as possible. This is possible because of the following corollary to Lemma 7.1.1 and Theorem 7.1.2.

Corollary 7.3.1 *Every order* $\leq$ *on a set* P *such that there are* $a, b \in P$ *with* $a \not\sim b$ *has a linear extension* $\leq_1$ *such that* $a \leq_1 b$ *and another linear extension* $\leq_2$ *such that* $b \leq_2 a$. *Thus for every order* $\leq$ *there is a family* $\{\leq_\alpha\}_{\alpha \in I}$ *of linear orders such that* $\leq = \bigcap_{\alpha \in I} \leq_\alpha$.

Proof. Exercise 7.   ∎

The above means any order can be realized as the intersection of linear orders, giving rise to the following definition.

Definition 7.3.2 *Let* $(P, \leq)$ *be an ordered set. Then a family* $\{\leq_i\}_{i \in I}$ *of linear orders* $\leq_i$ *on* P *is called a **realizer** of* $\leq$ *iff* $\leq = \bigcap_{i \in I} \leq_i$.

Example 7.3.3 Consider the ordered set in Figure 1.4, part e). One realizer for its order is

$$\leq_1 \; : \; a <_1 b <_1 c <_1 d <_1 e <_1 g <_1 f <_1 h <_1 k,$$
$$\leq_2 \; : \; c <_2 b <_2 a <_2 f <_2 e <_2 k <_2 d <_2 h <_2 g,$$
$$\leq_3 \; : \; a <_3 b <_3 c <_3 d <_3 f <_3 h <_3 e <_3 g <_3 k,$$
$$\leq_4 \; : \; b <_4 c <_4 f <_4 a <_4 d <_4 e <_4 g <_4 h <_4 k,$$
$$\leq_5 \; : \; b <_5 d <_5 a <_5 c <_5 e <_5 f <_5 g <_5 h <_5 k.$$

Another realizer is

$$\leq_1 \; : \; a <_1 b <_1 c <_1 d <_1 e <_1 g <_1 f <_1 h <_1 k,$$
$$\leq_2 \; : \; c <_2 b <_2 f <_2 a <_2 e <_2 k <_2 d <_2 h <_2 g,$$
$$\leq_3 \; : \; a <_3 b <_3 d <_3 c <_3 f <_3 h <_3 e <_3 g <_3 k.$$

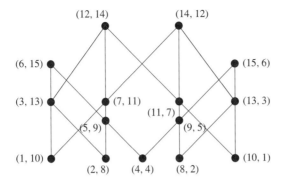

Figure 7.2: An ordered set of dimension 2. The linear orders whose intersection this order is are indicated in the picture. Alternatively, the up-set of every element is contained in the $90° + \varepsilon$ cone above the point (standard visualization for two-dimensional sets). The example is due to Rutkowski and it shows that there are ordered sets of dimension 2 that have the fixed point property (cf. Exercise 22) and no irreducible point.

Example 7.3.3 shows that realizers can have different numbers of orders in them. This automatically brings up the question what the smallest possible number of orders in a realizer would be. This is the definition of the dimension.

Definition 7.3.4 *Let* $(P, \leq)$ *be an ordered set. Then* $(P, \leq)$ *is of* **(linear) dimension** k *and we write* $\dim(P, \leq) = k$ *iff* $k \in \mathbb{N}$ *is the smallest natural number such that there is a realizer for* $\leq$ *that has* k *orders.*

Example 7.3.5

1. A four crown has dimension 2.

2. A six crown has dimension 3.

3. The ordered set in Figure 7.2 is of dimension 2.

Proof. We will only prove the claim in 1, since 2 will be proved in Example 7.3.7 and 3 is proved in Figure 7.2 itself.

Let $a, b \leq c, d$ be a four crown. Since a four crown is not a chain, it has dimension at least 2 and we are done if we can find a realizer with two orders. This realizer is given as follows. The first linear order is $a <_1 b <_1 c <_1 d$ and the second linear order is $b <_2 a <_2 d <_2 c$. ∎

Example 7.3.6 Let S be a set of points in $\mathbb{R}^k$, such that no two points are equal in any coordinate. Order S by $(x_1, \ldots, x_k) \leq (y_1, \ldots, y_k)$ iff $x_i \leq y_i$ for all $i = 1, \ldots, k$.

Then $(S, \leq)$ is an ordered set of dimension at most k. A realizer is given by the projections onto the coordinate axes. Indeed, any finite ordered set of dimension k can be embedded into $\mathbb{R}^k$ in this fashion. (This explains the terminology

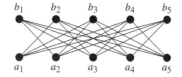

Figure 7.3: The standard example of a 5-dimensional ordered set.

"dimension". For a proof cf. Exercise 11 or Theorem 10.2.4.) Note, however, that not every ordered set that arises in this form is of dimension k. The set could be of lower dimension.   ∎

The visualization suggested in Example 7.3.6 is also given in Figure 7.2. The x-axis of $\mathbb{R}^2$ goes up at a $45° - \varepsilon$ angle versus the horizontal, while the y axis forms a $135° + \varepsilon$ degree angle with the horizontal. The indicated realizer is exactly what is obtained through projection on the x- and y-axes.

Example 7.3.7 The "standard example" St_n of an ordered set of dimension n.
As was mentioned in Example 7.3.6, we can geometrically construct ordered sets whose dimension is at most the dimension of the surrounding space. However, these sets could have dimension that is considerably less. For example, a chain has dimension 1, but it can be embedded into any $\mathbb{R}^n$ in the fashion of Example 7.3.6. The natural question is "Are there finite ordered sets of arbitrarily high dimension?" The following example answers this question in the affirmative.
Consider the ordered set $\{a_1, \ldots, a_n, b_1, \ldots, b_n\}$ ordered as follows.

1. $\{a_1, \ldots, a_n\}$ and $\{b_1, \ldots, b_n\}$ are both antichains.

2. No b_i is a lower bound of an a_j.

3. $a_i \leq b_j$ iff $i \neq j$.

Call an ordered set as above St_n. For a visualization of these sets, cf. Figure 7.3. Why is this set at least n-dimensional? The key is in the incomparable pairs $\{a_i, b_i\}$. Every realizer $\leq_1, \ldots, \leq_k$ of St_n must have one order $\leq_l$ in which $a_i >_l b_i$ and another order $\leq_{l'}$ in which $a_i <_{l'} b_i$.
Fix a realizer $\leq_1, \ldots, \leq_k$. For $i \in \{1, \ldots, n\}$ fixed, let $\leq_l$ be an element of the realizer with $a_i >_l b_i$. For $j \neq i$ we have $a_i < b_j$ and $b_i > a_j$. Thus for $j \neq i$ we have that $a_j <_l b_i <_l a_i <_l b_j$.
This means for every order $\leq_l$ in the realizer of $\leq$ there is at most one i with $a_i >_l b_i$. Therefore there must be at least n orders in the realizer of $\leq$, one for each $a_i >_l b_i$.
We leave the proof that $\dim(St_n) \leq n$ as Exercise 12.   ∎

Unlike other properties that we have investigated recently (lattice, completeness), dimension allows for some type of heredity. Specifically, we have monotonicity.

Proposition 7.3.8 *Let P be an ordered set and let $Q \subseteq P$ be an ordered subset. Then* $\dim(Q) \leq \dim(P)$.

Proof. Let $\leq_1, \ldots, \leq_k$ be a realizer of $\leq_P$. Then the intersection of the k orders $\leq_1 \mid_{Q \times Q}, \ldots, \leq_k \mid_{Q \times Q}$ is $\leq_Q$ and hence $\dim(Q) \leq k = \dim(P)$. $\blacksquare$

As a particular consequence we are able to give a lower bound on the dimension of a power set. The fact that the exact dimension of a power set $\mathcal{P}(\{1, \ldots, n\})$ is n is then a consequence of Theorem 10.2.5.

Corollary 7.3.9 $\dim(\mathcal{P}(\{1, \ldots, n\})) \geq n$.

Proof. Just note that the subset formed by the singleton sets together with the $(n-1)$-element sets is isomorphic to St_n. $\blacksquare$

7.3.1   A Characterization of Realizers

To obtain upper bounds on the dimension of a specific ordered set one would need to specify a realizer. The number of orders in the realizer would be an upper bound on the dimension. In order to make this process more feasible, it would be good to have a criterion to recognize a realizer that is more efficient than to compute the intersection of its orders. The notion of a critical pair is the key to such a criterion.

Definition 7.3.10 *Let P be an ordered set. Then the pair $(x, y) \in P^2$ is called a* **critical pair** *in P iff*

$$x \not\geq y, \; (\downarrow x) \setminus \{x\} \subseteq (\downarrow y) \setminus \{y\}, \text{ and } (\uparrow x) \setminus \{x\} \supseteq (\uparrow y) \setminus \{y\}.$$

Example 7.3.11 Note that in the standard example St_n the critical pairs are exactly the pairs of unrelated minimal and maximal elements.

Essentially in a critical pair (x, y) we have that there is more above x than there is above y and there is less below x than there is below y. Making $x > y$ ("**reversing**" the critical pair) will thus introduce more new comparabilities than making $x < y$. This seems to indicate that $x > y$ is harder to achieve in a given linear extension. Interestingly, once the above is achieved for all critical pairs, we have a realizer.

Proposition 7.3.12 (Compare [211].) *Let $(P, \leq)$ be a finite ordered set. Then $\leq_1, \ldots, \leq_d$ is a realizer of $\leq$ iff for each critical pair (x, y) of P there is a k such that $x \geq_k y$.*

Proof. Since for each critical pair (x, y) we have $x \not\geq y$, the part "$\Rightarrow$" is trivial.

To prove "$\Leftarrow$" let $\leq_1, \ldots, \leq_d$ be such that for each critical pair (x, y) of P there is a k such that $x \geq_k y$. Let $a, b \in P$ be such that $a \not\geq b$. We will need to show that there is a $k \in \{1, \ldots, d\}$ such that $a \geq_k b$. By symmetry this will prove the result.

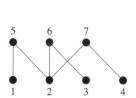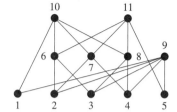

Figure 7.4: A three-dimensional set (left) and a four-dimensional ordered set with a critical pair whose removal decreases the dimension by 2 (right). The right example first occurred in [215].

The proof will be an induction on $c(a, b) := |({\downarrow} a) \setminus ({\downarrow} b)| + |({\uparrow} b) \setminus ({\uparrow} a)|$. For the basis step note that if $c(a, b) = 2$, then (a, b) is a critical pair. Thus by hypothesis there is a k with $a \geq_k b$.

Proceeding to the induction step we assume that $c(a, b) = n$ and that the result holds for all pairs a', b' with $a' \not> b'$ and $c(a', b') < n$. Suppose without loss of generality that $({\downarrow} a) \setminus ({\downarrow} b) \neq \emptyset$. Pick $a' \in ({\downarrow} a) \setminus ({\downarrow} b)$. Then $a' \not> b$ and $c(a', b) < c(a, b)$. Thus there is a k such that $a' \geq_k b$. This proves $a \geq_k a' \geq_k b$ and we are done.   ∎

Aside from giving a criterion for what a realizer is, Proposition 7.3.12 can be used to provide lower bounds on the dimension of an ordered set. This should not be surprising. For example, our computation of the dimension of the standard example heavily relied on the existence of critical pairs that could not be "reversed" simultaneously.

Example 7.3.13 To further demonstrate how Proposition 7.3.12 can be used to provide lower bounds on the dimension of an ordered set, consider Figure 7.4.

1. Consider the ordered set on the left in Figure 7.4. In Exercise 10 the reader will show that this ordered set has dimension at most 3. We will use Proposition 7.3.12 to show that the dimension of this ordered set is at least 3.

    To see this, first note that $(1, 6)$, $(2, 1)$, $(1, 7)$, $(4, 5)$, $(2, 4)$, $(4, 6)$, $(3, 7)$, $(2, 3)$, and $(3, 5)$ are all critical pairs. Moreover, no two pairs that are listed consecutively can be reversed by the same linear extension. Thus if the set had dimension 2, then the first and the last critical pair would have to be reversed by the same linear extension. This however is impossible. Thus any realizer of this ordered set must have at least three total orders.

2. Now consider the ordered set on the right in Figure 7.4. In Exercise 10 the reader will show that this ordered set has dimension at most 4. We will show that the dimension of this ordered set is at least 4.

For a contradiction, assume the set has dimension 3 and that a realizer is $\{\leq_1, \leq_2, \leq_3\}$. $(1, 11)$ and $(5, 10)$ are critical pairs that cannot be reversed in the same linear extension. We can assume $1 >_1 11$ and $5 >_2 10$. Now $(6, 9)$, $(7, 9)$, $(8, 9)$ are critical pairs that cannot be reversed in any linear extension that reverses $(1, 11)$ or $(5, 10)$. Therefore we must have $6 >_3 9$, $7 >_3 9$ and $8 >_3 9$.

No two of the critical pairs $(4, 6)$, $(3, 7)$ and $(2, 8)$ can be reversed in the same linear extension. Moreover, $(4, 6)$ and $(6, 9)$ cannot be reversed simultaneously and the same is true for $(3, 7)$ and $(7, 9)$ and for $(2, 8)$ and $(8, 9)$. Thus none of $(4, 6)$, $(3, 7)$ and $(2, 8)$ can be reversed in $\leq_3$, contradiction.

The reader versed in graph theory will notice that the above arguments rely on a connection to vertex coloring. Consider the graph whose vertices are the critical pairs and two pairs are joined by an edge iff they cannot occur in the same linear extension. In each of the above arguments we showed that this graph cannot be colored with less than a certain number of colors. (The reader is advised to sketch the graphs we argue about and verify the connection.) The connection between dimension and coloring will be made more precise in Remark 2.

7.4   Bounds on the Dimension

From what we have seen so far, it appears that computing the dimension of an ordered set is a challenging task. Indeed this is true in an algorithmic sense. At the end of Section 12.2 we mention that there is an efficient algorithm to determine if a given arbitrary ordered set has dimension ≤ 2. Remark 9 in Chapter 12 shows that for every $t \geq 3$ it is "hard" to decide if a given finite ordered set has dimension at most t. We will delay making the above notions of efficient and "hard" more precise to Chapter 12. Because determining the dimension is hard in general, upper or lower bounds are important tools. In this section we consider two upper bounds on the dimension in terms of easier order-theoretical parameters. Though the bounds appear weak, they are both sharp.

We start by considering the connection between the dimension and the width.

Theorem 7.4.1 *Let P be a finite ordered set. Then $\dim(P) \leq w(P)$.*

Proof. (Adapted from [263]; the argument is essentially that for Theorem 3.3 in [121], combined with Dilworth's Theorem. The result appears first in [57], though the connection would have to be made through Theorem 10.2.5.) First we shall show that if C is a chain in $(P, \leq)$, then there is a linear extension $\sqsubseteq$ of $\leq$ such that for all $c \in C$ and $p \in P$ such that $c \not> p$ we have $c \sqsupseteq p$.

Enumerate the elements of C as $c_1 < c_2 < \cdots < c_k$. Let A_k be the set of all elements $p \in P \setminus C$ such that c_k is the largest element of C that is below p. Let $A_0 := P \setminus (C \cup A_1 \cup \cdots \cup A_k)$. Note that if $p \in A_i$, $q \in A_j$ and $i < j$, then $p \not> q$. For each A_i $(i = 0, \ldots, k)$ find a linear extension $\sqsubseteq$ so that

$A_i = \{a_i^1 \sqsubset a_i^2 \sqsubset \cdots \sqsubset a_i^{n_i}\}$. The desired linear extension then is (elements listed in increasing order) $a_0^1, a_0^2, \ldots, a_0^{n_0}, c_1, a_1^1, a_1^2, \ldots, a_1^{n_1}, c_2, a_2^1, \ldots, a_{k-1}^{n_{k-1}}, c_k, a_k^1, a_k^2, \ldots, a_k^{n_k}$.

Now let $P = C_1 \cup \cdots \cup C_{w(P)}$ be a decomposition of P into chains as guaranteed by Dilworth's theorem. For each i let $\leq_i$ be a linear extension of P such that for all $c \in C_i$ and $p \in P \setminus C_i$ with $c \not> p$ we have $c >_i p$. Let $p, q \in P$ be incomparable elements. Find i, j such that $p \in C_i$ and $q \in C_j$. Then $p >_i q$ and $q >_j p$. This shows that $\leq$ is the intersection of the $\leq_i$.   ∎

The standard example St_n shows that the above inequality cannot be improved in general. In order to relate the dimension also to the size of the ordered set we need the following, somewhat technical result.

Lemma 7.4.2 (Cf. [138, 261].) *Let P be an ordered set and let $A \subseteq X$ be an antichain. Then* $\dim(P) \leq \max\{2, |P \setminus A|\}$.

Proof. The result is proved if we can prove it in the case that A is a maximal antichain. Thus in this proof we will assume A is a maximal antichain. Since A is maximal, for each $x \in P \setminus A$, there is an $a \in A$ such that $a \leq x$ or $a \geq x$. Note also that there is no $x \in P \setminus A$ such that there are $a, b \in A$ with $a \leq x \leq b$. We define

$$
\begin{aligned}
D &:= \{x \in P \setminus A : (\exists a \in A) x \leq a\}, \\
U &:= \{y \in P \setminus A : (\exists a \in A) y \geq a\}.
\end{aligned}
$$

The proof will be by induction on $|P \setminus A|$. Our basis step covers the cases $|P \setminus A| \leq 2$.

For $|P \setminus A| = 0$ a realizer with two orders consists of an extension $\leq_1$ of A and of its dual $\leq_2 := \leq_1^d$.

For $|P \setminus A| = 1$ assume without loss of generality that the element p of $P \setminus A$ is in D. A realizer with two orders consists of an extension $\leq_1$ of A such that $(A \setminus \uparrow p) \leq_1 p \leq_1 (\uparrow p) \setminus \{p\}$ and of the extension $\leq_2$ for which p is the smallest element and $\leq_2 |_{A \times A}$ is the dual of $\leq_1 |_{A \times A}$.

For $|P \setminus A| = 2$ assume first that the elements p, q of $P \setminus A$ are in D. Also assume that $q \not\geq p$. Find an extension $\leq_1$ of A such that

$$
\begin{aligned}
(A \setminus (\uparrow p \cup \uparrow q)) &\leq_1 q \leq_1 (\uparrow q) \setminus (\{q\} \cup \uparrow p) \leq_1 p \\
&\leq_1 (\uparrow p \cap \uparrow q) \setminus \{p\} \leq_1 (\uparrow p \setminus (\{p\} \cup \uparrow q)).
\end{aligned}
$$

In case $p \not> q$ find an extension $\leq_2$ for which p is the smallest element, such that

$$
\begin{aligned}
p &\leq_2 (\uparrow p \setminus (\{p\} \cup \uparrow q)) \leq_2 q \leq_2 (\uparrow p \cap \uparrow q) \setminus \{p\} \\
&\leq_2 (\uparrow q) \setminus (\{q\} \cup \uparrow p) \leq_2 (A \setminus (\uparrow p \cup \uparrow q))
\end{aligned}
$$

and such that $\leq_2 |_{A \times A}$ is the dual of $\leq_1 |_{A \times A}$. In case $q < p$, we must put q below p in $\leq_2$ above.

The case $p, q \in U$ is dual to the above. The case $p \in D$ and $q \in U$ is handled similarly and left to the reader as Exercise 24.

This leaves the induction step for $|P \setminus A| = k \geq 3$. The induction hypothesis is that the result holds for all ordered sets P and maximal antichains $A \subseteq P$ with $|P \setminus A| = k - 1$. Without loss of generality we can assume that $D \neq \emptyset$. Let $x \in D$ be a minimal element. Apply the induction hypothesis to $P \setminus \{x\}$ and A to obtain a realizer of $\leq |_{P \setminus \{x\} \times P \setminus \{x\}}$ with $k - 1$ linear orders. Turn each of these orders into an extension of $\leq$ by making x the smallest element. Find an extension $\leq_x$ of $\leq$ so that $x \leq_x y$ only if $x \leq y$. The thus obtained k linear orders are a realizer of P. ∎

The above allows us to connect the dimension with the size of the ordered set.

Theorem 7.4.3 (Hiraguchi's Theorem, cf. [121].) *For $|P| \geq 4$ the dimension of P is bounded by half its size. More precisely,* $\dim(P) \leq \left\lfloor \dfrac{|P|}{2} \right\rfloor$.

Proof. (Also cf. [262].) If $w(P) \leq \left\lfloor \dfrac{|P|}{2} \right\rfloor$, this follows from Theorem 7.4.1. If $w(P) > \left\lfloor \dfrac{|P|}{2} \right\rfloor$, then P has an antichain A with $|P \setminus A| \leq \left\lfloor \dfrac{|P|}{2} \right\rfloor$. Thus by Lemma 7.4.2 we have $\dim(P) \leq \max\{2, |P \setminus A|\}$, which is bounded by $\left\lfloor \dfrac{|P|}{2} \right\rfloor$ for $|P| \geq 4$. ∎

Again the standard example St_n shows that the above inequality cannot be improved in general. The original proof in [121] uses theorems that involve removals of four points which reduce the dimension by at most two each and removals of two points which reduce the dimension by at most 1 each. An open question is if there is a way to prove the result using just removals of 2 points at every stage.

> **Open Question 7.4.4** (Cf. [133].) *Does every ordered set P (with ≥ 3 elements) contain a pair of elements a, b whose removal decreases the dimension by at most 1? That is, are there a, b such that $dim(P) - dim(P \setminus \{a, b\}) \leq 1$?*

If the answer to the above is affirmative, then Theorem 7.4.3 would be an easy corollary. It was conjectured for a while, that critical pairs would be pairs as asked for in Open Question 7.4.4. The ordered set on the right in Figure 7.4, due to Reuter, shows that this is not the case. In Example 7.3.13 we have seen that the set has dimension 4. It is easy to check that $(8, 9)$ is a critical pair. Finally the set obtained by removing 8 and 9 has dimension 2. Thus removal of critical pairs can reduce the dimension by more than 1. Examples of larger size and any dimension > 4 are given in [137].

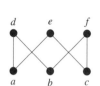 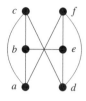

Figure 7.5: A six crown and the complement of its comparability graph. Note that the left is a diagram of an ordered set, while the right is merely a graph. The graph is drawn in this form to assist in the visualization of the argument in Example 7.5.2.

7.5 Ordered Sets of Dimension 2

Ordered sets of dimension 2 are the simplest higher-dimensional ordered sets. In this section we shall discuss an interesting structural property of ordered sets of dimension 2 that relates to graph theory. We leave a discussion of the fixed point property for ordered sets of dimension 2 to Corollary 8.4.6.

A standard tool in graph theory is the complement of a graph, which is obtained by erasing all original edges and by connecting any two vertices that were not originally connected by an edge.

Definition 7.5.1 *Let $G = (V, E)$ be a graph. The* **complement (graph)** *of G is defined to be the graph with vertex set V and edge set*

$$E_C := \{\{v, w\} : v, w \in V, v \neq w, \{v, w\} \notin E\}.$$

Clearly the complement of a graph is again a graph. It should be equally clear that the complement of a comparability graph need not be a comparability graph.

Example 7.5.2 Let P be a six crown. Then the complement graph of the comparability graph of P is itself not a comparability graph.

Proof. Let $(P, \leq)$ be a six crown as shown in Figure 7.5. The complement $\overline{C}$ of the comparability graph of $(P, \leq)$ is shown to the right of P. Assume $\overline{C}$ was the comparability graph of an ordered set $(P, \leq')$. Since $\{a, b, c\}$ is a clique in $\overline{C}$, $\{a, b, c\}$ has to be a chain with respect to $\leq'$. Without loss of generality we can assume that $a \leq' b \leq' c$. But then $b \leq' e$ would imply $e \geq' a$, which is not possible. On the other hand, $b \geq' e$ would imply $e \leq' c$, which is also not possible. Thus $\overline{C}$ cannot be a comparability graph.   ∎

The natural question to ask is for which comparability graphs the complement graph is again a comparability graph. This is answered in the next theorem.

Theorem 7.5.3 *Let P be an ordered set. Then the complement graph of the comparability graph $G_C(P)$ is a comparability graph iff P is of dimension ≤ 2.*

Proof. To prove "$\Leftarrow$", we let $(P, \leq)$ be an ordered set of dimension 2 and we let $\{\leq_1, \leq_2\}$ be a realizer for $\leq$. (In case P is one-dimensional, that is, a chain, choose $\leq_1 = \leq_2 = \leq$.) Define $\leq'$ by $p \leq' q$ iff $p \leq_1 q$ and $p \geq_2 q$. Then $\leq'$ is another order of dimension at most 2. We shall show now that the comparability graph of $(P, \leq')$ is the complement of the comparability graph of $(P, \leq)$.

First suppose $p \sim q$ and $p \neq q$. Then $p <_1 q$ and $p <_2 q$ and hence $p \not\sim' q$. Now suppose $p \not\sim q$. Then $p <_1 q$ and $p >_2 q$ or $p >_1 q$ and $p <_2 q$. This means that $p <' q$ or $p >' q$, that is $p \sim' q$. Thus we have proved $p \not\sim q$ iff $p \sim' q$.

To prove "$\Rightarrow$" let $(P, \leq)$ be an ordered set such that the complement graph H of the comparability graph $G_C(P)$ is also a comparability graph. Let $\leq'$ be an order relation on P such that $G_C(P, \leq') = H$. Define $p \leq_1 q$ iff $p \leq q$ or $p \leq' q$. To show that $\leq_1$ is a total order on P we proceed as follows. Reflexivity is trivial. Since any two elements of P are comparable via either $\leq$ or $\leq'$, it is also clear that any two elements in P are comparable via $\leq_1$. To prove antisymmetry assume $p \leq_1 q$ and $q \leq_1 p$. Assume without loss of generality that $p \sim q$ and $p \not\sim' q$. Then $p \leq q$ and $q \leq p$, which implies $p = q$. Finally to prove transitivity assume $p <_1 q <_1 r$. In case $p \sim q, q \sim r$ or $p \sim' q, q \sim' r$ there is nothing to prove. Now consider the case $p \sim q$ and $q \sim' r$. This means that $p \leq q$ and $q \leq' r$. Without loss of generality we can assume that $p \sim r$. Since $q \leq' r$, we have $q \not\sim r$, which means $r \not\leq p$. Thus $p \leq r$, which means $p \leq_1 r$. Thus $\leq_1$ is a total order on P.

We define $p \leq_2 q$ iff $p \leq q$ or $p \geq' q$. Then (with the same argument as above) $\leq_2$ is also a linear order on P. Now note that $\leq = \leq_1 \cap \leq_2$. This finishes the proof. ∎

The above is the foundation for the only known reconstruction result that relates to dimension. We will only present the idea of the proof.

Theorem 7.5.4 *Ordered sets of dimension 2 are recognizable from the deck.*

Idea of the Proof. Recognizability of ordered sets of dimension 2 is a corollary to the recognizability of comparability graphs in [268]. By Theorem 7.5.3 an ordered set is of dimension 2 iff the complement of its comparability graph is also a comparability graph. For the deck of P, one can form the deck of the complement of the comparability graph of P by forming the complement of the comparability graph of every card. P is of dimension 2 iff this is the deck of a comparability graph.

The idea for recognizability of comparability graphs in [268] is to prove that the forbidden subgraphs for comparability graphs identified in [93] are all reconstructible. This done, comparability graphs are recognizable as those graphs that are not equal to any of the forbidden subgraphs and whose cards are all comparability graphs. Since the proof of [93] is too lengthy to be included in this text, we shall not elaborate further.

Exercises

1. Let P be a finite ordered set of height h and let $r_i(P)$ be the number of elements of rank i in P. Prove that P has at least $\Pi_{i=0}^{h}[r_i(P)!]$ linear extensions.

2. Let $(P, \leq)$ be an ordered set and let $\leq'$ be a linear extension of $\leq$. Prove that the identity from $(P, \leq)$ to $(P, \leq')$ is an injective, surjective order-preserving map.

3. Compute all linear extensions of the following sets.

    (a) A four crown.

    (b) An antichain.

    (c) A chain.

    (d) A five fence.

4. An **alternating cycle** in an ordered set $(P, \leq)$ is a set $\{(x_i, y_i) : i = 1, \ldots, k\}$ so that $x_i \not\leq y_i$ and $y_i \leq x_{i+1}$ for $i = 1, \ldots, k$ modulo k. (That is, it's true for $i = 1, \ldots, k - 1$ and $y_k \leq x_1$.)

    (a) Prove that no order that extends $\leq$ contains an alternating cycle.

    (b) A **strict alternating cycle** is an alternating cycle such that the only comparabilities $y_i \leq x_j$ are the ones indicated above. Prove that a transitive relation that contains $\leq$ and fails to be antisymmetric must contain a strict alternating cycle.

5. Prove that the number of linear extensions is reconstructible from the maximal deck $\mathcal{M}_P$.

6. Prove that if $\leq$ and $\sqsubseteq$ are orders on P and $\leq$ is strictly contained in $\sqsubseteq$, then $\sqsubseteq$ has strictly fewer linear extensions than $\leq$.

7. Prove Corollary 7.3.1.

8. Compute the dimension of the following.

    (a) An antichain.

    (b) A chain.

    (c) The ordered set in part b) of Figure 1.1.

    (d) The ordered set in Figure 1.3.

    (e) The ordered set in Figure 1.4, part e).

9. (Crowns and fences again)

(a) Prove that every fence (including the one-way and the two-way infinite fence) has dimension 2.

(b) Prove that every $2n$-crown with $n \geq 3$ has dimension 3.

10. Consider the ordered sets in Figure 7.4.

(a) Prove that the ordered set on the left in Figure 7.4 has dimension at most 3. (Together with the argument in Example 7.3.13 this shows that its dimension is exactly 3.)

(b) Prove that the ordered set on the right in Figure 7.4 has dimension at most 4. (Together with the argument in Example 7.3.13 this shows that its dimension is exactly 4.)

(c) Prove that if the points 8 and 9 are removed from the ordered set on the right in Figure 7.4, then the remaining ordered set has dimension 2.

11. Prove that any finite ordered set of dimension $\leq n$ can be embedded into $\mathbb{R}^n$ as given in Example 7.3.6.

12. Prove that the standard example really has dimension at most n. Then prove that removal of any point from the standard example reduces the dimension by 1.

13. Compute the Dedekind–MacNeille completion of the standard example.

14. We have seen ways in which the adding of a relation leads to new ordered sets. Now consider removal.

(a) Give an example showing that in general removal of one comparability from an order relation will lead to a relation that need not be an order.

(b) Give an example that shows that in general there is no unique largest order relation that is contained in the relation obtained when deleting one comparability from an order relation.

(c) The edge reconstruction problem in graph theory is the problem of reconstructing a graph from all its one-edge-deleted subgraphs. Prove that ordered sets of height greater than 1 are reconstructible from their one-edge-deleted subgraphs

15. Prove that if P can be embedded in Q, then $\dim(P) \leq \dim(Q)$.

16. Prove Theorem 7.4.1 for infinite ordered sets.

17. (Cf. [121].) Prove that removal of a single point cannot decrease the dimension of an ordered set by more than 1. That is, prove that for an ordered set P and $x \in P$ we have $\dim(P \setminus \{x\}) \in \{\dim(P), \dim(P) - 1\}$.

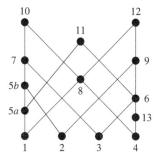

Figure 7.6: The ordered set for Exercise 19.

18. Let P be a finite ordered set. Prove that if p is doubly irreducible in P, that is, if p has a unique upper and a unique lower cover, then $\dim(P) = \dim(P \setminus \{p\})$.

19. Find the dimension of the ordered set in Figure 7.6.

    Then prove that the pair $(8, 11)$ is a critical pair such that its removal reduces the dimension to 2. (This set occurred in [215].)

20. Prove that an ordered set of height 1 is of dimension 2 iff it does not contain any crowns with more than four elements.

21. Prove that every two-dimensional ordered set has a retractable point.

22. Prove that the ordered set in Figure 7.2 has the fixed point property.

23. Dedekind's problem for two-dimensional ordered sets. Let P be a finite ordered set of dimension 2.

    (a) Prove there is a bijection $f : \{1, \ldots, n\} \to P$ and a permutation σ of $\{1, \ldots, n\}$ such that $a \not\sim b$ iff $\sigma(f^{-1}(a)) > \sigma(f^{-1}(b))$.

    (b) Prove that the number of antichains of P is the number of decreasing subsequences of $(\sigma(1), \ldots, \sigma(n))$ with σ as above.

24. Finish the proof of Lemma 7.4.2 by proving that if A is a maximal antichain with $P \setminus A = \{p, q\}$, $p \in D$ and $q \in U$, then P is two-dimensional. (Hint. Distinguish the cases $p < q$ and $p \not< q$. For an alternative proof, cf. [263], Theorem 11.1.)

25. (Covers in the set of extensions of an order, also cf. [34].) Let $(P, \leq)$ be a finite ordered set. Call an order $\sqsubseteq$ that contains $\leq$ an **immediate extension** of $\leq$ iff the only extensions of $\leq$ that are contained in $\sqsubseteq$ are $\leq$ and $\sqsubseteq$.

    (a) Prove that all immediate extensions are obtained by adding a (non-reversed) critical pair to the order relation.

(b) Prove that the dimension can increase or decrease when going from an order to an immediate extension.

(c) If $\leq$ is not a total order, is there always an immediate extension $\sqsubseteq$ of $\leq$ such that $\dim(P, \sqsubseteq) \leq \dim(P, \leq)$?

26. Show that for every even $n \in \mathbb{N}$ there is a sequence of n orders $\leq_1 \sqsubseteq \leq_2 \sqsubseteq \cdots \sqsubseteq \leq_n$ on $\{1, \ldots, n\}$ such that the even-indexed orders have the fixed point property and the odd-indexed orders do not. (Also cf. Open Problem 9.)

27. (Private communication with I. Zaguia.) A **weak order** is a union of antichains $A_1, \ldots, A_n$ such that $A_1 \leq A_2 \leq \cdots \leq A_n$. Let $(P, \leq)$ be a finite ordered set and let $f : P \to P$ be an order-preserving map. We will try to construct a weak order $\leq_f$ that extends $\leq$ such that f is still an order-preserving map on $(P, \leq_f)$. So let us assume such a weak order $\leq_f$ exists.

(a) Prove that if $a = f^k(a)$ for some $k \geq 1$, then $\{a, \ldots, f^k(a)\}$ must be an antichain in $\leq_f$.

(b) Prove that if ($\leq_f$ exists and) $f(a) \neq a = f^k(a)$ for some $k > 1$, then for all x such that there are i, j with $f^i(x) = f^j(a)$ we have that no $f^u(x)$ is strictly $\leq$-comparable to any $f^v(a)$.

(c) Prove that for any map as in 27b a weak order as desired exists. In particular, show that if $a = f^k(a)$ for some $k \geq 1$ implies $a = f(a)$, then the weak order can be chosen to be a linear order.

28. (Inspired by [75], also cf. [83].) Call a map $f : P \to Q$ strictly order-preserving iff $x < y$ implies $f(x) < f(y)$.

(a) Prove that if there is a strictly order-preserving map $f : P \to \mathbb{Z}$, then the intervals $[a, b] = \uparrow a \cap \downarrow b$ of P must all be of finite height.

(b) Prove that if P is countable, then the above finiteness condition is sufficient for existence of a strictly order-preserving map into the integers.

(c) Consider the following ordered set P.

- The minimal elements are the (unordered) natural numbers.
- The maximal elements are the (unordered) functions $f : \mathbb{N} \to \mathbb{N}$.
- Between any $n \in \mathbb{N}$ and any $f : \mathbb{N} \to \mathbb{N}$ there is a chain of length $f(n)$ such that n is the bottom element, f is the top, and the chain elements are not comparable to any other natural numbers or maps.

Show that all intervals in P are finite and yet there is no strictly order-preserving map of P into $\mathbb{Z}$. (Hint. Suppose $\alpha : P \to \mathbb{Z}$ is as desired and show that $h(n) := (\alpha(n))^2$ could not be mapped into $\mathbb{Z}$.)

(d) Show that there is a lattice that satisfies the interval finiteness condition which cannot be mapped into $\mathbb{Z}$ with a strictly order-preserving map.

Remarks and Open Problems

1. For more on dimension theory, cf. [135, 263]. Major open problems are presented within [263] and in the appendix.

2. Given a finite ordered set P, construct the following hypergraph $\mathcal{K}_P$. The vertices of $\mathcal{K}_P$ are the critical pairs of P and the hyperedges of $\mathcal{K}_P$ are those sets $\{(x_i, y_i) : i = 1, \ldots, k\}$ of critical pairs so that $\{(y_i, x_i) : i = 1, \ldots, k\}$ forms an alternating cycle. Using Exercise 4 it can be shown that the dimension of P is the chromatic number of $\mathcal{K}_P$. (That is, the smallest number k such that k colors can be used to color the vertices of $\mathcal{K}_P$ so that no hyperedge connects vertices that are all of the same color.)

   For more on the connection between coloring and dimension, cf. [84].

3. We have seen in Exercise 14c that the naive order analogue of the edge reconstruction problem is trivial except in the case of height 1.

   Consider the following enhanced deck $\mathcal{EN}_P$ of an ordered set P defined as follows. For every non-maximal element $x \in P$ let $(P, \leq \setminus U(x))$ be the ordered set obtained by erasing all comparabilities of the form (x, p) from P. Let $\mathcal{EN}_P([C])$ be the number of non-maximal points $x \in P$ such that $(P, \leq \setminus U(x)) \in [C]$. Is P reconstructible from the maximal deck plus $\mathcal{EN}_P$? To get a start at reconstruction "from the other end", that is, to find a set of nontrivial parameters that completely determine the order, one could modify this problem and add any other order parameters that are reconstructible from the deck $\mathcal{D}_P$ to the hypothesis.

4. How many indecomposable ordered sets of a given width, with the same number of comparabilities and with the same number of edges in the diagram, can have the same number of linear extensions? This will likely depend on the number of elements. The idea is of course that (if there is little ambiguity or none, which would mean only the set and its dual fit the description) in this fashion the given (reconstructible) parameters would allow us to (almost) reconstruct the ordered set up to duality. We shall revisit this idea briefly in Remark 9 in Chapter 11.

5. Linear extensions may be most natural in order, but they can also be applied to graphs. One application of such extensions is the notion of the width of a constraint graph of a constraint satisfaction problem (cf. Definition 12.5.9).

6. Let $\leq_l$ be a linear extension of $\leq$. Then the **jump number** of $(\leq, \leq_l)$ is defined to be

$$s_{(\leq,\leq_l)} := \#\{(x, y) : x \prec_l y \text{ and } x \not\geq y\}.$$

The **jump number** $s(\leq)$ of $(P, \leq)$ is the minimum of all $s_{(\leq,\leq_l)}$.

This parameter can be understood as follows. Suppose several jobs need to be scheduled on one machine and there is a (partial) order $\leq$ that must not be violated. Then one would expect the jobs to run most smoothly if there are few interruptions to the "natural" order $\leq$. This means we would want to minimize the number of jumps. The jump number exhibits to what extent this is possible.

For more on the jump number, cf. [258] and Chapter 9 in [263].

7. How hard is it to reconstruct ordered sets of dimension 2?

This has not been done yet, but the canonical visualization of ordered sets of dimension 2 suggests a deceptively simple approach. If we just knew the right way to draw the cards, reconstruction should be trivial ... (?)

8. For a connection between dimension and homology, cf. [214].

9. In Exercise 26 we found a chain of orders on n elements such that going from one order to the next one always changes the status of the fixed point property. What is the longest chain of orders $\leq_1 \subseteq \leq_2 \subseteq \cdots \subseteq \leq_k$ on $\{1, \ldots, n\}$ such that the even-indexed orders have the fixed point property and the odd-indexed orders do not?

8

Interval Orders

Consider the job of scheduling talks at a conference or allocating processor time to several concurrently running programs. These types of problems are what is handled in scheduling theory. The tasks involved each take a certain amount of time. Thus, abstractly, each task can be represented as an interval on the real line. Intervals can be ordered in a natural fashion (for scheduling and otherwise). An interval I is before another interval I' iff I is completely to the left of I'. This is essentially the idea that two tasks can only be related if one is finished before the other.

Studying interval orders means studying the properties of ordered sets that arise as above. Just as the chapters on lattices and on dimension, this chapter also can only be seen as a brief introduction. For further work on interval orders, the reader is referred to [85] and recent papers such as [20].

8.1 Definition and Examples

The real line is not the only structure in which intervals can be defined. Indeed, intervals are order-theoretical entities that can arise in any ordered set. For this chapter we will look at intervals in arbitrary chains.

Definition 8.1.1 *Let P be an ordered set. Then for $a \leq b$ in P the* **interval** $[a, b]$ *is defined to be*

$$[a, b] := (\uparrow a) \cap (\downarrow b).$$

Definition 8.1.2 *Let $(P, \leq)$ be an ordered set. Then $\leq$ is called an* **interval order** *iff there is a set J of intervals in a chain C such that $(P, \leq)$ is isomorphic to $(J, \leq_{\text{int}})$, where $[a, b] \leq_{\text{int}} [c, d]$ iff $b \leq c$ or $[a, b] = [c, d]$ (as in Example 1.1.2, part 7). The ordered set $(J, \leq_{\text{int}})$ is called the* **interval representation** *of P. We will also call P an* **interval ordered set**.

The most natural examples of interval ordered sets are of course sets of intervals on the real line. An interesting property of interval ordered sets is that all interval ordered sets with up to n elements are contained in a certain interval ordered set with $\frac{n}{2}(n + 1)$ elements. Thus in a sense Example 8.1.3, part 4 gives all finite examples of interval ordered sets.

Example 8.1.3

1. Every finite chain carries an interval order.

2. Any set of intervals in $\mathbb{R}$ ordered by $\leq_{\text{int}}$ is an interval ordered set.

3. Every subset of an interval ordered set carries an interval order.

4. Let P_n be the set of all intervals with more than one point and integer endpoints in $\{0, \ldots, n\}$, ordered by $[a, b] \sqsubseteq [c, d]$ iff $b \leq c$ or $[a, b] = [c, d]$. Then every interval ordered set with up to n elements is isomorphic to a subset of P_n in which no two intervals have the same left endpoint.

Proof. All claims except for part 4 are straightforward. To prove part 4 we proceed by induction on n. Since there is nothing to prove for $n = 1$, consider the induction step.

Let $Q := \{[a_1, b_1], \ldots, [a_n, b_n], [a_{n+1}, b_{n+1}]\}$ be a set of $n + 1$ intervals ordered by $\leq_{\text{int}}$. Assume without loss of generality that $[a_{n+1}, b_{n+1}]$ is an interval such that no other interval has a larger left endpoint. Then for every non-maximal (with respect to the order $\leq_{\text{int}}$) interval $[a_k, b_k]$ we have $b_k \leq a_{n+1}$.

By induction hypothesis $H := \{[a_1, b_1], \ldots, [a_n, b_n]\}$ is already isomorphic to a subset $H' = \{[a'_1, b'_1], \ldots, [a'_n, b'_n]\}$ (such that no two a'_i are equal) of P_n via an isomorphism Φ that maps $[a_k, b_k]$ to $[a'_k, b'_k]$. Replace every interval $[a'_k, b'_k] \in H'$ that is the Φ-image of a $(Q, \leq_{\text{int}})$-maximal element, with $[a''_k, b''_k] := [a'_k, n + 1]$. For all other intervals, let $[a''_k, b''_k] := [a'_k, b'_k]$. The thus obtained set $H'' = \{[a''_1, b''_1], \ldots, [a''_n, b''_n]\}$ is still isomorphic to H. (Recall that no two intervals in H' have the same left endpoint.)

We claim that $\Phi^{-1}[H'' \cap P_n]$ is the set of intervals whose right boundary is at most a_{n+1}. To see this, let $[a_k, b_k] \in Q$ be an interval with $k \leq n$ and $b_k \leq a_{n+1}$. Then by definition $[a''_k, b''_k] = [a'_k, b'_k]$ and $b'_k \leq n$. This means that $[a_k, b_k] \in \Phi^{-1}[H'' \cap P_n]$. Conversely, if $[a_k, b_k] \in \Phi^{-1}[H'' \cap P_n]$, then $\Phi([a_k, b_k]) = [a'_k, b'_k] = [a''_k, b''_k]$. This means that $[a_k, b_k]$ cannot have been maximal in Q, which means that $b_k \leq a_{n+1}$.

We thus conclude that $H'' \cup \{[n, n + 1]\}$ is isomorphic to Q. The isomorphism is the map that for $k \leq n$ maps $[a_k, b_k]$ to $[a''_k, b''_k]$ and which maps $[a_{n+1}, b_{n+1}]$ to $[n, n + 1]$. ∎

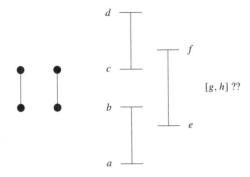

Figure 8.1: The ordered set $\mathbf{2} + \mathbf{2}$ and an illustration why it cannot be contained in an interval ordered set.

Example 8.1.3, part 4 provides us with essentially all finite examples of interval ordered sets. This allows us to prove that, while the standard example St_n is not interval ordered, there are interval ordered sets of any dimension.

Example 8.1.4 *The examples P_n in part 4 in Example 8.1.3 show that there are interval orders of any dimension.*

Assume for a contradiction that the dimension of the P_n is uniformly bounded by d. Then for all $n \in \mathbb{N}$ there are linear extensions $\leq_1^n, \ldots, \leq_d^n$ whose intersection is the order on P_n. For $n \in \mathbb{N}$ consider P_n. For any three numbers $a < b < c$ in $\{0, \ldots, n\}$ note that $[a, b]$ and $[b-1, c]$ are incomparable elements in P_n. Thus there must be a $k_n(a, b, c) \in \{1, \ldots, d\}$ such that $[a, b] \geq_{k_n(a,b,c)}^n [b-1, c]$.

It can be concluded from Ramsey's theorem (cf. [212]) that for large enough n there will be a quadruple $a < b < c < d$ such that $k_n(a, b, c) = k_n(b-1, c, d) = k$. (This is because there will be a quintuple $a < b-1 < b < c < d$ such that k_n is constant on all its sub-triples.) But then $[a, b] \geq_k^n [b-1, c] \geq_k^n [c-1, d]$, which is impossible because $[a, b] \leq [c-1, d]$.

Thus the sequence of the dimensions of the P_n must be unbounded.   ∎

It would be quite difficult to check if a given ordered set with n elements carries an interval order by testing if it is isomorphic to an ordered subset of P_n. We will now consider a powerful graphical characterization for interval orders. It is a characterization via forbidden suborders. We have seen a trivial characterization of this kind for chains in Proposition 2.5.3.

Theorem 8.1.5 *Let $(P, \leq)$ be an ordered set. Then $\leq$ is an interval order iff P does not contain the ordered set $\mathbf{2} + \mathbf{2}$ depicted in Figure 8.1 as an ordered subset.*

Proof. (Adapted from the introduction to [181].) First suppose that P has an interval order and assume P contains the ordered set $\mathbf{2} + \mathbf{2}$, with the first 2-chain being $A < C$ and the second being $E < G$. In the interval representation of P, let $A = [a, b]$ and $C = [c, d]$. Then $b \leq c$. Let $E =: [e, f]$ in the interval representation of P. Then, since E is not comparable to either A or C, we have

$e < b$ and $f > c$. However now for $G = [g, h]$ we must have $g \geq f > c \geq b$, a contradiction to the assumption that G and A were incomparable.

Conversely, suppose P is an ordered set that contains no copy of $\mathbf{2} + \mathbf{2}$. We need to find an interval representation for P. Recall that when embedding ordered sets into lattices via Proposition 3.3.7, Theorem 5.3.4 or Proposition 5.5.5 the points p generally were mapped to their principal ideals $\downarrow p$. A similar idea will be used here.

We claim that for any two elements $x, y \in P$ we have

$$(\downarrow x) \setminus \{x\} \subseteq (\downarrow y) \setminus \{y\} \qquad \text{or} \qquad (\downarrow y) \setminus \{y\} \subseteq (\downarrow x) \setminus \{x\}.$$

For $x \sim y$, this is clear. For $x \not\sim y$ assume the result was not true. Then there would be an $x' < x$ such that $x' \not< y$ and a $y' < y$ such that $y' \not< x$. Then $x' \not\sim y'$ and since $x \not\sim y$ we also have $x' \not\sim y$ and $y' \not\sim x$. However this means that $\{x, y, x', y'\}$ is isomorphic to $\mathbf{2} + \mathbf{2}$, contradicting our assumption. This means that $\{(\downarrow x) \setminus \{x\} : x \in P\}$ is linearly ordered. This insight is the key to constructing the ground set for the interval representation.

For $x \in P$ we define

$$L(x) := (\downarrow x) \setminus \{x\} \qquad \text{and} \qquad R(x) := \bigcap \{(\downarrow p) \setminus \{p\} : p \in P, p > x\}.$$

Clearly, $L(x) \subseteq R(x)$. These points will of course become the left and right endpoints in our interval representation. However first we need to prove that $C := \{L(x) : x \in P\} \cup \{R(x) : x \in P\}$ is linearly ordered by inclusion. Let a, b be elements of C. If $a = L(x)$ and $b = L(y)$, there is nothing to prove. If $a = R(x)$ and $b = R(y)$, by the dual of the previous claim, we can assume without loss of generality that $(\uparrow x) \setminus \{x\} \subseteq (\uparrow y) \setminus \{y\}$. However then $R(x) \supseteq R(y)$. Finally in case $a = L(x)$ and $b = R(y)$ there is nothing to prove if $x \sim y$. In case $x \not\sim y$, every strict lower bound l of x is below every strict upper bound u of y (otherwise $\{x, y, l, u\}$ is isomorphic to $\mathbf{2} + \mathbf{2}$). However then $a = L(x) \subseteq R(y) = b$.

Now let $\Phi(x) := [L(x), R(x)]$. We claim that $x < y$ iff $R(x) \subseteq L(y)$. First, if $x < y$, then $(\downarrow y) \setminus \{y\}$ is one of the sets over which we take the intersection when computing $R(x)$. Thus $R(x) \subseteq (\downarrow y) \setminus \{y\} = L(y)$. Second, if $R(x) \subseteq L(y)$, then $x \in R(x) \subseteq L(y) = (\downarrow y) \setminus \{y\}$. This means $x < y$. We have proved that P carries an interval order. ∎

Theorem 8.1.5 allows for a quick visual identification of interval ordered sets. The proof of Theorem 8.1.5 has several consequences. First note that in the construction of the interval representation $x \in R(x)$ and $x \notin L(x)$. Therefore

Scholium 8.1.6 *Every interval ordered set has an interval representation in which none of the intervals is a singleton.*

So even if we start with an interval order in which some of the intervals are singletons, we can find another representation in which this is not the case. Another important part of the proof of Theorem 8.1.5 is the insight that in an interval ordered set the sets of strict lower bounds (and by duality the sets of strict upper bounds) are linearly ordered. We record this fact for future reference.

Scholium 8.1.7 *Let P be an ordered set with an interval order and let $x \not\sim y$ in P. Then $(\uparrow x) \setminus \{x\} \subseteq (\uparrow y) \setminus \{y\}$ or $(\uparrow y) \setminus \{y\} \subseteq (\uparrow x) \setminus \{x\}$.* ∎

8.2   The Fixed Point Property for Interval Orders

Scholium 8.1.7 immediately leads to another structural insight into interval orders, which is in turn related to the fixed point property.

Corollary 8.2.1 *Finite interval ordered sets are collapsible.*

Proof. This proof is an induction on the size n of the ordered set. For $n = 0, 1$ there is nothing to prove, so assume P is an interval ordered set with $|P| = n > 1$ and that all interval ordered sets with fewer than n elements are collapsible.

If P has a smallest element, there is nothing to prove, since in this case P is $\mathcal{I}$-dismantlable to a singleton. So suppose P has at least two minimal elements. Let $m_1, \ldots, m_k$ be the minimal elements of P. By Scholium 8.1.7, we can assume that $(\uparrow m_i) \setminus \{m_i\} \supseteq (\uparrow m_{i+1}) \setminus \{m_{i+1}\}$ for all $i = 1, \ldots, k - 1$. This implies that m_2 is retractable to m_1. Since subsets of interval ordered sets are again interval ordered, the induction hypothesis applies to $P \setminus \{m_2\}$ as well as to $(\uparrow m_2) \setminus \{m_2\}$. Both these sets are collapsible. Thus P must be collapsible also. ∎

We trivially conclude that a finite interval ordered set has the fixed point property iff it is connectedly collapsible. However, even more is the case.

Proposition 8.2.2 *A finite interval ordered set has the fixed point property iff it is $\mathcal{I}$-dismantlable to a singleton.*

Proof. All we need to prove is "⇒". Let P be a finite interval ordered set with no irreducible points and the fixed point property. Since a finite ordered set has the fixed point property iff its $\mathcal{I}$-core has the fixed point property, we are done if we can show that P must be a singleton.

Assume the contrary. Then $P_0 := P$ has at least two minimal elements a_0 and b_0. By Scholium 8.1.7 we can assume without loss of generality that a_0 is retractable to b_0. The set $P_1 := (\uparrow a_0) \setminus \{a_0\}$ is not empty, as otherwise P_0 would be disconnected and thus not have the fixed point property. Moreover P_1 is also interval ordered, has the fixed point property (cf. Theorem 4.2.6), and no point in P_1 has a unique upper cover in P_1 (otherwise the unique upper cover in P_1 would also be a unique upper cover in P_0).

We now proceed inductively. Assume

$$a_0, b_0, \ldots, a_n, b_n, \quad \text{and} \quad P_{k+1} := (\uparrow a_k) \setminus \{a_k\} \neq \emptyset \ (k \in \{0, \ldots, n\}),$$

have already been chosen and are such that

1. For $k \in \{0, \ldots, n\}$

   (a) a_k, b_k are minimal in P_k,

    (b) a_k is retractable to b_k in P_k,

2. For $k \in \{0, \ldots, n + 1\}$ the set P_k

    (a) Has an interval order,

    (b) Has the fixed point property and

    (c) Has no points with a unique upper cover.

Then a_n has at least two upper covers, which means P_{n+1} has at least two minimal elements. Choose a_{n+1} and b_{n+1} minimal in P_{n+1} such that a_{n+1} is retractable to b_{n+1} in P_{n+1}. Then $P_{n+2} := (\uparrow a_{n+1}) \setminus \{a_{n+1}\} \neq \emptyset$ (otherwise P_{n+1} is disconnected). Moreover P_{n+2} has an interval order, the fixed point property and no points with a unique upper cover in P_{n+2} (otherwise P_{n+1} has a point with a unique upper cover).

The above inductive process never stops, contradicting the finiteness of P. Thus P must have been a singleton and we are done. ∎

8.3 Dedekind's Problem for Interval Orders and Reconstruction

Dedekind's problem of finding the number of antichains in an ordered set has a particularly easy solution for interval orders.

Lemma 8.3.1 *Let P be a finite interval ordered set. Let m be a minimal element such that $(\uparrow m) \setminus \{m\} \supseteq (\uparrow m') \setminus \{m'\}$ for all other minimal elements m' of P. Then $\#\mathrm{A}(P) = \#\mathrm{A}(P \setminus \{m\}) + 2^{|\mathrm{Min}(P)|-1}$.*

Proof. Every antichain in P either contains m or it does not. The antichains that do not contain m are the antichains of $P \setminus \{m\}$. The antichains that contain m are the antichains of $P \setminus (\uparrow m)$ with m added to each antichain. Thus $\#\mathrm{A}(P) = \#\mathrm{A}(P \setminus \{m\}) + \#\mathrm{A}(P \setminus (\uparrow m))$. By choice of m the set $P \setminus (\uparrow m)$ is the set of all minimal elements of P except m. Since the number of antichains in a $(|\mathrm{Min}(P)| - 1)$-element antichain is $2^{|\mathrm{Min}(P)|-1}$ the result follows. ∎

The recursive formula is now easily translated into a closed formula.

Theorem 8.3.2 *Let P be a finite interval ordered set and let $u_1, \ldots, u_n$ be the sequence of the filter sizes $| \uparrow p|$ of P in nonincreasing order and with multiplicity. Then*

$$\#\mathrm{A}(P) = 1 + \sum_{i=1}^{n} 2^{n-u_i-i+1}.$$

Proof. This proof is an induction on n. For $n = 1$ there is nothing to prove. Now assume the result holds for interval ordered sets of size $n - 1$ and let P be an interval ordered set with n elements.

Let $x \in P$ be an element with $|\uparrow x| = u_1$. Then x must be a minimal element such that $(\uparrow x) \setminus \{x\} \supseteq (\uparrow m') \setminus \{m'\}$ for all other minimal elements m' of P. By Lemma 8.3.1 we have

$$\#A(P) = \#A(P \setminus \{x\}) + 2^{|\mathrm{Min}(P)|-1}.$$

Now with $i = 1$ we have $|\mathrm{Min}(P)| - 1 = n - u_1 = n - u_1 + i - 1$.

By induction hypothesis we have

$$\#A(P \setminus \{x\}) = 1 + \sum_{i=2}^{n} 2^{(n-1)-u_i-(i-1)+1},$$

which finishes the proof. ∎

The above indicates that one knows a lot about an interval ordered set once one knows the filters or ideals of the set. Indeed finite interval ordered sets are determined up to isomorphism by their ideal decks. For convenience of notation in the proof we introduce the ideal size sequence.

Definition 8.3.3 *Let P be a finite ordered set. The **ideal size sequence** of P is the function $I_P : \mathbb{N} \to \mathbb{N}$ that assigns to each natural number k the number of ideals of size k in P.*

Theorem 8.3.4 *Let P and Q be finite interval ordered sets. Then P is isomorphic to Q iff $\mathcal{I}_P = \mathcal{I}_Q$, that is, iff the ideal decks are equal.*

Proof. The direction "$\Rightarrow$" is trivial. This proof of the direction "$\Leftarrow$" is by induction on n. For $n = 1$ there is nothing to prove.

For the induction step, assume that the result holds for interval ordered sets of size $n - 1$ and let P, Q be interval ordered sets of size n such that $\mathcal{I}_P = \mathcal{I}_Q$. Let k be the largest number such that P (and hence Q) has an ideal of size k. Since $\mathcal{I}_P = \mathcal{I}_Q$ there are elements $m^P \in P$ and $m^Q \in Q$ with $|\downarrow m^P| = |\downarrow m^Q| = k$ such that $\downarrow m^P$ is isomorphic to $\downarrow m^Q$. Then m^P and m^Q are maximal and each of m^P and m^Q is an upper bound of all non-maximal elements of $P \setminus \{m^P\}$ and $Q \setminus \{m^Q\}$ respectively. Moreover $\mathcal{I}_{P \setminus \{m^P\}} = \mathcal{I}_{Q \setminus \{m^Q\}}$. Therefore by induction hypothesis there is an isomorphism $\Phi : P \setminus \{m^P\} \to Q \setminus \{m^Q\}$. If $\Phi[(\downarrow m^P) \setminus \{m^P\}] = (\downarrow m^Q) \setminus \{m^Q\}$, then we are done, as Φ can be extended to an isomorphism of P and Q by mapping m^P to m^Q. If this is not the case, the following argument shows how to adjust Φ to obtain an isomorphism from $P \setminus \{m^P\}$ to $Q \setminus \{m^Q\}$ such that strict lower bounds of m^P go to strict lower bounds of m^Q.

Let P' be the set obtained from $P \setminus \{m^P\}$ by removing all its maximal elements and let Q' be the set obtained from $Q \setminus \{m^Q\}$ by removing all its maximal elements. Let $J_P := I_{(\downarrow m^P) \setminus \{m^P\}} - I_{P'}$ and let $J_Q := I_{(\downarrow m^Q) \setminus \{m^Q\}} - I_{Q'}$. Since $\downarrow m^P$

is isomorphic to $\downarrow m^Q$ and P' is isomorphic to Q', we have $J_P = J_Q$. $J_P(i)$ is the number of lower covers of m^P that have i lower bounds and are maximal in $P \setminus \{m^P\}$.

Let $p \in P$ be a lower cover of m^P in P. If $\Phi(p)$ is not a lower bound of m^Q, then $\Phi(p)$ must be maximal in $Q \setminus \{m^Q\}$. Maximal elements in $Q \setminus \{m^Q\}$ that are not below m^Q in Q are also maximal in Q. Thus $\Phi(p)$ is maximal in Q. Moreover p is maximal in $P \setminus \{m^P\}$. With $l := | \downarrow p|$ we must have $J_P(l) > 0$. Then m^Q has $J_Q(l) = J_P(l) > 0$ lower covers that are maximal in $Q \setminus \{m^Q\}$. The set M_l^Q of maximal elements of $Q \setminus \{m^Q\}$ with l lower bounds is thus partitioned into two nonempty sets. These sets are comprised of those elements of M_l^Q that are maximal in Q and those that are not maximal in Q. The same holds for the set M_l^P of maximal elements of $P \setminus \{m^P\}$ with l lower bounds. Corresponding subsets are of equal size because $J_P(l) = J_Q(l)$.

Since Φ maps maximal elements of $P \setminus \{m^P\}$ with l lower bounds to maximal elements of $Q \setminus \{m^Q\}$ with l lower bounds, there must be a maximal element q in $Q \setminus \{m^Q\}$ that is below m^Q in Q, has l lower bounds and which is not the Φ-image of a lower cover of m^P. Note that $(\downarrow q) \setminus \{q\} = (\downarrow \Phi(p)) \setminus \{\Phi(p)\}$, since both have equally many lower bounds. Define

$$\Phi'(x) := \begin{cases} \Phi(x); & \text{if } x \notin \{\Phi^{-1}(q), p\}, \\ \Phi(p); & \text{if } x = \Phi^{-1}(q), \\ q; & \text{if } x = p. \end{cases}$$

Then Φ' is an isomorphism from $P \setminus \{m^P\}$ to $Q \setminus \{m^Q\}$ and $\Phi'(p)$ is below m^Q.

If $\Phi'[(\downarrow m^P) \setminus \{m^P\}] = (\downarrow m^Q) \setminus \{m^Q\}$, then we are done. Otherwise note that Φ' maps one more strict lower bound of m^P to a point below m^Q than Φ. Thus continuing in this fashion will eventually produce a map with the desired property. This proves that P and Q are isomorphic.   ■

The reader will use a similar proof in Exercise 9 to prove the more common and computationally simpler characterization of isomorphism of interval orders. Aside from being a result about isomorphism, the above is a result on reconstructibility. If we know that a given deck stems from an interval ordered set, then we can reconstruct the set. The typical reconstruction proof for a class of ordered sets consists of proving recognizability and subsequently reconstructibility. If, like here, only the second part of the proof is available, one speaks of the following.

Definition 8.3.5 *A class $\mathcal{O}$ of ordered sets is called* **weakly reconstructible** *iff for any two P, Q in $\mathcal{O}$ with $\mathcal{D}_P = \mathcal{D}_Q$ we have that P is isomorphic to Q.*

With this definition we have the following immediate corollary.

Corollary 8.3.6 *Let P be an interval ordered set with > 3 points. Then P is weakly reconstructible from its ideal deck $\mathcal{I}_P$.*   ■

Of course, recognizability of interval ordered sets is quite trivial. Thus we can conclude that interval ordered sets are reconstructible.

Corollary 8.3.7 *Let P be an interval ordered set with > 3 points. Then P is reconstructible.*

Proof. First note that by Theorem 8.1.5 interval ordered sets with at least five elements are recognizable from their decks. Since ordered sets with four elements are reconstructible, this means interval ordered sets are recognizable.

Since the ideal deck is reconstructible by Theorem 3.1.6 the result follows from Corollary 8.3.6.   ∎

Reconstructibility of interval orders can also trivially be concluded from Exercise 9 and Theorem 3.5 in [247]. The presentation here was chosen to make the proof self-contained.

8.4   Interval Dimension

The idea of dimension can be defined using more general classes of orders than just linear orders. Indeed, any class of orders that contains the chains will contain a realizer for any ordered set. The availability of more orders however might allow for smaller realizers. In this section we will briefly examine how interval orders can be used to define a notion of dimension. Our focus will be on the fixed point property for interval dimension 2.

Definition 8.4.1 *Let $(P, \leq)$ be an ordered set. Then P is **of interval dimension** k iff $k \in \mathbb{N}$ is the smallest natural number such that there are k interval orders $\leq_1, \ldots, \leq_k$ on P such that $\leq = \bigcap_{i=1}^{k} \leq_i$.*

Remark 8.4.2 Clearly the interval dimension of an ordered set is less than or equal to its linear dimension.   ∎

Example 8.4.3 Let A be a set of products of intervals $[a_1, b_1] \times \cdots \times [a_n, b_n]$ ordered by $[a_1^1, b_1^1] \times \cdots \times [a_n^1, b_n^1] \leq [a_1^2, b_1^2] \times \cdots \times [a_n^2, b_n^2]$ iff $b_i^1 \leq a_i^2$ for all $i = 1, \ldots, n$ or the two products are equal. Then $(A, \leq)$ is of interval dimension at most n.

Comparing this example with Example 7.3.6 we find the ordering that we are familiar with from linear dimension. The change is that instead of considering sets of points we now consider n-dimensional boxes.

While Example 8.4.3 shows that interval dimension is similar to linear dimension there are of course important differences. The standard example shows that the interval dimension of a set can be 1 while the linear dimension is arbitrarily large. Thus there is no companion inequality for Remark 8.4.2 that goes in the other direction. For our following investigation of the structure of sets of interval dimension 2, we use a feature of interval orders that is not available for chains. Namely, it is possible to erase comparabilities in an interval order and obtain another interval order. This freedom is of course not given for chains.

Lemma 8.4.4 *Let $(P, \leq)$ be a finite ordered set of interval dimension k. Let the orders $\leq_1, \ldots, \leq_k$ be k interval orders such that $\leq = \bigcap_{i=1}^{k} \leq_i$. Then there are interval orders $\leq_1^m, \ldots, \leq_k^m$ with $\leq = \bigcap_{i=1}^{k} \leq_i^m$ that have the same minimal elements as $\leq$.*

Proof. Let M be the set of minimal elements of $(P, \leq)$ and for each $i \in \{1, \ldots, k\}$ let M_i be the set of minimal elements of $(P, \leq_i)$. Clearly for each i, we have $M_i \subseteq M$. Our proof is an induction on $n := \sum_{i=1}^{k} |M \setminus M_i|$.

Since the case $n = 0$ is trivial, assume now that $n > 0$ and the result has been proved for all $j < n$. Let $m \in P$ be $\leq$-minimal and not $\leq_i$-minimal for some i. Define $\leq_i' := \leq_i \setminus \{(x, m) : x <_i m\}$. Clearly $\leq_i'$ is still an order. To see that $\leq_i'$ is still an interval order, assume there are distinct points $a, b, c, d \in P$ with $a <_i' c, b <_i' d, b \not\leq_i' c$ and $a \not\leq_i' d$. Since $\leq_i$ is an interval order that only differs from $\leq_i'$ in the lower bounds of m, we infer $m \in \{a, b, c, d\}$. Since m is minimal in $(P, \leq_i')$ we can assume without loss of generality $m = a$. But then $a <_i c, b <_i d, b \not\leq_i c$ and $a \not\leq_i d$, a contradiction. Thus $\leq_i'$ is still an interval order. Moreover

$$\{(x, m) : x <_i m\} \cap \leq = \emptyset,$$

so replacing $\leq_i$ with $\leq_i'$ does not remove any comparabilities in $\leq$. For $j \neq i$ let $\leq_j' := \leq_j$. Then $\leq = \bigcap_{j=1}^{k} \leq_j'$.

Applying the induction hypothesis to the above representation of $\leq$ yields the result.   ∎

We can now use the special form of the above interval realizer to show that ordered sets of interval dimension 2 are collapsible.

Theorem 8.4.5 (This generalizes [88], Theorems 2,6.) *Let P be a finite ordered set of interval dimension 2. Then P has a retractable point. Consequently every finite ordered set of interval dimension 2 is collapsible.*

Proof. Let $\leq$ be the order on P and let $\leq_1, \leq_2$ be two interval orders with $\leq_1 \cap \leq_2 = \leq$. Let M_i be the set of minimal elements of $(P, \leq_i)$ and let M be the set of minimal elements of $(P, \leq)$. By Lemma 8.4.4 we can assume that $M_1 = M_2 = M$.

If $|M| = 1$, there is nothing to prove, so assume $|M| \geq 2$. By Scholium 8.1.7 and finiteness of P there are elements $a, b \in M$ such that for all $m \in M$ we have $(\uparrow_1 m) \setminus \{m\} \subseteq (\uparrow_1 b) \setminus \{b\}$ and for all $m \in M \setminus \{b\}$ we have $(\uparrow_1 m) \setminus \{m\} \subseteq (\uparrow_1 a) \setminus \{a\}$. Assume neither a is retractable to b, nor b is retractable to a in P. Then by definition of interval orders we must have $(\uparrow_2 b) \setminus \{b\} \subseteq (\uparrow_2 a) \setminus \{a\}$. Moreover there is a $d \in P$ with $d > b$ and $d \not\geq a$. Since $d \geq_2 a$ we must have $d \not\geq_1 a$. Thus the only minimal $\leq_1$-lower bound of d is b. Hence d is $\leq$-above only one $\leq$-minimal element, namely b. This implies that $(P, \leq)$ must have an irreducible point (cf. Chapter 4, Exercise 13).

Collapsibility is now proved via an easy induction.   ∎

The above immediately gives us the characterization of the fixed point property for interval dimension 2 in terms of connected collapsibility. Unlike for interval ordered sets, for interval dimension 2 it is not possible to characterize the fixed point property via $\mathcal{I}$-dismantlability. In Figure 7.2 we have an ordered set of linear and interval dimension 2, which has no irreducible point. Yet this set still has the fixed point property.

Corollary 8.4.6 *Let P be a finite ordered set of interval dimension 2. Then P has the fixed point property iff P is connectedly collapsible.*

Proof. By Theorem 8.4.5 every finite ordered set of interval dimension 2 is collapsible. Thus by Theorem 4.3.13 any such set has the fixed point property iff it is connectedly collapsible.   ∎

Exercises

1. Prove that if we demand strict inequality between endpoints in the definition of interval orders, we arrive at the same class of orders. That is, if $[a, b] \leq_{int} [c, d]$ iff $b < c$ or $[a, b] = [c, d]$, then every interval ordered set in this sense is interval ordered in the sense of Definition 8.1.2 and vice versa.

   We have seen in Scholium 8.1.6 that the same is true if we demand that all intervals we work with are nontrivial (i.e., contain more than one point). Prove that it remains true if we demand both above conditions.

2. Let P_n be the set in Example 8.1.3, part 4. Prove that no proper subset of P_n contains all interval ordered sets with up to n elements. (Hint: Stack chains and antichains.)

3. Let P be an ordered set. Prove that the following are equivalent.

   (a) P has an interval order.
   (b) For all $x, y \in P$ we have $(\uparrow x) \setminus \{x\} \subseteq (\uparrow y) \setminus \{y\}$ or $(\uparrow y) \setminus \{y\} \subseteq (\uparrow x) \setminus \{x\}$.
   (c) For all $x, y \in P$ we have $(\downarrow x) \setminus \{x\} \subseteq (\downarrow y) \setminus \{y\}$ or $(\downarrow y) \setminus \{y\} \subseteq (\downarrow x) \setminus \{x\}$.

4. Interval orders and dismantlability.

   (a) Give an example of an interval ordered set that is not dismantlable,

   (b) Give an example of a dismantlable ordered set that does not carry an interval order.

5. In this exercise the proofs of Example 8.1.3, part 4 and Theorem 8.1.5 are unified and the results are strengthened a bit. This exercise shows very nicely that with deep enough insight to define the right auxiliary quantities one can achieve elegant proofs. This result can be found in [85] or in [181].

   Let P be a finite ordered set. For $x \in P$ define

$$
\begin{aligned}
L(x) &:= (\downarrow x) \setminus \{x\}, \\
R(x) &:= (\uparrow x) \setminus \{x\}, \\
L^* &:= \{L(x) : x \in X\}, \\
R^* &:= \{R(x) : x \in X\}, \\
l(x) &:= |\{L \in L^* : L \subseteq L(x)\}|, \\
r(x) &:= |\{R \in R^* : R \supseteq R(x)\}|.
\end{aligned}
$$

   Then for all x we have $l(x) \le r(x)$ and the following are equivalent.

   (a) P carries an interval order.

   (b) P carries an interval order with endpoints of comparable intervals being distinct

   (c) P does not have a suborder isomorphic to $\mathbf{2} + \mathbf{2}$.

   (d) R^* is linearly ordered by inclusion.

   (e) L^* is linearly ordered by inclusion,

   (f) $x < y$ is equivalent to $r(x) < l(y)$.

6. Prove that the Dedekind–MacNeille completion of an interval ordered set is again interval ordered.

7. Let P be a finite interval ordered set. Let $a, b \in P$ be distinct elements such that $|\downarrow a| = |\downarrow b|$. Define the order $\le'$ on P as follows.

   (a) If $p \notin \{a, b\}$ or if q is not a strict upper bound of a or b, then $p \le' q$ iff $p \le q$.

   (b) For all $q \in P \setminus \{a, b\}$ we set

      • $a \le' q$ iff $b \le q$,
      • $b \le' q$ iff $a \le q$.

Prove that $(P, \leq)$ is isomorphic to $(P, \leq')$.

Also give an example that shows we need $\leq$ to be an interval order.

8. Let P be a finite interval ordered set and let $f : P \to P$ be an automorphism. Prove that for all $p \in P$ we have $(\uparrow p) \setminus \{p\} = (\uparrow f(p)) \setminus \{f(p)\}$ and $(\downarrow p) \setminus \{p\} = (\downarrow f(p)) \setminus \{f(p)\}$.

9. Let P be a finite ordered set. Let $IF_P(m, n) : \mathbb{N} \times \mathbb{N} \to \mathbb{N}$ be the number of elements $x \in P$ such that $| \uparrow x| = m$ and $| \downarrow x| = n$. Prove that if P and Q are interval ordered sets, then P is isomorphic to Q iff $IF_P = IF_Q$. Conclude that interval ordered sets are weakly reconstructible from IF_P.

10. Revisiting the relation $\sqsubseteq$ from Proposition 1.1.3, which also features prominently in Section 4.6.

   (a) Show that if $\sqsubseteq$ is considered as a relation on a set $\mathcal{J}$ of intervals in an ordered set, then $\sqsubseteq$ is an order relation.

   (b) Let $\mathcal{J}$ be a set of intervals on the real line. Show that a finite ordered set P is isomorphic to a set $(\mathcal{J}, \sqsubseteq)$ iff P is two-dimensional.

   (c) Let $\mathcal{J}$ be a set of intervals in a totally ordered set. Show that an ordered set P is isomorphic to a set $(\mathcal{J}, \sqsubseteq)$ iff P is two-dimensional.

11. **Semi-orders.** An ordered set P is called **semi-ordered** iff there is a set J of intervals of unit length on the real line ordered by $[a, b] \leq [c, d]$ iff $b \leq c$ or $[a, b] = [c, d]$. Such orders are also called **unit interval orders**.

   (a) (The Scott–Suppes theorem, cf. [251].) Prove that a countable ordered set P is semi-ordered iff P does not contain a subset of the form $\omega+\omega$, $\mathbf{2}+\mathbf{2}$ or $\mathbf{3}+\mathbf{1}$, where $\mathbf{n}$ is a chain with n elements and the "+" indicates that the sets are components of the union.

   (b) Verify that the ordered set in Figure 7.1 is semi-ordered.

   (c) Prove that every semi-ordered set with up to n elements is isomorphic to an ordered subset of the set P_n^1 of all intervals of unit length with rational endpoints in

$$\left\{ \frac{p}{q} : p = 0, \ldots, n^2, q = 0, \ldots, n \right\},$$

ordered by $[a, b] \sqsubseteq [c, d]$ iff $b \leq c$ or $[a, b] = [c, d]$.

Remarks and Open Problems

1. For more on interval orders cf. [20, 85]. For the connection to dimension, cf. [263], Chapter 8 and [209, 210].

2. Interval ordered sets can be generalized in various ways. Consider

   (a) By Theorem 8.1.5 interval ordered sets are characterized by the forbidden subset $\mathbf{2} + \mathbf{2}$. One could also investigate classes of ordered sets for which other types of subsets are forbidden. The semi-orders in Exercise 11 are an example.

   (b) The notion of dimension can be generalized by considering a class less restrictive than chains for the realizers. Similarly, interval orders also can be generalized by looking at intervals in less restrictive classes than chains. For some work in this direction cf. [182]. One has to be careful with this generalization, though. Every ordered set can be embedded into a lattice. Thus, for example, replacing the totally ordered set in the definition of an interval order with a lattice simply gives us all ordered sets. (In [182] the class of sets in which the intervals are formed has a forbidden subset.)

   One can also go in the opposite direction and pose more conditions on the intervals as done in Exercise 11.

   Parts 2a and 2b are addressed for example in [20].

3. Ordered sets can also be defined using shapes in $\mathbb{R}^n$ ordered via geometric containment. If we don't demand that the set is closed under union and intersection, we can indeed represent all ordered sets in such a fashion. For more on these geometric containment orders cf. [86].

4. Can Theorem 8.4.5 be used to efficiently recognize if an ordered set of interval dimension 2 has the fixed point property?

5. Is Lemma 8.4.4 extendable? That is, is the interval dimension of an ordered set $(P, \leq)$ the smallest number k for which there are k interval orders (denoted $\leq_1, \ldots, \leq_k$) such that for all $p \in P$ and all $j \in \{1, \ldots, k\}$ we have

$$\mathrm{rank}_{\leq}(p) = \mathrm{rank}_{\leq_j}(p) \text{ and } \leq = \bigcap_{j=1}^{k} \leq_j?$$

6. It is proved in [102] that the interval dimension of an ordered set and of its Dedekind–MacNeille completion are equal. The same holds for the linear dimension, cf. Exercise 8 in Chapter 10.

7. (Inspired by Example 8.1.3, part 4 and Exercise 2.) Let a_n be the smallest natural number such that there is an ordered set P of size a_n that contains

an isomorphic copy of every ordered set Q of size $\leq n$. Compute a_n or find estimates.

Clearly a_n is a finite number, as one can always pick one representative of each of the finitely many isomorphism classes of n-element ordered sets and form an ordered set whose connected components are the thus picked sets. It is equally clear that this construction is much too inefficient to get close to a_n.

More generally, for any class $\mathcal{C}$ of ordered sets, let $a_n^{\mathcal{C}}$ be defined as follows. $a_n^{\mathcal{C}}$ is the smallest number such that there is an ordered set $P \in \mathcal{C}$ of size $|P| = a_n^{\mathcal{C}}$ that contains isomorphic copies of all ordered sets $Q \in \mathcal{C}$ with $|Q| \leq n$. Compute $a_n^{\mathcal{C}}$ or find estimates.

8. (Variation on 7.) Let b_n be the smallest natural number such that there is an ordered set P of size b_n that contains an isomorphic copy of every ordered set Q of size $\leq n$ as a **covering subset**. That is, Q is an ordered subset and the diagram of Q is contained in the diagram of P. Compute b_n or find estimates. $b_n^{\mathcal{C}}$ can be defined analogously to $a_n^{\mathcal{C}}$ and the same question can be asked.

9. In [28] it was proved that the $\dfrac{1}{3} - \dfrac{2}{3}$ conjecture holds for semi-orders. Can it be proved for interval orders?

9
Lexicographic Sums

So far, we have introduced basic concepts of ordered sets and several important classes of ordered sets. In this chapter we will ask ourselves for the first time how we can use existing ordered sets to build new ordered sets. The construction we exhibit here, lexicographic sums, comes from a very simple pictorial idea. Take an ordered set T and replace each of its points t with an ordered set P_t. The resulting structure will be a new, larger ordered set. Natural questions to ask are of course how various order-theoretical properties and parameters behave under lexicographic constructions.

9.1   Definition and Examples

There are two ways to look at lexicographic sums. On one hand we can consider the construction as a tool to build larger sets from smaller sets. On the other hand we can analyze a given ordered set and see if we can represent it as a lexicographic sum that is made up of smaller sets. The "building" point-of-view is reflected in the definition.

Definition 9.1.1 *Let T be a nonempty ordered set considered as an index set. Let $\{P_t\}_{t \in T}$ be a family of pairwise disjoint nonempty ordered sets that are all disjoint from T. We define the* **lexicographic sum** *$L\{P_t | t \in T\}$ to be $\bigcup_{t \in T} P_t$ ordered by*

$p_1 \leq p_2$ *iff*

    1. $p_i \in P_{t_i}$, $t_1 \neq t_2$ and $t_1 < t_2$, or

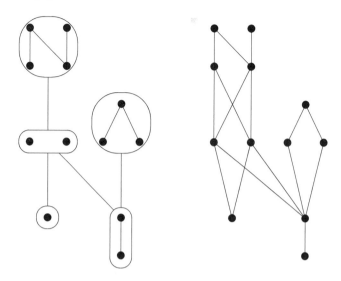

Figure 9.1: Visualization of a lexicographic sum. To the right, the diagram of the ordered set, to the left the ordered set with pieces and their comparabilities indicated.

2. $p_1, p_2 \in P_t$ and $p_1 \leq_{P_t} p_2$.

The P_t will be called the **pieces** *of the lexicographic sum and T will be called the* **index set**. *For $p \in L\{P_t | t \in T\}$ define $I(p)$ to be the unique $t \in T$ such that $p \in P_t$. We will also say that $L\{P_t | t \in T\}$ is the* **lexicographic sum of the P_t over T**.

To avoid re-stating the obvious we will always assume that when we consider a lexicographic sum that the pieces P_t are all mutually disjoint and also disjoint from T.

The definition is illustrated pictorially in the left ordered set in Figure 9.1. The index set T is the ordered set made up of the large ovals and the connections between the large ovals. The pieces are the ordered sets within the large circles. The comparabilities that arise are shown in the diagram on the right.

The simplest types of lexicographic sums are the lexicographic sums in which we "stack" ordered sets and those in which we put the pieces "side by side".

Definition 9.1.2 *Let $L\{P_t | t \in T\}$ be a lexicographic sum.*

1. *If the index set T is a chain, we will call the lexicographic sum* **linear**.

2. *A linear lexicographic sum with finitely many pieces $P_1, \ldots, P_n$ will also be denoted $P_1 \oplus \cdots \oplus P_n$, with the lowest piece at the beginning of the sum.*

3. *A lexicographic sum of* $P_1, \ldots, P_n$ *where the index set is the n-element antichain will be denoted* $P_1 + \cdots + P_n$. *It will be called the* **disjoint sum** *of* $P_1, \ldots, P_n$.

4. *If all pieces are singletons or if there is only one piece, the lexicographic sum will be called* **trivial**.

The point-of-view of breaking down an ordered set into smaller parts that are put together in a lexicographic sum is reflected in the following notation.

Definition 9.1.3 *Let P be an ordered set.*

1. *If P is isomorphic to a nontrivial lexicographic sum* $L\{P_t | t \in T\}$, *P will be called* **decomposable**. $L\{P_t | t \in T\}$ *will be called a* **lexicographic sum decomposition** *of P.*

   Otherwise P will be called **indecomposable**.

2. *If P is isomorphic to a nontrivial linear lexicographic sum, P will be called* **series decomposable**. *The linear lexicographic sum will be called a* **series decomposition** *of P.*

With this language in mind, consider Figure 9.1 once more. Going from left to right, the set on the right has been built out of five pieces. Formally, these pieces are the one and two element antichains, the 2-chain, a 3-fence and a 4-fence. The pieces were put into a lexicographic sum over an index set that is a 4-fence with an additional upper cover attached to the maximal element with two lower covers. Going from right to left, we see that, starting with the set on the right, we can decompose it into the lexicographic sum as indicated above.

Example 9.1.4

1. The four crown is a lexicographic sum of two two-element antichains over a 2-chain as the index set.

2. Every disconnected ordered set is decomposable.

3. Every chain with more than two elements is decomposable.

4. The set $\mathcal{P}(X) \setminus \{\emptyset, X\}$ with $\mathcal{P}(X)$ being the power set of the set X is not decomposable.

5. The lexicographic order on $\mathbb{N}^n$ as described in Example 1.1.2, part 9 can be described inductively as $\mathbb{N}^n = L\{\mathbb{N}_j^{n-1} | j \in \mathbb{N}\}$ with all $\mathbb{N}_n^{n-1}$ isomorphic to $\mathbb{N}^{n-1}$.

Let us now investigate how lexicographic constructions affect previously introduced notions.

Lemma 9.1.5 *Let* $P := L\{P_t | t \in T\}$ *be a lexicographic sum. Then P has a retract that is isomorphic to T.*

Proof. For each $t \in T$ choose an element $p_t \in P_t$. The map that maps each piece P_t to the chosen element p_t is a retraction of P to an ordered set that is isomorphic to T. ∎

Proposition 9.1.6 *If T and all P_t are complete lattices, then the lexicographic sum $L\{P_t | t \in T\}$ is a complete lattice. Conversely, if $L\{P_t | t \in T\}$ is a (complete) lattice, then T is a (complete) lattice.*

Proof. Let T and all P_t be complete lattices. Let A be a subset of $L\{P_t | t \in T\}$. Then $I[A] \subseteq T$ has a supremum $\bigvee_T I[A]$. The supremum of A in P now is

$$\bigvee A = \bigvee_{P_{\bigvee_T I[A]}} (A \cap P_{\bigvee_T I[A]}).$$

The mentioned partial converse is an easy consequence of Lemma 9.1.5 and Exercise 4 in Chapter 5. ∎

Note that even if the lexicographic sum is a complete lattice it is not necessary for all the pieces to be complete lattices. As an example consider the linear lexicographic sum of a one-point ordered set, a two-point antichain and another one-point ordered set. This set is a lattice, and yet the middle piece is not a lattice. We further explore this problem in Exercise 2.

Proposition 9.1.7 *The dimension of a lexicographic sum is*

$$\dim(L\{P_t | t \in T\}) = \max\{\dim(T), \dim(P_t) : t \in T\},$$

provided the right side exists as a finite number. If the right side is not finite, then the set $L\{P_t | t \in T\}$ is not finite dimensional.

Proof. First recall that the dimension of a subset can never exceed the dimension of the whole set. Therefore by Lemma 9.1.5 the inequality

$$\dim(L\{P_t | t \in T\}) \geq \max\{\dim(T), \dim(P_t) : t \in T\}$$

is trivial. In particular, if the right side is not finite, then the dimension of P is not finite either.

Now assume that the right side is finite, say k. For each ordered set $X \in \{T\} \cup \{P_t : t \in T\}$ denote the order on X by $\leq^X$ and find a realizer $\leq_1^X, \ldots, \leq_{l_X}^X$, with $l_X \leq k$. For $i \in \{l_X + 1, \ldots, k\}$ let $\leq_i^X := \leq_{l_X}^X$. For $i = 1, \ldots, k$, let P_t^i be the set P_t ordered by $\leq_i^{P_t}$ and let T^i be the set T ordered by $\leq_i^T$.

We claim the orders $\leq_i$ of the lexicographic sums $P_i := L\{P_t^i | t \in T^i\}$ have the order $\leq$ of $L\{P_t | t \in T\}$ as their intersection. We first note that $\leq_i |_{P_t \times P_t} = \leq_i^{P_t}$.

This means that $\bigcap_{i=1}^{k} \leq_i |_{P_t \times P_t} = \bigcap_{i=1}^{k} \leq_i^{P_t} = \leq^{P_t}$, so the pieces are given their original order by the intersection of these linear extensions. Thus if $p, q \in P_s \subseteq$

$L\{P_t | t \in T\}$, then $p \le q$ iff $p \le^{P_s} q$ iff for all i we have $p \le_i q$. Now consider $p \in P_r$ and $q \in P_s$ with $r \ne s$. Then $p < q$ iff $r <^T s$ iff $r <_i^T s$ for all i iff $p <_i q$ for all i.

This proves $\le$ is the intersection of the $\le_i$. Hence

$$\dim(L\{P_t | t \in T\}) \le \max\{\dim(T), \dim(P_t) : t \in T\}$$

and we are done.   ∎

Proposition 9.1.8 *Even if T and all P_t carry interval orders, the lexicographic sum $L\{P_t | t \in T\}$ need not have an interval order.*

Proof. A trivial example will do. The two-element chain and the two-element antichain both carry interval orders. Yet the lexicographic sum of two two-element chains over a two-element antichain is the forbidden set $\mathbf{2} + \mathbf{2}$.   ∎

We conclude our first exposure to lexicographic sums by investigating the fixed point property. The first step is to consider linear lexicographic sums.

Lemma 9.1.9 *Let $Q := L\{P_c | c \in C\}$ be a chain-complete lexicographic sum with a chain C as index set. Then P has the fixed point property iff one piece P_c has the fixed point property.*

Proof. If none of the pieces have the fixed point property, it is easy to see that Q does not have the fixed point property.

Now consider the case in which one P_c has the fixed point property. Let $f : Q \to Q$ be order-preserving. If f maps P_c to itself, then f has a fixed point. Otherwise for some $p_c \in P_c$ we have $f(p_c) \sim p_c$ and hence by the Abian–Brown theorem, f must have a fixed point.

Thus chain-complete linear lexicographic sums have the fixed point property iff they have a piece with the fixed point property.   ∎

Proposition 9.1.10 (Compare with [122, 123, 124].) *Let $L\{P_t | t \in T\}$ be a chain-complete lexicographic sum. Then $L\{P_t | t \in T\}$ has the fixed point property iff*

1. *T has the fixed point property, and*

2. *For all $t_0 \in T$ one of P_{t_0}, $L\{P_t | t < t_0\}$ and $L\{P_t | t > t_0\}$ has the fixed point property.*

Proof. Let $P := L\{P_t | t \in T\}$ be a chain-complete lexicographic sum with arbitrary index set T. First suppose P has the fixed point property. By Lemma 9.1.5, part 1 must hold. To prove part 2 we proceed as follows. Fix a point $p_{t_0} \in P_{t_0}$. The map that maps all points that are not comparable to any $p \in P_{t_0}$ to p_{t_0} is a retraction of P onto $L\{P_t | t < t_0\} \oplus P_{t_0} \oplus L\{P_t | t > t_0\}$. Thus $L\{P_t | t < t_0\} \oplus P_{t_0} \oplus L\{P_t | t > t_0\}$ must have the fixed point property. By Lemma 9.1.9 this is the case iff one of its three pieces has the fixed point property. Thus 2 must hold.

Conversely, suppose $P := L\{P_t | t \in T\}$ is a chain-complete lexicographic sum that satisfies conditions 1 and 2. Let $f : P \to P$ be an order-preserving map. Define $F : T \to \mathcal{P}(T) \setminus \{\emptyset\}$ by $F(t) := I[P_t]$. Then for some $t \in T$ we must have $F(t) = \{t\}$. Indeed, otherwise for each $t \in T$, choose $g(t) \in F(t) \setminus \{t\}$. This map would be an order-preserving fixed point free self-map of T, a contradiction to 1. Find a $t_0 \in T$ with $F(t_0) = \{t_0\}$. Then f maps $L\{P_t | t < t_0\} \oplus P_{t_0} \oplus L\{P_t | t > t_0\}$ to itself. Since by 2 one of the three pieces must have the fixed point property, we conclude via Lemma 9.1.9 that f has a fixed point.   ∎

Note that a characterization of exactly which pieces in a lexicographic sum could be fixed point free ordered sets is unattainable. The reason is that in a lexicographic sum over a two-point chain, either piece could have the fixed point property in order for the whole set to have the fixed point property. This means, there can be pairs of points $\{s, t\} \subseteq T$ such that either one could be assigned a fixed point free piece without the sum loosing the fixed point property. However if both are assigned a fixed point free piece, the sum also becomes fixed point free. This situation occurs in other index sets also (cf. Exercise 3).

9.2   The Canonical Decomposition

Consider the ordered set in Figure 9.1 once more. There are indeed several ways to write this set as a lexicographic sum. On one hand, one could write the piece that is a 3-fence as a linear lexicographic sum to obtain a "finer" partition of the ordered set. On the other hand one could merge the 2-antichain and the 4-fence into a larger piece to obtain a lexicographic sum with a smaller index set. Is there a canonical way to write lexicographic sums? For finite ordered sets there is a natural idea, which is to "make the pieces as large as possible".

Definition 9.2.1 (Cf., e.g., [60, 134].) *Let P be an ordered set and let $S \subseteq P$ be nonempty. Then S is called **order-autonomous** iff for all $p \in P \setminus S$ we have that*

1. *If there is an $s \in S$ with $p \leq s$, then $p \leq S$, and*

2. *If there is an $s \in S$ with $p \geq s$, then $p \geq S$.*

Note that every piece in any lexicographic sum decomposition of an ordered set is order-autonomous. Conversely, any order-autonomous set in P can be a piece in a lexicographic sum decomposition of P. Moreover we have

Lemma 9.2.2 *Let P be an ordered set and let S_1 and S_2 be order-autonomous subsets with $S_1 \cap S_2 \neq \emptyset$. Then the union $S_1 \cup S_2$ is order-autonomous.*

Proof. Exercise 8.   ∎

Taking pieces that are as large as possible of course is nothing else but finding subsets that are order-autonomous and maximal with respect to inclusion. We just have to be careful not to take the whole set, which is vacuously order-autonomous in itself.

Definition 9.2.3 *An order-autonomous subset S of the ordered set P is called* **maximal** *iff* $S \neq P$ *and for all order-autonomous subsets* $Q \subseteq P$ *with* $S \subseteq Q$ *we have* $Q \in \{S, P\}$.

The canonical decomposition for an ordered set is now a decomposition into disjoint maximal order-autonomous subsets. This idea is important for connected, not series decomposable, ordered sets. (Disconnected ordered sets decompose naturally into their components and for series decomposable ordered sets a similar idea is discussed in Exercise 9.)

Proposition 9.2.4 (The canonical decomposition of finite ordered sets.) *Let P be a finite ordered set. Then every* $p \in P$ *is contained in a maximal order-autonomous subset S of P. Moreover, if P is not series decomposable with more than two summands and not disconnected with more than two components, then the set S is uniquely determined by p.*

Proof. First consider the case in which P is series decomposable. Decompose P into the lexicographic sum $L\{P_c | c \in C\}$ with C as large as possible. Then $B := L\{P_c | c \in C \setminus \{\bigvee C\}\}$ and $T := L\{P_c | c \in C \setminus \{\bigwedge C\}\}$ are both maximal order-autonomous subsets of P and each $p \in P$ is in at least one of these two sets. If $|C| = 2$, then B and T are disjoint and every order-autonomous subset not equal to P is contained in B or T. This proves the result for series decomposable sets. The proof in case P is disconnected is similar.

This leaves us with the case in which P is connected and not series decomposable. Since P is finite and every singleton subset of an ordered set is order-autonomous, every $p \in P$ is contained in a maximal order-autonomous subset. To prove uniqueness, let S_1 and S_2 be two maximal order-autonomous subsets of P with $S_1 \cap S_2 \neq \emptyset$. Suppose $S_1 \setminus S_2$ is not empty. Then, since P is connected and $S_1 \cup S_2 = P$, there is a point $p \in S_1 \setminus S_2$ such that p is comparable to a point in S_2. Say $p \leq s_2$ for some $s_2 \in S_2$. Then $p \leq S_2$. However this implies that every element of S_2 is an upper bound of S_1 (and hence of $S_1 \setminus S_2$). This contradicts our assumption that P was not series decomposable. Thus $S_1 \setminus S_2 = \emptyset$ and since the argument is entirely symmetric we have $S_1 = S_2$. Therefore any two maximal order-autonomous subsets in P are either disjoint or equal, which finishes the proof. ∎

The canonical decomposition of a finite ordered set will be useful whenever a "standardization" of a lexicographic sum is necessary. This will be helpful in the following when we talk about comparability invariance and about reconstruction of lexicographic sums. Note that the ambiguity observed in the lexicographic decompositions for series decomposable finite ordered sets gets worse for infinite ordered sets (cf. Exercise 10). For more on order-autonomous sets consider for example [134] and [245].

Since the canonical decomposition only applies in full strength to connected, non-series decomposable ordered sets we introduce the following terminology.

Definition 9.2.5 *We shall call an ordered set P* **co-connected** *iff P is not series decomposable.*

The language is motivated by the fact that if P is not series-decomposable, then the complement of the comparability graph of P must be connected.[1]

9.3 Comparability Invariance

With ordered sets being closely related to graphs, it is natural to ask if a certain property of ordered sets is of graph-theoretical nature or not. How would one define "of graph-theoretical nature", though? A property should be considered of graph-theoretical nature iff it is a property of the comparability graph. That is, two ordered sets with isomorphic comparability graphs either both have the property or neither does. The definition we give formalizes this idea by considering "parameters" of an ordered set. Note that while quantities like the dimension of an ordered set are obvious parameters, properties can also be encoded as a parameter via indicator functions. For example, we can define $\alpha_{FPP}(P)$ to be one iff P has the fixed point property and zero otherwise. In this fashion the parameter α_{FPP} encodes the fixed point property.

Definition 9.3.1 *Let C be a class of ordered sets and let α be a parameter for ordered sets. Then α is called a C-**comparability invariant** iff for any ordered sets $P, Q \in C$ we have that isomorphism of $G_C(P)$ and $G_C(Q)$ implies $\alpha(P) = \alpha(Q)$.*

The definition of comparability invariants with respect to certain classes of ordered sets is important. Indeed, there are properties that are comparability invariants for large classes of ordered sets (say for the class of finite ordered sets), but which are not comparability invariants in the class of all ordered sets. For example, the fixed point property is not a comparability invariant in general. Indeed $\mathbb{N}\cup\{\infty\}$ and $\mathbb{N}$ have isomorphic comparability graphs, but the first set has the fixed point property, while the second does not. In the following we will concentrate on finite ordered sets, where there is a strong connection between comparability invariance and a certain type of theorem for lexicographic sums. In particular we will see that the fixed point property is a comparability invariant for finite ordered sets.

Note that if P is an ordered set, then P and P^d, the dual of P have the same comparability graph. Thus comparability invariants must be invariant under dualization. Theorem 9.3.4 shows that for finite ordered sets, invariance under dualization is close to comparability invariance. (For comparability invariance we must

---

[1] Terminology suggested by J.-X. Rampon.

be able to dualize any piece of a lexicographic decomposition without changing the parameter.) We shall start with a result on the structure of comparability graphs. The following theorem shows that if P is indecomposable, its comparability graph has exactly two transitive orientations.

Theorem 9.3.2 *Let $G = (V, E)$ be a comparability graph. Then one of the following holds for G.*

1. *G has exactly two transitive orientations.*

2. *There is a proper subset $S \subset V$ with at least two elements such that S is order-autonomous in every transitive orientation of G.*

3. *The complement graph $\overline{G}$ of G is not connected.*

This result, which has a surprisingly rich proof, was reported in [60], Lemma on p.270 as a consequence of [134] Theorems 3.1 and 3.4 and Corollary 4.2. A proof can be constructed in the very instructive sequence of Exercises 26–29. The theorem is used to prove the following crucial lemma.

Lemma 9.3.3 *Let P and Q be finite ordered sets with isomorphic comparability graphs. Let Φ be the isomorphism between the comparability graphs $G_C(P)$ and $G_C(Q)$. If P has no nontrivial lexicographic sum decomposition, then the map $\Phi : P \to Q$ is also an order-isomorphism between P and Q or between P and Q^d.*

Proof. Suppose Φ is not an order-isomorphism between P and Q or between P and Q^d. Then there are four ways to give $G_C(P)$ a transitive orientation. They are the orders of P, P^d and those induced by $\Phi^{-1}[Q]$ and $\Phi^{-1}[Q^d]$. Theorem 9.3.2 now leads to a contradiction to the indecomposability of P. ∎

Theorem 9.3.4 (Cf. [60].) *Let α be a parameter of ordered sets. Then α is a comparability invariant in the class of finite ordered sets iff*

For all finite lexicographic sums $L\{P_t | t \in T\}$ such that for at most one $t \in T$ the set P_t has more than one element, we have that

$$\alpha(L\{P_t | t \in T\}) = \alpha(L\{P_t^d | t \in T\}).$$

Proof. The direction "⇒" is trivial, since the two sets involved have the same comparability graph.

To prove "⇐", let P and Q be two ordered sets with isomorphic comparability graphs. Call this graph isomorphism between the comparability graphs Φ. We will prove there is a finite sequence $P = P_0, P_1, \ldots, P_N$ such that P_N is order-isomorphic to Q and P_k is obtained from P_{k-1} by dualization of exactly one order-autonomous subset. (This includes the possibility of dualizing the set itself, since the whole set is vacuously order-autonomous.) Since the parameter α is by assumption not affected by going from P_{k-1} to P_k, we have $\alpha(P) = \alpha(Q)$, establishing that α is a comparability invariant for finite ordered sets.

The proof is an induction on the size n of the set P. Since for $n = 1$ there is nothing to prove, let $|P| = n$ and assume the result has been proved for ordered sets of size $< n$. We will first consider the case in which P is disconnected (which immediately implies that Q is disconnected). The idea in this case will then merely be transferred to the other cases.

If P is disconnected, then the isomorphism Φ between $G_C(P)$ and $G_C(Q)$ is comprised of isomorphisms of the comparability graphs of the components. Let $C^1, \ldots, C^m$ be the components of P and let $K^1, \ldots, K^m$ be the components of Q enumerated such that $\Phi|_{C^i}$ is an isomorphism from $G_C(P)[C^i]$ to $G_C(Q)[K^i]$. By induction hypothesis, for each pair $(C^i, \Phi[C^i] = K^i)$ (with orders induced by P and Q, respectively) there is a finite sequence $C^i = C_0^i, C_1^i, \ldots, C_{N_i}^i$ such that $C_{N_i}^i$ is order-isomorphic to K^i and C_j^i is obtained from C_{j-1}^i by dualization of exactly one order-autonomous subset. The desired sequence from P to Q is now obtained by replacing each C^i in P (starting with C^1 and ending with C^m) successively with the sets in the sequence $C^i = C_0^i, C_1^i, \ldots, C_{N_i}^i$.

If P is series decomposable, then the isomorphism between $G_C(P)$ and $G_C(Q)$ is comprised of isomorphisms of the components of the complements of the comparability graphs. Let $C^1, \ldots, C^m$ be the components of the complement of $G_C(P)$ and let $K^1, \ldots, K^m$ be the components of the complement of $G_C(Q)$ enumerated so that $\Phi|_{C^i}$ is an isomorphism between $G_C(P)[C_i]$ and $G_C(Q)[K_i]$. From now on we will interpret the C_i and K_i as ordered subsets of P and Q respectively.

The set P is a linear lexicographic sum of the C^i and Q is a linear lexicographic sum of the K^i. Without loss of generality we can assume $Q = K^1 \oplus K^2 \oplus \cdots \oplus K^m$. Then there is a permutation σ on $\{1, \ldots, m\}$ such that $P = C^{\sigma(1)} \oplus C^{\sigma(2)} \oplus \cdots \oplus C^{\sigma(m)}$. Recall that every permutation can be represented as a composition of permutations that switch two adjacent elements. Successively switch two adjacent pieces C_i in P until the set $P' = C^1 \oplus C^2 \oplus \cdots \oplus C^m$ is reached. Then continue with a successive replacement of the pieces as in the proof for disconnected ordered sets.

Now consider the case that P is connected and co-connected. The above two cases have shown the main ingredients of the proof already. Once we have turned P into a set whose index set is order-isomorphic to the index set of Q via Φ, we can use the induction hypothesis and a replacement process as for the components of disconnected ordered sets to finish the proof. Our main tool here will be the canonical decomposition.

In case P is connected and co-connected, Q is connected and co-connected also. Let $P = L\{P_t | t \in T\}$ and $Q = L\{Q_u | u \in U\}$ be the canonical decompositions of P and Q. We claim that for all $t \in T$, there is a unique $u \in U$ such that $\Phi[P_t] = Q_u$. Indeed, suppose there was a $t \in T$ such that there is no $u \in U$ with $\Phi[P_t] \subseteq Q_u$.

First consider a $q \in Q \setminus \Phi[P_t]$ such that $q > b$ for some $b \in \Phi[P_t]$. This means that $\Phi^{-1}(q)$ is related to $\Phi^{-1}(b)$ in P, say without loss of generality $\Phi^{-1}(q) > \Phi^{-1}(b)$. Then $\Phi^{-1}(q) > P_t$. This means that q is related to all elements of $\Phi[P_t]$. This fact will be used frequently.

Now let $B \subseteq Q$ be the union of all Q_u that intersect $\Phi[P_t]$. Since Q is connected and co-connected, we have $|U| \geq 4$. Moreover, since we work with the canonical decomposition, no union of two pieces Q_u is order-autonomous in Q. We claim that for any two distinct $Q_{u_1}, Q_{u_2} \subseteq B$, we must have $Q_{u_1} \leq Q_{u_2}$ or $Q_{u_2} \leq Q_{u_1}$. Indeed, suppose no element of Q_{u_1} is related to any element of Q_{u_2}. If $q \in Q \setminus (Q_{u_1} \cup Q_{u_2})$ satisfies $q > b'$ for some $b' \in Q_{u_1} \cup Q_{u_2}$, then $q > b$ for some $b \in \Phi[P_t]$. The above shows that then q is related to every element of $\Phi[P_t]$. Thus, in particular q is related to an element of $\Phi[P_t] \cap Q_{u_1}$ and to an element of $\Phi[P_t] \cap Q_{u_2}$. Since we assumed Q_{u_1} and Q_{u_2} had no related elements, we conclude $q > Q_{u_1} \cup Q_{u_2}$. This and the dual argument show $Q_{u_1} \cup Q_{u_2}$ would be order-autonomous in Q. Since $Q_{u_1} \cup Q_{u_2} \neq Q$, this is a contradiction. This means that B is a linear lexicographic sum of pieces Q_u.

Now let u_{max} be the largest element of the set $I_Q[B]$ and let u_{min} be the smallest element of the set $I_Q[B]$. Consider the set

$$C := Q_{u_{min}} \oplus L\{Q_u | u_{min} < u < u_{max}\} \oplus Q_{u_{max}}.$$

Since C is series-decomposable, we have $C \neq Q$. We claim that C is order-autonomous in Q. Indeed, let $q \in Q \setminus C$ be such that $q > c$ for some $c \in C$. Then $q > c'$ for some $c' \in Q_{u_{min}}$. This means that $q > Q_{u_{min}}$. Let $b \in \Phi[P_t] \cap Q_{u_{min}}$. Then $q > b$ and hence q is related to all elements of $\Phi[P_t]$, which implies q is related to all elements of B. Since $q \notin C$ this means that $q > B$. This finally implies $q > C$. Hence the set C is order-autonomous in Q and not equal to Q, contradiction to the maximality of the Q_u. Thus for all $t \in T$, there must be a unique $u \in U$ such that $\Phi[P_t] \subseteq Q_u$. By symmetry we infer $\Phi[P_t] = Q_u$.

The above shows that Φ induces a natural isomorphism Ψ between $G_C(T)$ and $G_C(U)$ by making $\Psi(t)$ the unique element $u \in U$ so that $\Phi[P_t] = Q_u$. Moreover $\Phi|_{P_t}$ is an isomorphism between $G_C(P_t)$ and $G_C(\Phi[P_t])$. T and U are both connected and co-connected ordered sets with no nontrivial order-autonomous subsets. Thus by Lemma 9.3.3, T is order-isomorphic via Ψ to U or to U^d. If T is isomorphic to U, let $P_1 := P$, otherwise let $P_1 := P^d$.

Now continue with a successive replacement of the pieces as in the proof for disconnected ordered sets.   ∎

With this tool in hand we are able to easily prove comparability invariance or non-invariance of several parameters. It is interesting that while the fixed point property turns out to be a comparability invariant in the class of finite ordered sets, the related property of not having a fixed point free automorphism is not a comparability invariant. To prove this fact, we introduce here the concept of rigidity. A **rigid** ordered set has only one automorphism, the identity. As an example consider a fence with an even number of elements. Rigidity essentially says that the ordered set lacks internal symmetry of a certain kind. It is thus somewhat opposite of having a fixed point free automorphism, which essentially says that the ordered set has a certain type of symmetry.

Corollary 9.3.5 *In the class of finite ordered sets:*

1. *The fixed point property is a comparability invariant.*

2. *Dimension is a comparability invariant.*

3. *Being an interval order is a comparability invariant.*

4. *Being a lattice is not a comparability invariant.*

5. *"f is rigid", that is, "f has no nontrivial automorphism", is not a comparability invariant.*

6. *"f has a fixed point free automorphism" is not a comparability invariant.*

Proof. To prove part 1 we prove that the fixed point property is a comparability invariant for finite ordered sets by proving the condition of Theorem 9.3.4 holds for ordered sets of any size. To do this we proceed by induction on $|P|$. For $|P| = 1$ there is nothing to prove. We proceed to the induction step assuming that the condition of Theorem 9.3.4 holds for all ordered sets Q with $|Q| < n$.

Let $P = L\{P_t | t \in T\}$ be an ordered set of size n that has the fixed point property and is such that only for $t' \in T$ has the piece $P_{t'}$ more than one element. Then by Proposition 9.1.10, T has the fixed point property, and for all $t_0 \in T$ one of P_{t_0}, $L\{P_t | t < t_0\}$ and $L\{P_t | t > t_0\}$ has the fixed point property. Then for all $t_0 \in T$ that are incomparable to t', one of $P_{t_0}^d$, $L\{P_t^d | t < t_0\}$ and $L\{P_t^d | t > t_0\}$ has the fixed point property (all pieces are singletons and thus their own duals). Since an ordered set has the fixed point property iff its dual has the fixed point property, by Lemma 9.1.9 one of $P_{t'}^d$, $L\{P_t^d | t < t'\}$ and $L\{P_t^d | t > t'\}$ has the fixed point property. For $t_0 < t'$ note that the induction hypothesis applies to $L\{P_t | t > t_0\}$ and thus one of $P_{t_0}^d$, $L\{P_t^d | t < t_0\}$ and $L\{P_t^d | t > t_0\}$ has the fixed point property. The case $t_0 < t'$ is handled similarly. Thus by Proposition 9.1.10 the ordered set $L\{P_t^d | t \in T\}$ also has the fixed point property.

We have proved the condition in Theorem 9.3.4 for the fixed point property. Thus the fixed point property is a comparability invariant in the class of finite ordered sets.

Part 2 follows directly from Proposition 9.1.7, Theorem 9.3.4 and the fact that $\dim(P) = \dim(P^d)$.

Part 3 is almost vacuously true. Let $P = L\{P_t | t \in T\}$ be a finite ordered set that carries an interval order and is such that only for $t' \in T$ has the piece $P_{t'}$ more than one element. Then $P_{t'}$ is an antichain or $\{p \in P : (\forall x \in P_{t'})x \not> p\}$ is an antichain (which could possibly be empty). Thus dualizing $P_{t'}$ will not create a set of the form $\mathbf{2} + \mathbf{2}$ and hence $P = L\{P_t^d | t \in T\}$ carries an interval order. By Theorem 9.3.4 being an interval order is thus a comparability invariant in the class of finite ordered sets.

To prove 4 note that every finite lattice L is series decomposable into the top element and the rest of the lattice. Obtain P from L as follows. Remove the top element of L and add a new smallest element. Then $G_C(P)$ is isomorphic to $G_C(L)$. If L is such that the top element has more than one lower cover, then P is not a lattice.

To prove 5 and 6 consider the following example. Take a non-self-dual rigid set P (for example, a linear lexicographic sum of a singleton and a fence with an even number of elements) and look at the disjoint union $P + P^d$ of it and its dual. This set is rigid and thus has in particular no fixed point free automorphism. Yet $P + P^d$ has the same comparability graph as $P + P$, the disjoint union of the set P with itself. However $P + P$ has a fixed point free automorphism (and is thus also not rigid). ∎

It should be noted that dimension is a comparability invariant for all ordered sets, cf. [8], as is being an interval ordered set, cf. Exercise 13. Comparability invariance of the dimension for finite sets was first established (independent of each other) in the papers [7, 99, 264].

9.4 Lexicographic Sums and Reconstruction

While it is still an open question if nontrivial lexicographic sums are reconstructible, we will prove in this section that "many" lexicographic sums are reconstructible. To be completely correct, in this text we will only prove weak reconstructibility of many lexicographic sums. The proof of recognizability of decomposable ordered sets can be found in [127]. By quoting said paper we are able to establish the reconstruction result in Theorem 9.4.6. Since disconnected ordered sets are reconstructible by Proposition 2.8.9 and since series decomposable sets are reconstructible by Exercise 16, we will be able to concentrate on connected and co-connected ordered sets. This allows us to make extensive use of the canonical decomposition.

Definition 9.4.1 *For every connected and co-connected card C^k of an ordered set we let $L\{C_s^k | s \in S^k\}$ stand for the canonical decomposition of C^k.*

Lemma 9.4.2 *Let P be a finite, nontrivially decomposable connected and co-connected ordered set. Let $L\{P_t | t \in T\}$ be the canonical decomposition of P and let $x \in P$.*

1. *If $A \subseteq P$ is order-autonomous in P and $x \in P$, then $A \setminus \{x\}$ is order-autonomous in $P \setminus \{x\}$ or empty.*

2. *For every connected and co-connected card C^k, the index set S^k of the canonical decomposition $L\{C_s^k | s \in S^k\}$ of C^k has at most as many elements as T. Let*

$$m := \max\{|S^k| : C^k \text{ is a connected and co-connected card}\}.$$

3. *There is a connected and co-connected card C^k of P such that the index set S^k of the canonical decomposition $L\{C_s^k | s \in S^k\}$ of C^k is isomorphic to T (hence $m = |T|$).*

4. *We will call connected and co-connected cards with $|S^k| = m$ the **crucial cards**. Let $N \subseteq P$ be the set of all points that are in a nontrivial maximal order-autonomous subset of P. The crucial cards are exactly the cards of P that are obtained by removal of an element of N.*

5. *Let C^k be a connected and co-connected card of P. If $|S^k| = m$, then S^k is isomorphic to T.*

Proof. To prove 1 let $A \subseteq P$ be order-autonomous in P and let $x \in P$. If $A \setminus \{x\}$ is empty, there is nothing to prove. If $A \setminus \{x\} \neq \emptyset$, let $p \in P \setminus (A \cup \{x\})$ be comparable to $y \in A \setminus \{x\}$, say $p \geq y$. Then $p \geq A$ and hence we have $p \geq A \setminus \{x\}$. The other comparability is handled dually. This proves that $A \setminus \{x\}$ is order-autonomous in $P \setminus \{x\}$.

For 2 let $C^k := P \setminus \{x\}$ be a connected and co-connected card. Then for each maximal order-autonomous set $P_t \subseteq P$, the set $P_t \setminus \{x\}$ is order-autonomous in C^k or empty. If $P_t \setminus \{x\}$ is nonempty, it is contained in a unique maximal order-autonomous subset of C^k. Thus C^k has at most as many maximal order-autonomous subsets as P and hence $|S^k| \leq |T|$.

For the proof of 3 let $P_r \subseteq P$ be a maximal order-autonomous subset with $|P_r| \geq 2$. Let $x \in P_r$. Then all $Q_t := P_t \setminus \{x\}$ are nonempty order-autonomous in $P \setminus \{x\}$ and $P \setminus \{x\} = L\{Q_t | t \in T\}$. Since T is neither disconnected, nor series decomposable, $P \setminus \{x\}$ is connected and co-connected. Suppose there is a $t \in T$ such that Q_t is not maximal order-autonomous in $P \setminus \{x\}$. Then there is a maximal order-autonomous $A \subset P \setminus \{x\}$ with $A \supset Q_t$ (proper containments, maximality in $P \setminus \{x\}$). We will show that A or $A \cup \{x\}$ (neither of which is equal to P_t or P) would have to be order-autonomous in P, thus arriving at a contradiction.

First consider the case $Q_r \not\subset A$. If $Q_r \not\subset A$, then $Q_r \cap A = \emptyset$ by maximality of A in $P \setminus \{x\}$. For all $p \in P \setminus (A \cup \{x\})$ the fact $p \leq a$ ($p \geq a$) for some $a \in A$ implies $p \leq A$ ($p \geq A$). This leaves us with x. $x \geq a$ for some $a \in A$ implies that $a \leq P_r$, since P_r is order-autonomous in P. Hence, since A is order-autonomous in $P \setminus \{x\}$, this means that $A \leq Q_r$, which again by order-autonomy of P_r in P implies $A \leq P_r$ and in particular $A \leq x$. The other comparability is handled dually. Thus $A \neq P$ is order-autonomous in P and properly contains one of the maximal order-autonomous subsets of P, a contradiction.

Now consider the case $Q_r \subset A$. If $Q_r \subset A$, let $p \in P \setminus (A \cup \{x\})$. If $p \geq a$ for some $a \in A$, then $p \geq A$, hence $p \geq Q_r$, which implies $p \geq x$ and hence $p \geq (A \cup \{x\})$. If $p \geq x$, then $p \geq P_r$, that is, in particular $p \geq Q_r$, which implies $p \geq A$ and hence $p \geq (A \cup \{x\})$. The reversed comparabilities are handled dually. Thus in this case we also have that $A \cup \{x\} \neq P$ is order-autonomous in P and properly contains P_r, a contradiction.

Thus all Q_t are maximal order-autonomous in $P \setminus \{x\}$ and the decomposition $P \setminus \{x\} = L\{Q_t | t \in T\}$ is the canonical decomposition of $P \setminus \{x\}$.

For the proof of 4 note that by the proof of 3 if C^k is obtained from P by removing a point out of a maximal order-autonomous set with ≥ 2 elements, C^k is crucial. On the other hand if C^k is obtained from P by removing a point p such that $\{p\}$ is maximal order-autonomous, then by 1 all $P_t \setminus \{p\}$ are order-autonomous in $P \setminus \{p\}$ or empty. There are $m - 1$ nonempty $P_t \setminus \{p\}$. Thus if $P \setminus \{p\}$ is connected and co-connected, then $|S^k| \leq m - 1 < m = |T|$ and C^k is not crucial.

For part 5 note that crucial cards are exactly those that were obtained by removing an element of a nontrivial autonomous subset. The proof of 3 shows that the index set of the canonical decomposition of a crucial card must be isomorphic to T.   ∎

Lemma 9.4.3 *m and the crucial cards and hence T are weakly reconstructible. That is, they can be identified from $\mathcal{D}_P$ if it is known that P is nontrivially decomposable, connected and co-connected.*

Proof. The number m can be determined from the formula in Lemma 9.4.2 part 2 (the right side can be determined from the deck). Through m, we obtain the crucial cards. Finally we obtain T from the crucial cards via part 5 of Lemma 9.4.2.   ∎

Proposition 9.4.4 *Let P be a finite connected and co-connected ordered set with > 3 elements. Then the maximal order-autonomous subsets of the canonical decomposition of P are weakly reconstructible. That is, if Q is another finite, connected and co-connected ordered set with $\mathcal{D}_P = \mathcal{D}_Q$, then there is a bijection between the maximal order-autonomous subsets of P and Q such that each set is isomorphic to its image.*

Proof. Since the number of maximal order-autonomous subsets of P is known to be m, which is weakly reconstructible from the deck, we only need to reconstruct the non-singleton maximal order-autonomous subsets. Let $C^1, \ldots, C^l$ be the crucial cards. For each crucial card C^k let g_k be the number of elements of C^k that are contained in non-singleton maximal order-autonomous subsets of C^k. Let g be the number of elements of P that are contained in non-singleton maximal order-autonomous subsets of P. By Lemma 9.4.2, part 4 we know that $g = l$, so g can be determined. A crucial card C^k is obtained by removing an element from a maximal order-autonomous subset with ≥ 3 elements iff $g_k = g - 1$. C^k is a crucial card obtained by removing an element from a maximal order-autonomous subset with 2 elements iff $g_k = g - 2$.

For each crucial card C^k of P create a new card D^k as follows. If $g - g_k = 1$ let D^k be the disjoint sum of the non-singleton maximal order-autonomous subsets of C^k. If $g - g_k = 2$ let D^k be the disjoint sum of the non-singleton maximal order-autonomous subsets of C^k and a singleton. The $D^1, \ldots, D^l$ are the deck of an ordered set D. This ordered set D is the disjoint sum of the non-singleton maximal order-autonomous subsets of P.

If D is disconnected with ≥ 3 elements (a fact that can be recognized from $D^1, \ldots, D^l$), we reconstruct D from $D^1, \ldots, D^l$. Since not every component of D is necessarily maximal order-autonomous in P, we proceed as follows. There is a $k_0 \in \{1, \ldots, l\}$ such that either no other crucial card C^k has fewer non-singleton maximal order-autonomous subsets than C^{k_0}, or no other card crucial C^k has a smaller non-singleton maximal order-autonomous subset. C^{k_0} is obtained by removing an element of a smallest possible non-singleton maximal order-autonomous subset of P. If $g - g_{k_0} = 1$, then all non-singleton maximal order-autonomous subsets of C^{k_0} except for the smallest one are in fact non-singleton maximal order-autonomous subsets of P. If $g - g_{k_0} = 2$, then all non-singleton maximal order-autonomous subsets of C^{k_0} are non-singleton maximal order-autonomous subsets of P, and P has exactly one more two-element maximal order-autonomous subset. In either case we obtain the number j of non-singleton maximal order-autonomous subsets of P. If $j = 1$, then the disconnected ordered set D is the only non-singleton maximal order-autonomous subset of P. If $j \geq 2$, then in both cases above, all but one non-singleton maximal order-autonomous subset of P have been found. Call these sets $Q_1, \ldots, Q_{j-1}$. There is a unique (up to isomorphism) ordered set Q_j such that D is the disjoint sum of the sets $Q_1, \ldots, Q_j$. Q_j is the last non-singleton maximal order-autonomous subset of P.

This leaves the cases in which $D^1, \ldots, D^l$ is the deck of a connected ordered set D with ≥ 3 elements or of a 2-element set (which are not necessarily known to be reconstructible from the deck). In this case we know that P has exactly one non-singleton maximal order-autonomous subset with $|D|$ elements, namely the set D.

Let C^1 be a crucial card. If $|D| = 2$, then D is a 2-antichain iff $e_H(C^1) < e_H(P) - 1$ or C^1 has one less minimal or one less maximal element than P. This leaves the case $|D| > 2$. In this case there is a unique maximal order-autonomous subset D' on C^1 that is not a singleton. Naturally D' was obtained from D by removal of one element. Let U be the set of upper bounds of D' on C^1 and let L be the set of lower bounds of D' on C^1. Then U is also the set of upper bounds of D in P and L is also the set of lower bounds of D in P. P was not series decomposable. Thus for each ordered set of the form $L \oplus \overline{D} \oplus U$ with $|\overline{D}| = |D|$ we can use Kelly's Lemma (cf. Proposition 1.5.14) to find out how many copies of $L \oplus \overline{D} \oplus U$ are contained in P. D will be the unique ordered set for which C^1 contains one less copy of $L \oplus D \oplus U$ than P.   ∎

To state the main result of this section in full generality we need the recognizability of decomposable ordered sets. For a proof of this result we refer the reader to [127]. Thus the conscientious reader may want to replace the word "reconstructible" in Theorem 9.4.6 with "weakly reconstructible" until (s)he has verified the results in [127].

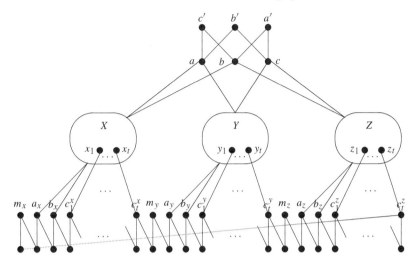

Figure 9.2: The ordered set described in Proposition 9.5.1.

Lemma 9.4.5 (For the proof, cf. [127], Theorem 2.) *If P is a finite ordered set with $|P| \geq 12$, then P is recognizable as decomposable or indecomposable from the deck.* ∎

Theorem 9.4.6 *Let P be a finite, decomposable ordered set with ≥ 12 elements such that there is a piece P_i of P such that a card C_i of P_i is not isomorphic to any pieces of P. Then P is reconstructible.*

Proof. Find a crucial card that contains a maximal order-autonomous set that is isomorphic to C_i. Replace C_i with P_i to obtain an ordered set isomorphic to P. ∎

9.5   An Almost Lexicographic Construction

The construction presented in this section is needed to prove Theorem 12.4.5 in Section 12.4. We present it here, since it shows very nicely how a certain overall lexicographic type structure influences the structure of the set of fixed point free order-preserving maps. Note that from the proof of Lemma 9.1.9 it is easy to see that fixed point free order-preserving maps in series-decomposable finite ordered sets must have a special structure. They must map all pieces to themselves.

The following construction exploits a similar idea to generate a type of ordered set that can only have a fixed point free order-preserving map under certain circumstances. Indeed the construction is essentially lexicographic, except for part 5 in Proposition 9.5.1. This structure will be the framework that will allow us to embed the problem 3SAT into the problem if a given ordered set has a fixed point free order-preserving self-map. (The proof presented in this text follows the presentation of the original in [62].)

Proposition 9.5.1 (Cf. [62], Section 3.) *Let X, Y, and Z be finite ordered sets. An ordered set P as depicted in Figure 9.2 is defined as follows.*

1. *The underlying set is the union of the six crown $\{a, b, c, a', b', c'\}$, the ordered sets X, Y, Z and the $2n$-crown C_{2n}.*

2. *$\{a, b, c, a', b', c'\}$ is ordered as depicted, that is $a < b', c'; b < a', c';$ and $c < a', b'$.*

3. *X, Y, Z, and C_{2n} carry their original orders.*

4. *$X \leq \{a, b\}, Y \leq \{a, c\}, Z \leq \{b, c\},$ and $C_{2n} \leq \{a, b, c\}$.*

5. *Each element of X, Y, Z is above at least two maximal elements of C_{2n}; no element of X, Y, Z is an upper bound of C_{2n} (it is intentional that this does not completely specify the comparabilities involving X, Y and Z).*

6. *The only further comparabilities are those added to the above by transitivity.*

If P has a fixed point free order-preserving self map, then P has a fixed point free order-preserving self map f such that:

1. *$f|_{\{a,b,c,a',b',c'\}}$ and $f|_{C_{2n}}$ are automorphisms.*

2. *$f[X] \subseteq Y, f[Y] \subseteq Z, f[Z] \subseteq X,$ or $f[X] \subseteq Z, f[Z] \subseteq Y, f[Y] \subseteq X$.*

Proof. Suppose P has a fixed point free order-preserving self map. Then by Proposition 3.1.1 there is a fixed point free order-preserving self map $f : P \to P$ such that f maps minimal elements to minimal elements and maximal elements to maximal elements.

First we shall prove that $f[\{a', b', c'\}] = \{a', b', c'\}$. Indeed suppose $a' \notin f[\{a', b', c'\}]$. Then we must have $f(b') = c'$ and $f(c') = b'$. This implies that $f(a) \leq \{f(c'), f(b')\} = \{b', c'\}$ and since $f(a) \not> a$, this means $f(a) \in Z$ and in particular $f(a) \leq \{b, c\}$. Without loss of generality assume $f(a') = c'$. Then $f(b) \leq \{f(c'), f(a')\} = \{b', c'\}$ and $f(b) \not> b$, which implies $f(b) \leq a$. Therefore $f[C_{2n}] \leq \{f(a), f(b)\}$ implies $f[C_{2n}] \leq \{a, b, c\}$. But this means that $f[C_{2n}] \subseteq C_{2n}$, which means in particular by Exercise 32 in Chapter 2 that $f|_{C_{2n}}$ is an automorphism. However then $f(a) \in Z$ is not an upper bound of $f[C_{2n}]$, contradicting the fact that f was order-preserving.

Thus $f[\{a', b', c'\}] = \{a', b', c'\}$, which implies that $f[C_{2n}] \leq \{a', b', c'\}$, meaning that $f[C_{2n}] \subseteq C_{2n}$. However this implies

$$f[\{a, b, c, a', b', c'\}] \subseteq \{a, b, c, a', b', c'\}.$$

Using Exercise 32 in Chapter 2 once more, we conclude that 1 must be true.

In particular, since every element of $X \cup Y \cup Z$ is below two distinct minimal elements of $\{a, b, c, a', b', c'\}$ and above two distinct maximal elements of C_{2n}, this implies that $f[X \cup Y \cup Z] \subseteq X \cup Y \cup Z$. What remains to be proved is part 2. Assume without loss of generality that $f(a) = b$, $f(b) = c$ and $f(c) = a$ (the other permutation is handled similarly). Then $f(X) \leq \{f(a), f(b)\} = \{b, c\}$, which forces $f[X] \subseteq Z$. $f[Z] \subseteq Y$ and $f[Y] \subseteq X$ are proved similarly. ∎

Proposition 9.5.2 (Cf. [62], Section 5.) *With assumptions as in Proposition 9.5.1, let $t > 1$, and let $\{x_1, \ldots, x_t\} \subseteq X$, $\{y_1, \ldots, y_t\} \subseteq Y$, and $\{z_1, \ldots, z_t\} \subseteq Z$ be antichains. Let $n = 3t + 9$ and let the maximal elements of C_{2n} in cyclic order be*

$$m_x, a_x, b_x, c_1^x, \ldots, c_t^x, m_y, a_y, b_y, c_1^y, \ldots, c_t^y, m_z, a_z, b_z, c_1^z, \ldots, c_t^z.$$

To specify the set we only need to specify the comparabilities in part 5 in Proposition 9.5.1.

1. *Let $X > \{a_x, b_x\}$, $Y > \{a_y, b_y\}$, and $Z > \{a_z, b_z\}$.*

2. *For $i \in \{1, \ldots, t\}$ let $x_i > c_i^x$, $y_i > c_i^y$, and $z_i > c_i^z$.*

If the thus defined ordered set P has a fixed point free order-preserving map, then there is a fixed point free order-preserving map $f : P \to P$ such that for all $i \in \{1, \ldots, t\}$ we have $f(x_i) \geq y_i$, $f(y_i) \geq z_i$, and $f(z_i) \geq x_i$, or there is a fixed point free order-preserving map $f : P \to P$ such that for all $i \in \{1, \ldots, t\}$ we have $f(x_i) \geq z_i$, $f(z_i) \geq y_i$, and $f(y_i) \geq x_i$.

Proof. Let $f : P \to P$ be a fixed point free order-preserving map as guaranteed by Proposition 9.5.1 and assume without loss of generality that $f[X] \subseteq Y$, $f[Y] \subseteq Z$, $f[Z] \subseteq X$. Then we have that $f[\{a_x, b_x\}] = \{a_y, b_y\}$, $f[\{a_y, b_y\}] = \{a_z, b_z\}$, and $f[\{a_z, b_z\}] = \{a_x, b_x\}$ with a's mapped to a's and b's mapped to b's. Thus, since $f|_{C_{2n}}$ is an automorphism, for all $i \in \{1, \ldots, t\}$ we have $f(c_i^x) = c_i^y$, $f(c_i^y) = c_i^z$, and $f(c_i^z) = c_i^x$.

Now since $f(x_i) \geq f[\{a_x, b_x, c_i^x\}] = \{a_y, b_y, c_i^y\}$ and since all elements in Y that are above this set are also above y_i we must have $f(x_i) \geq y_i$. Similarly we prove $f(y_i) \geq z_i$ and $f(z_i) \geq x_i$. ∎

Exercises

1. Investigating Proposition 9.1.6.

   (a) Show that even if T is a complete lattice and all P_t are lattices, the lexicographic sum $L\{P_t | t \in T\}$ need not be a lattice.

   (b) Show that if T is a lattice and all P_t are lattices with a smallest and a largest element, then $L\{P_t | t \in T\}$ is a lattice.

2. Let the lexicographic sum $L\{P_t | t \in T\}$ be a complete lattice. Prove that:

(a) Every piece P_t is conditionally complete.

(b) If $t_0 = \bigvee_T \{t \in T : t < t_0\} = \bigwedge_T \{t \in T : t > t_0\}$, then P_{t_0} must be a complete lattice.

State and prove a similar result for lattices.

3. Find an index set T that is not a chain and that contains elements $s, s' \in T$ such that

   - Any finite lexicographic sum $L\{P_t | t \in T\}$ with $|P_t| = 1$ for all $t \in P \setminus \{s\}$ or with $|P_t| = 1$ for all $t \in P \setminus \{s'\}$ has the fixed point property, and

   - Any finite lexicographic sum $L\{P_t | t \in T\}$ with P_s and $P_{s'}$ being two element antichains does not have the fixed point property.

4. Prove a result analogous to Proposition 9.1.10 for the connected relational fixed point property for finite ordered sets. (Hint. Use retractable sets.)

   Conclude that there are ordered sets with the relational fixed point property that are not connectedly collapsible.

5. Let $P = L\{P_t | t \in T\}$ be a lexicographic sum that is

   - The decomposition into components if P is disconnected,

   - Any linear decomposition if P is series decomposable,

   - The canonical decomposition if P is connected and co-connected.

   Let $\Phi : P \to P$ be an automorphism. For $t \in T$ let $\tau(t)$ be a (fixed) element of P_t. Prove that:

   (a) $I \circ \Phi \circ \tau$ is an automorphism of T.

   (b) For all $t \in T$, $\Phi|_{P_t}$ is an isomorphism from P_t to $P_{I \circ \Phi \circ \tau(t)}$.

6. A finite ordered set P is called **series parallel** iff P is a singleton, or $P = P_1 + P_2$ or $P = P_1 \oplus P_2$ with P_1, P_2 being series parallel ordered sets. A finite ordered set is called **N-free** iff for all $a \prec b \succ c \prec d$, we have that $a \leq d$. Prove that every series parallel ordered set is N-free.

7. Prove that an ordered set is series-parallel iff it does not contain a 4-fence.

8. Prove Lemma 9.2.2.

9. In series decomposable ordered sets, maximal order-autonomous subsets can overlap. It thus makes sense to look for some type of minimal order-autonomous subset that reflects the fact that the set is series decomposable.

   An order-autonomous proper subset $S \neq P$ of the ordered set P is called

- A **chain-link** iff every element of S is related to all elements in $P \setminus S$,
- A **minimal chain-link** iff S is a chain-link and every proper order-autonomous subset of S is not a chain-link.

Prove

(a) P contains a chain-link iff P is series decomposable.

(b) Let P be a series decomposable ordered set. Then every element $p \in P$ is contained in a unique minimal chain-link.

(c) Let $\overline{G}$ be the complement of the comparability graph of the series-decomposable ordered set P. Prove that $C \subseteq P$ is a minimal chain-link iff $\overline{G}[C]$ is a component of $\overline{G}$.

10. Construct an infinite ordered set P such that:

(a) P is not series decomposable.

(b) No element of P is contained in a maximal order-autonomous subset.

11. An ordered set is called **weakly ordered** iff it does not contain the disjoint sum of a singleton and a two element chain. Prove that every weakly ordered set is a linear sum of antichains.

12. Find the number of linear extensions of a linear lexicographic sum of antichains. (That is, the number of linear extensions of a weak order.)

13. Prove that being an interval ordered set is a comparability invariant in the class of all ordered sets.

14. Extend the example in the proof of Corollary 9.3.5, part 4 by showing that being a lattice also is not a comparability invariant for infinite ordered sets.

15. Let $P_1, \ldots, P_n$ be ordered sets. Formulate the definition of $P_1 \odot \cdots \odot P_n$ (cf. Exercise 15 in Chapter 1) in terms of lexicographic sums. Use Example 9.1.4, part 5 as a model.

16. (Cf. [151], Theorem 4.7.) Let P be a series decomposable ordered set. Prove that P is reconstructible.

17. Prove that the following are comparability invariants for finite ordered sets. Are they comparability invariants for all ordered sets?

(a) The width of an ordered set,

(b) The height of an ordered set,

(c) The number of linear extensions of an ordered set.

18. Construct an indecomposable ordered set P such that all minimal cards of P are connected and decomposable.

19. Let P be a finite ordered set with ≥ 12 elements such that there is a $b > 0$ such that

- There is a maximal order-autonomous subset of P with $> b$ elements, and
- No maximal order-autonomous subset of P is of size b.

Prove that P is reconstructible.

20. Prove that any truncated lattice that contains a four crown as an ordered subset is reconstructible.

21. Let P be a finite ordered set with ≥ 12 elements such that every maximal element of P is contained in a nontrivial order-autonomous subset of P. Prove that P is reconstructible.

22. Let P be a finite ordered set with ≥ 12 elements that has only one non-singleton maximal order-autonomous subset. Assume this subset has at least three elements. Prove that P is reconstructible.

23. Prove that all fixed point free order-preserving maps of the set given in Proposition 9.5.1 must be as described in Proposition 9.5.1. (Hint: Prove that $f[C_{2n}] \subseteq C_{2n}$ for all fixed point free order-preserving maps of P.)

24. (Convex subsets.) Let P be an ordered set and let $C \subseteq P$. Then C is called **convex** iff for all $x \leq y$ in C we have that $[x, y] \subseteq C$.

    (a) Show that every piece in a lexicographic sum decomposition is convex.

    (b) Show that the intersection of convex subsets of an ordered set is again convex.

    (c) Show that the union of (even intersecting) convex subsets need not be convex.

25. (More intersections of convex sets.) This exercise shows how a natural result for intervals on the real line (cf. part 25a) has analogues for convex sets in ordered sets that are not chains.

    (a) Prove that if $\{C_i\}_{i \in I}$ is a family of convex subsets of a finite chain such that $C_i \cap C_j \neq \emptyset$ for all $i, j \in I$, then $\bigcap_{i \in I} C_i \neq \emptyset$.

    (b) Prove that if $\{C_i\}_{i \in I}$ is a family of convex subsets of a finite ordered set of width w such that any subfamily of $\leq 2w$ sets has nonempty intersection, then $\bigcap_{i \in I} C_i \neq \emptyset$.

(c) For $w > 0$ find an ordered set of width w and a family $\{C_i\}_{i=1,\ldots,2w}$ with empty intersection such that any $(2w-1)$-element subfamily has nonempty intersection.

26. Let P be an ordered set and let $K \subseteq P$ be an ordered subset. Define the **convex hull** of K to be

$$\mathrm{con}_P(K) := \bigcap \{C \subseteq P : K \subseteq C, \text{ and } C \text{ is convex}\}.$$

(a) Show that the convex hull of an ordered subset K is the smallest convex subset of P that contains K.

(b) Prove that if P and Q are ordered sets with the same comparability graph and S is order-autonomous in P, then $\mathrm{con}_P(S)$ is order-autonomous in Q.

(c) We have encountered a convex hull in the proof of Theorem 9.3.4. Where?

27. Prove that if P is a finite, decomposable, co-connected ordered set, then P has an order-autonomous subset that is order-autonomous in any transitive orientation of the comparability graph $G_C(P)$ of P.

Hint. Use Exercise 26b to show that maximal order-autonomous sets are order-autonomous in any transitive orientation of $G_C(P)$.

28. Let P be a finite ordered set. For any two edges $\{x, y\}$ and $\{x, z\}$ of $G_C(P)$ with $x \neq y$ and $x \neq z$ say $\{x, y\} \wedge \{x, z\}$ iff $y \not\sim z$ or $y = z$ (cf. [93], p.25; [134], p.5).

(a) Prove that the transitive closure of the relation $\wedge$ is an equivalence relation on the nontrivial edges of the comparability graph.

The equivalence classes of $\wedge$ are called **edge classes** (cf. [93], p25; [134], p.5).

(b) Let C be an edge class. Prove that any transitive orientation of $G_C(P)$ induces one of two dual transitive orientations on the edges in C.

(c) For any edge class C let $V(C)$ be the set of points $p \in P$ such that there is an $x \in P$ with $\{p, x\} \in C$. Prove that $V(C)$ is order-autonomous in any transitive orientation of $G_C(P)$.

(d) Let C be an edge class. Prove that the graph $(V(C), C)$ is connected.

(e) (Cf. [93], Hilfssatz 2.3.) Let P be connected and co-connected. Prove that $G_C(P)$ has exactly one edge class C such that $V(C) = P$.

Hint. Let L be a smallest possible set such that the graph induced on $P \setminus L$ by the complement $\overline{G_C(P)}$ of the comparability graph is disconnected. Let $K_1, \ldots, K_n$ be the vertex sets of the components of this graph.

- Note that for all $i \neq j$ and $k_i \in K_i$, $k_j \in K_j$ we have $k_i \sim k_j$.
- Prove that all edges that go from a K_i to a K_j with $i \neq j$ are in the same edge class. Call this edge class D.
- Prove that all edges from L to a K_i are in D.
- Conclude that $V(D) = P$.
- Prove that for any further edge class D' the set $V(D')$ must be completely contained in a K_i.

29. Use Exercises 26 through 28 (more precisely, Exercises 27, 28b, 28c and 28e) to prove Theorem 9.3.2.

30. Let P be an ordered set and let $\mathcal{C}(P)$ be the set of its convex subsets ordered by inclusion. Show that $\mathcal{C}(P)$ is a complete lattice.

Remarks and Open Problems

1. Is it possible to prove an analogue of Proposition 9.1.10 without the assumption of chain-completeness?

2. A natural graph theoretical question to ask is which graphs actually are comparability graphs. This question was answered in [93]. For an in-depth treatment of comparability graphs the reader should consider the survey [134]. Also the reader will have to decide for him/herself how much comparability invariance says about ordered sets. True, many properties only depend on the comparability graph, which may say they are "graph-theoretical". Yet, as for example the fixed point property shows (cf. Section 4.4.4) the only sensible approach to many of these properties is through the ordered structure. Moreover, as Theorem 9.3.4 shows, in the setting of finite ordered sets, comparability invariance is equivalent to a simple result about lexicographic sums.

   A natural follow-up question is if there are infinitary analogues for Theorem 9.3.4. A step in this direction is made in [245].

3. Another graph that is associated with ordered sets is the covering graph. The **covering graph** has the points of the ordered set as vertices and there is an edge between p and q iff $p \prec q$ or $q \prec p$. All questions that one can ask about comparability graphs can also be asked for covering graphs. However, characterization of covering graphs is problematic.

   > **Open Question 9.5.3** (Cf. [222].) *Characterize the undirected graphs that are covering graphs of ordered sets.*

   Certainly not every graph is a covering graph. The simplest example is the triangle. In general it is difficult, namely NP-complete (cf. Definition 12.4.3), to determine if a given graph is a covering graph (cf. [29, 189]). This remains so even for lattices (cf. [231]), but can become "easy" (that is, polynomially solvable, cf. Definition 12.2.1) for some classes of ordered sets (such as distributive and modular lattices, cf. [6]).

What NP-completeness means is that a computational characterization is considered "hard". This would however still allow for structural character- izations, which in turn are also computationally "hard", but which could lead to deeper understanding. The only **covering graph invariant** found so far seems to be the genus of an ordered set (cf. [77]). The **genus** of an ordered set is the smallest possible genus of a surface on which the cover- ing graph can be drawn as a planar graph such that a given "up"-direction on the surface makes the covering graph a diagram.

4. The proof of Lemma 9.4.5 in [127] is quite technical. Is there a short proof of Lemma 9.4.5?

5. Is the class of decomposable ordered sets reconstructible? Considering The- orem 9.4.6 one would conjecture that the answer is "yes" and that recon- struction should be possible with a reasonably short argument. However, a proof is not available so far. The main problem seems to be with lexico- graphic sums that have only one nontrivial piece of size 2.

6. N-free ordered sets have been referred to in a few exercises throughout (the latest being Exercise 6 in this chapter). Using results from [151] (or [247]), Lemma 9.4.3 and Proposition 9.4.4 it was proved in [248] that N- free ordered sets are reconstructible.

7. The natural generalization of an interval is the notion of a convex subset of an ordered set. Thus, another possible generalization of interval ordered sets is to represent ordered sets as sets of convex subsets of some type of ordered set. One can order convex sets in various ways. These questions are addressed for example in [184]

10

Sets $P^Q = \text{Hom}(Q, P)$ and Products

Order-preserving mappings from one ordered set to another form a natural ordered set under the pointwise order. Since products in general are defined similar to homomorphism sets, we shall investigate homomorphism sets and products of ordered sets closely together. In this chapter we introduce some of the salient results on these sets such as the fixed point theorem for products of two finite ordered sets (cf. Theorem 10.2.11), Hashimoto's refinement theorem (cf. Theorem 10.4.4) and the cancelation property for exponents (cf. Theorem 10.5.9). The automorphism conjecture (cf. Open Question 11.5.1) as well as the open problems at the end of this chapter show that there are interesting problems related to homomorphism sets and products that remain open.

10.1   Sets $P^Q = \text{Hom}(Q, P)$

We shall start by defining the homomorphism sets and by investigating their relationship to the fixed point property.

Definition 10.1.1 *Let P, Q be ordered sets. We define the set of* (**order**) **homomorphisms** *from P to Q to be*

$$\text{Hom}(Q, P) := \{f : Q \to P | f \text{ is order−preserving }\},$$

ordered by the pointwise order $f \leq g$ iff $f(q) \leq g(q)$ for all $q \in Q$. If $Q = P$ we will call $\text{End}(P) := \text{Hom}(P, P)$ the set of (**order**) **endomorphisms** *of P.*

Notation 10.1.2 *Due to the formal laws associated with forming sets of homomorphisms (cf. Section 10.5) which are similar to exponentiation, we also denote* $\text{Hom}(Q, P)$ *by* P^Q.

Homomorphism sets appear to be new structures for us. Yet there are examples of homomorphism sets that we have already encountered. They were just represented in a slightly different way.

Example 10.1.3 Let **2** denote the 2-element chain $\{0, 1\}$ and let n denote the n-element antichain. Then the set $\mathbf{2}^n$ is isomorphic to $\mathcal{P}(n)$, the power set of an n-element set ordered by inclusion. The isomorphism maps each f to the set $\{k : f(k) = 1\}$.

There is no guarantee that we can find a copy of the domain Q in $\text{Hom}(Q, P)$. Indeed, if P is a singleton, then so is $\text{Hom}(Q, P)$. However, we can find a copy of P.

Lemma 10.1.4 *Let P be an ordered set. Then P is a retract of* $\text{Hom}(Q, P)$.

Proof. Fix $q_0 \in Q$ and define $R_{q_0} : \text{Hom}(Q, P) \rightarrow P$ by $R_{q_0}(f) := f(q_0)$. For any $p \in P$ define f_p to be the constant function that returns p, that is, $f_p(q) := p$ for all $q \in Q$. Let $I : P \rightarrow \text{Hom}(Q, P)$ be defined by $I(p) := f_p$. Then R_{q_0} and I are a retraction-coretraction pair as in the alternative definition of retractions given in Exercise 1 in Chapter 4. ∎

With regards to the fixed point property we can now record an extensive characterization. Indeed, especially for endomorphism sets many notions related to the fixed point property are equivalent.

Theorem 10.1.5 (Cf. [12], Theorem 4.5, [67], Theorem 6.13, and also [257].) *Let P be a chain-complete ordered set. Then the following are equivalent.*

1. *P is $\mathcal{C}$-dismantlable to a singleton.*

2. *$\text{Hom}(Q, P)$ is $\mathcal{C}$-dismantlable to a singleton for all ordered sets Q.*

3. *$\text{Hom}(Q, P)$ is connected for all ordered sets Q.*

4. *$\text{Hom}(Q, P)$ has the fixed point property for all ordered sets Q.*

5. *$\text{End}(P)$ has the fixed point property.*

6. *$\text{End}(P)$ is $\mathcal{C}$-dismantlable to a singleton.*

7. *$\text{End}(P)$ is connected.*

Proof. For the implication "1⇒2" let $n \in \mathbb{N}$ be such that for $i = 1, \ldots, n$ there are retractions $r_i : P_{i-1} \rightarrow P_i$ in $\mathcal{U} \cup \mathcal{D}$ (recall Exercise 14 in Chapter 4) with $P_0 = P$ and P_n a singleton. Define $R_i : \text{Hom}(Q, P_{i-1}) \rightarrow \text{Hom}(Q, P_i)$

by $R_i(f) := r_i \circ f$. Clearly the R_i are retractions in $\mathcal{U} \cup \mathcal{D}$ and $\mathrm{Hom}(Q, P_n) \subseteq \mathrm{Hom}(Q, P)$ is a singleton.

For part "2$\Rightarrow$4" by Theorem 4.2.5 we only need to prove that $\mathrm{Hom}(Q, P)$ is chain-complete. To do this let $C \subseteq \mathrm{Hom}(Q, P)$ be a chain. For each $q \in Q$ the set $\{c(q) : c \in C\}$ is a chain. Define $s(q) := \bigvee \{c(q) : c \in C\}$. Then s is the supremum of C in $\mathrm{Hom}(Q, P)$.

The implications "2$\Rightarrow$3$\Rightarrow$7", "2$\Rightarrow$6$\Rightarrow$7", and the implications "4$\Rightarrow$5$\Rightarrow$7" are trivial.

This leaves us with the implication "7$\Rightarrow$1". To prove that the needed retractions exist, we will proceed by induction on the length n of the fence from id_P to a constant function. In case $n = 0$ there is nothing to prove, since P is a singleton. Now assume there is a fence of length n from id_P to a constant function and that the result is proved for all sets P' for which there is a shorter fence from $\mathrm{id}_{P'}$ to a constant function. Without loss of generality let $\mathrm{id}_P = F_0 > F_1 < F_2 > \cdots F_n$ be a fence in $\mathrm{End}(P)$ such that F_n is a constant function. By an easy adaptation of Lemma 4.3.7 there is a retraction $R_1 \leq F_1 < F_0$ such that $R_1 = R_1 \circ F_1$. Since $R_1 < \mathrm{id}_P$ we have that $R_1 \in C$. Moreover

$$\mathrm{id}_{R_1[P]} = R_1|_{R_1[P]} = R_1 \circ F_1|_{R_1[P]} < R_1 \circ F_2|_{R_1[P]} > \cdots R_1 \circ F_n|_{R_1[P]}$$

is a fence in $\mathrm{End}(R_1[P])$ such that $R_1 \circ F_n|_{R_1[P]}$ is a constant function. Thus $R_1[P]$ is C-dismantlable to a singleton by induction hypothesis. This implies that P is C-dismantlable to a singleton. ∎

With the fixed point property for sets $\mathrm{End}(P) = P^P$ characterized in many ways (as long as we assume chain-completeness) one could now consider sets $\mathrm{Hom}(Q, P) = P^Q$. One manifestation of this problem is the famous product problem (cf. Open Question 10.2.6), which handles exponents Q that are finite antichains. We will explore this problem in more detail later on. For connected exponents little is known. All we can record is that P^Q needs not be disconnected, even when both base and exponent are not dismantlable.

Example 10.1.6 Let P be an arbitrary ordered set. Let Q be an ordered set with n elements and let $nA^2 := L\{A_k | k \in \{1, \ldots, n\}\}$, where $\{1, \ldots, n\}$ is carrying its natural order and each A_k is a two-element antichain. Then the ordered set $(nA^2 \oplus P)^Q$ is connected.

Proof. Let m_1 and m_2 be the two minimal elements of $nA^2 \oplus P$. We will show that each order-preserving map $f : Q \to nA^2 \oplus P$ is connected to one of the maps $g_i(q) = m_i$ ($i = 1, 2$). If $|f[Q] \cap \{m_1, m_2\}| \leq 1$, we have $f \geq g_1$ or $f \geq g_2$, so there is nothing to prove. In case $\{m_1, m_2\} \subseteq f[Q]$ there is a $k \in \{1, \ldots, n\}$ such that $f[Q] \cap A_k = \emptyset$. Fix $a \in A_k$ and define

$$\tilde{f}(q) := \begin{cases} f(q); & \text{if } f(q) > A_k, \\ a; & \text{if } f(q) < A_k. \end{cases}$$

Then $\tilde{f}$ is order-preserving and $f < \tilde{f} > g_1, g_2$. ∎

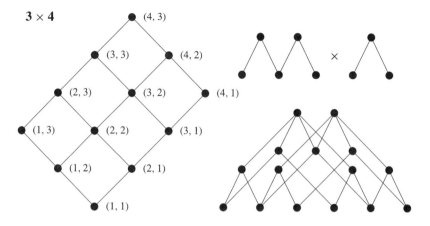

Figure 10.1: The product of a 3-chain with a 4-chain and the product of a 5-fence with a 3-fence.

The above example shows at least that a characterization of the fixed point property for P^Q with both P^Q and Q connected will not be as strong as the characterization for P^P in Theorem 10.1.5. Indeed, for chain-complete sets by Lemma 9.1.9 the ordered set $nA^2 \oplus P$ has the fixed point property iff P has the fixed point property. In particular $nA^2 \oplus P$ need not be dismantlable. Yet the set $(nA^2 \oplus P)^Q$ is connected for all P. Nothing nontrivial is known about the fixed point property for these sets or sets $\text{Hom}(Q, P)$ in which Q is neither P nor an antichain.

10.2   Finite Products

The possibly simplest sets of homomorphisms $\text{Hom}(Q, P)$ are those in which the domain Q is an antichain. These ordered sets are nothing but the $|Q|$-fold product of a set P with itself. While we will define products in full generality, this section will mainly focus on finite products. Infinite products will be investigated in Sections 10.3 and 10.4.

Definition 10.2.1 *Let A be an index set and let $\{(P_\alpha, \leq_\alpha)\}_{\alpha \in A}$ be an indexed family of ordered sets. We define the* **product** $\Pi_{\alpha \in A} P_\alpha$ *of the ordered sets P_α to be the (set-theoretical) product*

$$\Pi_{\alpha \in A} P_\alpha = \left\{ f : A \to \bigcup_{\alpha \in A} P_\alpha \,\middle|\, f(\alpha) \in P_\alpha \right\}$$

equipped with the **pointwise order**

$$f \leq g :\Leftrightarrow (\forall \alpha \in A) \, f(\alpha) \leq_\alpha g(\alpha).$$

In case the index set is finite, say $\{1, \ldots, n\}$, *we also denote the product by* $P_1 \times \cdots \times P_n$. *The elements of a product are often given componentwise as* $p = (p_\alpha)_{\alpha \in A}$ *or as* $p = (p_1, \ldots, p_n)$ *for finite index sets.*

*The **natural projection** $\pi_{P_\beta} : \Pi_{\alpha \in A} P_\alpha \to P_\beta$ is defined to be*

$$\pi_{P_\beta}((p_\alpha)_{\alpha \in A}) := p_\beta.$$

If there is no confusion possible, then π_{P_β} may also be abbreviated as π_β.

Note that the definition of products requires the existence of selection functions. This means that for infinite index sets we will need the Axiom of Choice. As previously, we will use the Axiom of Choice as needed. Finite products that are not too large can be visualized as shown in Figure 10.1.

The same task that led us to Lemma 10.1.4 can be posed for products. Find copies of the building blocks, which are of course the factors here, in the whole structure. The examples in Figure 10.1 show that such copies should abound. The next result (the only one on infinite products in this section) verifies this impression. Moreover all the embedded copies actually are retracts. These embedded copies will become very important when we prove Hashimoto's refinement theorem.

Lemma 10.2.2 *Let $\Pi_{\alpha \in A} X_\alpha$ be an infinite product of ordered sets X_α. For the elements $s^0 \in \Pi_{\alpha \in A} X_\alpha$, $\lambda \in A$ and $x_\lambda \in X_\lambda$ let*

$$(s^0, x_\lambda, \lambda)_\alpha := \begin{cases} s_\alpha^0; & \text{for } \alpha \neq \lambda, \\ x_\lambda; & \text{for } \alpha = \lambda. \end{cases}$$

Let $X_\lambda^{s^0} := \{(s^0, x_\lambda, \lambda) : x_\lambda \in X_\lambda\}$. Then $X_\lambda^{s^0}$ is isomorphic to X_λ and X_λ is a retract of $\Pi_{\alpha \in A} X_\alpha$ in the sense of Exercise 1 in Chapter 4.

Proof. Exercise 13.   ∎

Let us now consider how products connect with notions that were introduced earlier.

Proposition 10.2.3 *Let P, Q be ordered sets.*

1. *If P, Q are (complete) lattices, then $P \times Q$ is a (complete) lattice.*

2. *Let P, Q be ordered sets with more than one point. Then*

$$\max\{\dim(P), \dim(Q)\} \leq \dim(P \times Q) \leq \dim(P) + \dim(Q).$$

3. *The product of two interval orders need not be an interval order.*

Proof. For part 1 simply note that the needed pointwise suprema and infima will exist and that they will be the suprema and infima in the product.

To prove part 2 first note that the first inequality trivially follows from the fact that $P \times Q$ has retracts isomorphic to P and Q respectively. To prove the second inequality, let $(P, \leq_P)$ and $(Q, \leq_Q)$ be ordered sets. If one of P, Q is infinite dimensional, there is nothing to prove. Thus let $\leq_P^1, \ldots, \leq_P^n$ and $\leq_Q^1, \ldots, \leq_Q^m$ be realizers of P and Q respectively. For $i = 1, \ldots, n$ define $\leq_i$ on $P \times Q$ by $(p, q) \leq_i (p', q')$ iff $p <_P^i p'$ or $p = p'$ and $q \leq_Q^1 q'$. (This is the lexicographic sum of n copies of $(Q, \leq_Q^1)$ over $(P, \leq_P)$.) Similarly, for $j = 1, \ldots, m$ define $\leq^j$ on $P \times Q$ by $(p, q) \leq^j (p', q')$ iff $q <_Q^j q'$ or $q = q'$ and $p \leq_P^1 p'$.

We claim that $\{\leq_1, \ldots, \leq_n, \leq^1, \ldots, \leq^m\}$ is a realizer of the product order $\leq$. Indeed, it is trivial to see that $\leq$ is contained in the intersection of the above orders. Conversely, let $(p, q) \leq_i (p', q')$ for all $i \in \{1, \ldots, n\}$ and also $(p, q) \leq^j (p', q')$ for all $j \in \{1, \ldots, m\}$. If $p \neq p'$, we have $p <_P^i p'$ for all $i \in \{1, \ldots, n\}$ and thus $p <_P p'$. Similarly, if $q \neq q'$, we have $q <_Q^j q'$ for all $j \in \{1, \ldots, m\}$ and thus $q <_Q q'$. This means that $p \leq_P p'$ and $q \leq_Q q'$ which means $(p, q) \leq (p', q')$ in the product order.

Finally for part 3 note that the product of the two 3-chains $\{1 < 2 < 3\}$ and $\{a < b < c\}$ contains the subset $\{(3, a), (3, b), (1, c), (2, c)\}$. ∎

Products can now be used to give a different characterization of linear dimension. (Cf. Example 7.3.6.)

Theorem 10.2.4 (O. Ore) *The product of m nontrivial chains has dimension m. As a consequence, the linear dimension of an ordered set P is the smallest number k such that P can be embedded as an ordered subset into a product of k chains.*

Proof. Let $C_1, \ldots, C_m$ be m chains, each with at least two elements. Then $C_1 \times \cdots \times C_m$ contains a copy of $\mathbf{2}^m$, which is (cf. Example 10.1.3) isomorphic to $\mathcal{P}(m)$. By Corollary 7.3.9 we have $\dim(\mathcal{P}(m)) \geq m$ which means $\dim(C_1 \times \cdots \times C_m) \geq m$. By part 2 of Proposition 10.2.3 we have $\dim(C_1 \times \cdots \times C_m) \leq m$. Therefore $\dim(C_1 \times \cdots \times C_m) = m$.

For the second statement, let $(P, \leq)$ be an ordered set of dimension k and let $\{\leq_1, \ldots, \leq_k\}$ be a realizer of $\leq$. Let $(H, \sqsubseteq)$ be the product of the ordered sets $(P, \leq_i)$. Then $p \leq p'$ iff $p \leq_i p'$ for all $i \in \{1, \ldots, k\}$ which is the case iff $(p, \ldots, p) \sqsubseteq (p', \ldots, p')$ in H. Thus the map $e : P \rightarrow H$, $p \mapsto (p, \ldots, p)$ is an embedding of P into a product of k chains. Finally P cannot be embedded into a product of fewer than k chains, since then the dimension of P would be less than k. ∎

Note that the above gives a formal basis for our visualization of ordered sets of a given dimension as indicated in Example 7.3.6. The left set in Figure 10.1 shows for example how the visualization of products of two chains connects with our visualization of ordered sets of dimension 2. Moreover, the above gives a strong upper bound on the dimension of finite distributive lattices.

Theorem 10.2.5 (A product version of Dilworth's Chain Decomposition Theorem, cf. [57], Theorem 1.2.) *Let D be a finite distributive lattice and let k be the largest number of lower covers that any element of D can have. Then D is a sublattice of a product of k chains.*

Proof. By Birkhoff's characterization of finite distributive lattices, D is isomorphic to $D(\mathcal{J}(D))$. We shall first prove that $w(\mathcal{J}(D)) = k$.

Let $A = \{a_1, \ldots, a_m\} \subseteq \mathcal{J}(D)$ be an antichain. Let

$$A_D := (\downarrow_D a_1) \cup \cdots \cup (\downarrow_D a_m).$$

Then the sets $A_D^i := A_D \setminus \{a_i\}$ are all lower covers of A in $D(\mathcal{J}(D))$. Thus $m \leq k$, that is $w(\mathcal{J}(D)) \leq k$.

Conversely for every $S \in D(\mathcal{J}(D))$ there is an antichain $A = \{a_1, \ldots, a_m\} \subseteq \mathcal{J}(D)$ such that $S = A_D$. The lower covers of S are exactly the sets A_D^i as described above. Thus $k \leq w(\mathcal{J}(D))$ and hence the two quantities are equal.

By Dilworth's Chain Decomposition Theorem there are $k = w(\mathcal{J}(D))$ chains $C_1, \ldots, C_k$ such that $\mathcal{J}(D) = C_1 \cup \ldots \cup C_k$. Add a new smallest element 0 to each chain C_i. For every set $A \in D(\mathcal{J}(D))$ (which, as we recall, is a down-set in $\mathcal{J}(D)$) let a_i be the supremum of $(C_i \cup \{0\}) \cap A$. The map $\Phi : A \mapsto (a_1, \ldots, a_k)$ embeds $\mathcal{J}(D)$ into $C_1 \times \cdots \times C_k$. Moreover it is easy to see that $\Phi[\mathcal{J}(D)]$ is a sublattice of $C_1 \times \cdots \times C_k$. ∎

10.2.1 Finite Products and the Fixed Point Property

One of the most attractive problems in the theory of ordered sets is the question if the fixed point property is "productive". The problem was open even for the case of two finite factors for over 15 years. M. Roddy solved this (most important) case in [228]. We will present a slight refinement of his argument (due to himself) here.

> **Open Question 10.2.6** *Let P and Q be ordered sets with the fixed point property. Does $P \times Q$ necessarily have the fixed point property? (The most important special case was solved by Roddy in [228], cf. Theorem 10.2.11 here.)*

The key idea in the proof is to actually concentrate on one factor. The notion of a factor map is what allows us to do so.

Definition 10.2.7 *Let P, X be ordered sets and let $f : P \times X \to P \times X$ be order-preserving. For $x \in X$ we define the **factor map** $f_x : P \to P$ to be $f_x(\cdot) := \Pi_P f(\cdot, x)$. For $p \in P$ we define $f_p(\cdot) := \Pi_X f(p, \cdot)$.*

As we will be working with factor maps, it will be good for these factor maps to have as simple a structure as possible. Nice maps as defined below fit this description. We will see that for a finite ordered set the factor maps can be replaced with nice maps.

Definition 10.2.8 *An order-preserving map $f : P \to P$ is called **flat** iff for all $x \in f[P]$ we have that x is comparable to $f(x)$ iff $f(x) = x$. f will be called a **nice map** iff*

1. *f is flat,*

2. *f is injective on its range,*

3. *There is a retraction $r_f : P \to f[P]$ such that $f = f \circ r_f$.*

Essentially, nice maps are maps for which the Abian–Brown theorem is instantaneous and for which (in the finite setting) the range does not change under iterations. As we will see, both properties allow us to eliminate many technical difficulties. We shall first show that the fixed point property is productive for ordered sets $P \times X$ and order-preserving maps $f : P \times X \to P \times X$ with sufficiently structured factor maps. The proof of Theorem 10.2.11 is then a simple application of Proposition 4.1.4.

Theorem 10.2.9 *Let P, X be ordered sets such that X has the fixed point property. Assume P is connected and chain-complete. Let $f : P \times X \to P \times X$ be an order-preserving map such that:*

1. *The factor maps $f_x : P \to P$ are all nice.*

2. *Each factor map $f_x : P \to P$ has a fixed point.*

3. *$x \le y$ implies $r_x := r_{f_x} \le r_{f_y} =: r_y$.*

Then f has a fixed point.

The key notion for the proof is the notion of an s-fence below. Consider that it is easy to find a fixed point p for a factor map f_x. Thereafter, it is equally easy to find a fixed point x' for the factor map f_p. Unfortunately it can be that $x' \ne x$. Jumping back and forth between P and X in this fashion merely leads to a process in which one hand destroys what the other hand has just repaired. Using an s-fence, a fixed point in X is maintained throughout the argument. This allows us to avoid the problem indicated above. The actual proof is an argument by contradiction. However the construction that leads to the contradiction can be turned into an algorithm to find the fixed point of a product map.

Definition 10.2.10 *An $(n + 2)$-tuple $(a_0, \dots, a_n; x)$ is called an* **up s-fence** *of length n iff the following are satisfied.*

1. *$a_0 \le a_1 \ge \cdots a_n$ in P,*

2. *$x \in X$,*

3. *$f_{a_0}(x) = x$,*

4. *$r_x(a_j) = a_j$, for $j > 0$, and*

5. *$f_x(a_n) = a_n$.*

A **down s-fence** *is defined dually and an* **s-fence** *is one or the other.*

Proof of Theorem 10.2.9. First we shall show that P contains an s-fence. Fix $q \in P$ and consider the order-preserving map $x \mapsto f_{r_x(q)}(x)$. Let x_0 be one of its fixed points. Let $a_0 := r_{x_0}(q)$ and find a fixed point p of f_{x_0}. Since P is connected, there is a fence $(a_0, \ldots, a_n)$ from a_0 to $a_n := p$ in $r_{x_0}[P]$. Now $(a_0, \ldots, a_n; x_0)$ is the desired s-fence.

Let $(a_0, \ldots, a_n; x_0)$ be an s-fence of minimum length. If $n = 0$, then we are done, since $f_{x_0}(a_0) = a_0$ and $f_{a_0}(x_0) = x_0$ implies $f(a_0, x_0) = (a_0, x_0)$. We will lead the assumption $n > 0$ to a contradiction. Assume without loss of generality that $a_0 < a_1$. Define sequences $\{x_i\}_{i\in\mathbb{N}}$, $\{a_{0,i}\}_{i\in\mathbb{N}}$ as

$$a_{0,0} = a_0, \quad a_{0,1} = a_1, \quad x_i = f_{a_{0,i}}(x_{i-1}), \quad \text{and } a_{0,i+1} = r_{x_i}(a_{0,i}) \ (i \geq 1).$$

Then the sequences $\{x_i\}_{i\in\mathbb{N}}$ and $\{a_{0,i}\}_{i\in\mathbb{N}}$ are both increasing, which can be seen as follows.

$$
\begin{aligned}
a_{0,1} &= a_1 \geq a_0 = a_{0,0}, \\
x_1 &= f_{a_{0,1}}(x_0) = f_{a_1}(x_0) \geq f_{a_0}(x_0) = x_0, \\
a_{0,2} &= r_{x_1}(a_{0,1}) \geq r_{x_0}(a_{0,1}) = a_{0,1}, \\
a_{0,i+1} &= r_{x_i}(a_{0,i}) \geq r_{x_{i-1}}(a_{0,i}) \geq r_{x_{i-1}}(a_{0,i-1}) = a_{0,i} \quad (i \geq 2), \\
x_{i+1} &= f_{a_{0,i+1}}(x_i) \geq f_{a_{0,i}}(x_{i-1}) = x_i \quad (i \geq 1).
\end{aligned}
$$

Let a be any upper bound of the sequence $\{a_{0,i}\}_{i\in\mathbb{N}}$ and let

$$Y = \uparrow \{x_i : i \in \mathbb{N}\}.$$

Define $h : Y \mapsto X$ by $h(y) = f_{r_y(a)}(y)$. We claim that h maps Y to itself. To see this, it is enough to show that $h(y) \geq x_i$ for each $i \in \mathbb{N}$ and $y \in Y$. Indeed, for $y \in Y$, $r_y(a) \geq r_{x_{i-1}}(a_{0,i-1}) = a_{0,i}$ gives

$$h(y) \geq f_{a_{0,i+1}}(x_i) = x_{i+1} \geq x_i$$

for all i, hence $h(y) \in Y$.

By Theorem 4.2.11 $h : Y \to Y$ has a fixed point. Let y be any fixed point of h and define the $(n+1)$-tuple $(b_1, \ldots, b_n; y)$ by setting

$$b_1 = r_y(a), \quad b_j = r_y(a_j), \quad 1 < j \leq n.$$

We claim that $(b_1, \ldots, b_n; y)$ is a down s-fence of length $n - 1$, which is the contradiction we seek. To prove this we have to check all properties of an s-fence. Properties 2 and 4 are trivial. For property 1 note that

$$b_1 = r_y(a) \geq r_y(a_2) = b_2,$$

since $a \geq a_{0,1} = a_1 \geq a_2$. The other inequalities of 1 hold because r_y is order-preserving. For 3 note that $f_{b_1}(y) = f_{r_y(a)}(y) = h(y) = y$. Finally to prove 5 we establish three claims, A, B, C.

(A) $f_y(a_n)$ *is a fixed point of* f_y. Indeed, $f_y(a_n) \geq f_{x_0}(a_n) = a_n$, and the claim follows from the fact that f_y is nice.

(B) $f_y(a_n) = r_y(a_n)$. To see this, note that by (A) and the definition of r_y, $f_y(f_y(a_n)) = f_y(a_n) = f_y r_y(a_n)$ and $r_y(a_n) \in f_y[P]$. Since f_y is one-to-one on its range, we have $f_y(a_n) = r_y(a_n)$.

(C) $f_y(b_n) = b_n$ *and thus 5 holds.*

$$f_y(b_n) = f_y r_y(a_n) \overset{\text{(B)}}{=} f_y f_y(a_n) \overset{\text{(A)}}{=} f_y(a_n) \overset{\text{(B)}}{=} r_y(a_n) = b_n.$$

This completes the proof of Theorem 10.2.9.   ∎

The main scope of Theorem 10.2.9 is to prove that the product of two finite ordered sets with the fixed point property has the fixed point property, too. Below we state the most general result that can be derived from Theorem 10.2.9. The "remaining cases" to prove the fixed point property is productive have eluded researchers so far.

Theorem 10.2.11 (Also cf. [228], Theorem 1.1.) *Let* P *be a chain-complete ordered set with no infinite antichains and the fixed point property and let* Q *be an ordered set with the fixed point property. Then* $P \times Q$ *also has the fixed point property.*

Proof. We shall first prove this result in the case that P is finite. Let $g : P \times Q \to P \times Q$ be order-preserving. Define $f : P \times Q \to P \times Q$ by setting $f_q := g_q^{|P|!+1}$ and $f_p := g_p$. Then f is order-preserving. For all $q \in Q$ define $r_q := g_q^{|P|!}$. Then by Proposition 4.1.4, r_q is a retraction (onto $f_q[P]$) and $q_1 \leq q_2$ implies $r_{q_1} \leq r_{q_2}$.

Clearly every fixed point of g is also a fixed point of f. Conversely if (p, q) is a fixed point of f, then $g_p(q) = f_p(q) = q$ and

$$p = f_q(p) = g_q^{|P|!+1}(p) = g_q(r_q(p)) = g_q(r_q(r_q(p))) = r_q(f_q(p)) = f_q(p),$$

which implies $g_q(p) = g_q(r_q(p)) = f_q(p) = p$. Thus f and g have the same fixed points and we are done if we can show that f has a fixed point. To do this we show that f is a nice map, which will allow us to apply Theorem 10.2.9.

To see that each f_q is injective on its image, let $a, b \in f_q[P]$. Then $a = g_q^{|P|!+1}(x)$ and $b = g_q^{|P|!+1}(y)$. If $f_q(a) = f_q(b)$, then

$$
\begin{aligned}
a &= r_q^2(a) = g_q^{3|P|!+1}(x) = g_q^{|P|!-1} f_q(a) \\
&= g_q^{|P|!-1} f_q(b) = g_q^{3|P|!+1}(y) = r_q^2(b) = b.
\end{aligned}
$$

Thus each f_q is injective on its image. In particular this means that if $p \in f_q[P]$ is comparable to $f_q(p)$, then $f_q(p) = p$, so f_q is flat. With $r_{f_q} := r_q$ we have via Proposition 4.1.4 that $f_q = f_q \circ r_{f_q}$. Thus each f_q is nice.

Now f satisfies all hypotheses of Theorem 10.2.9. This means f has a fixed point and we are done in the case that P is finite. The rest follows from Li and Milner's structure theorem and Exercise 9.   ∎

10.3 Infinite Products

A natural generalization of Open Question 10.2.6 is to ask the same question for infinite families of ordered sets with the fixed point property. At first glance this does not appear promising. Even the product of nice families of ordered sets with the fixed point property needs not be connected (cf. Lemma 10.3.3). However, taking trivially disconnected products out of consideration, nothing much beyond the (simple) results of this section is known. We are thus led to Open Question 10.3.7. Progress in this area would have at least two consequences. These are better understanding of order-preserving maps on products and a better understanding of the fixed point property for infinite ordered sets.

We start with connectivity as a warm-up to infinite products. While the main scope of the following results are infinite products, they are of course equally valid for any finite products that fall into their scope.

Lemma 10.3.1 *Let* $\{P_\alpha\}_{\alpha \in A}$ *be a family of ordered sets. For any two elements* $(p_\alpha)_{\alpha \in A}, (q_\alpha)_{\alpha \in A} \in \Pi_{\alpha \in A} P_\alpha$ *we have*

$$\sup_{\alpha \in A} \text{dist}(p_\alpha, q_\alpha) \leq \text{dist}((p_\alpha)_{\alpha \in A}, (q_\alpha)_{\alpha \in A}) \leq \sup_{\alpha \in A} \text{dist}(p_\alpha, q_\alpha) + 1.$$

Proof. The first inequality follows since every P_α is a retract of $\Pi_{\alpha \in A} P_\alpha$. To see the other inequality let $n := \sup_{\alpha \in A} \text{dist}(p_\alpha, q_\alpha)$. If $n = \infty$ there is nothing to prove. If n is finite, then for each α one can find $x_\alpha^0, \dots, x_\alpha^{n+1}$ such that $p_\alpha = x_\alpha^0 \geq x_\alpha^1 \leq \cdots x_\alpha^{n+1} = q_\alpha$. (To see this, first find a fence that connects p_α and q_α. Then duplicate elements of that fence as necessary to get the above.) Then $(x_\alpha^0)_{\alpha \in A} \geq (x_\alpha^1)_{\alpha \in A} \leq \cdots (x_\alpha^{n+1})_{\alpha \in A}$ connects the two points in the product. ∎

Proposition 10.3.2 *Let* $\{P_\alpha\}_{\alpha \in A}$ *be a family of ordered sets. The following are equivalent.*

1. *$\Pi_{\alpha \in A} P_\alpha$ is connected and of finite diameter.*

2. *All P_α are connected and there is an $n \in \mathbb{N}$ such that all P_α are of diameter $\leq n$.*

Proof. Easy consequence of Lemma 10.3.1. ∎

Proposition 10.3.2 almost characterizes connected products, but easy examples show that ordered sets can be connected and of unbounded diameter. These will need to be considered separately to completely characterize connectivity of products.

Lemma 10.3.3 (Also cf. [67].) *Let* $\{P_\alpha\}_{\alpha \in A}$ *be a family of ordered sets. Assume for each $n \in \mathbb{N}$ there is an α_n such that $n \neq m$ implies $\alpha_n \neq \alpha_m$ and such that the diameter of P_{α_n} is larger than n. Then $\Pi_{\alpha \in A} P_\alpha$ is not connected.*

Proof. For each $n \in \mathbb{N}$ find $p_{\alpha_n}, q_{\alpha_n} \in P_{\alpha_n}$ whose distance is larger than n. For $\alpha \in A \setminus \{\alpha_n : n \in \mathbb{N}\}$ let $p_\alpha, q_\alpha \in P_\alpha$ be arbitrary. Then the distance $\text{dist}((p_\alpha)_{\alpha \in A}, (q_\alpha)_{\alpha \in A})$ cannot be finite, since for all $n \in \mathbb{N}$ we have $n < \text{dist}(p_{\alpha_n}, q_{\alpha_n}) \leq \text{dist}((p_\alpha)_{\alpha \in A}, (q_\alpha)_{\alpha \in A})$. ∎

Theorem 10.3.4 *Let $\{P_\alpha\}_{\alpha \in A}$ be a family of ordered sets. The following are equivalent.*

1. *$\Pi_{\alpha \in A} P_\alpha$ is connected.*

2. *All P_α are connected and there is an $n \in \mathbb{N}$ such that all but finitely many P_α are of diameter $\leq n$.*

Proof. "2⇒1": The product of the factors whose diameter is bounded by n is connected by Proposition 10.3.2. The product of the factors with diameter $> n$ is a finite product of connected sets, hence it is connected (cf. Exercise 12). Thus the product of all factors is connected.

"1⇒2": If 2 was false, then $\Pi_{\alpha \in A} P_\alpha$ would have to be disconnected by Lemma 10.3.3. ∎

While connectivity is a simple necessary condition for the fixed point property, $\mathcal{C}$-dismantlability was a strong sufficient condition. Moreover we have seen in Theorem 10.1.5 that $\mathcal{C}$-dismantlability to a singleton is closely related to connectivity of $\text{End}(P)$. These ideas are again present in the next result.

Proposition 10.3.5 *Let $\{P_\alpha\}_{\alpha \in A}$ be a family of ordered sets. The following are equivalent.*

1. *All P_α are chain-complete and there is an $n \in \mathbb{N}$ such that for all $\alpha \in A$ the diameter of $\text{End}(P_\alpha)$ is $\leq n$.*

2. *$\Pi_{\alpha \in A} P_\alpha$ is $\mathcal{C}$-dismantlable to a singleton and chain-complete.*

Proof. "1⇒2": For each α select a $p_\alpha \in P_\alpha$. Then for each α there are order-preserving maps $r_\alpha^0, r_\alpha^1, \ldots, r_\alpha^{n+1} \in \text{End}(P_\alpha)$ with $r_\alpha^0 = \text{id}_{P_\alpha}$, $r_\alpha^{n+1}(x) = p_\alpha$ for all $x \in P_\alpha$ and such that $r_\alpha^0 \leq_\alpha r_\alpha^1 \geq_\alpha \cdots r_\alpha^{n+1}$. Define $R^j((q_\alpha)_{\alpha \in A}) := (r_\alpha^j(q_\alpha))_{\alpha \in A}$ for $j = 0, \ldots, n+1$. Then for every mapping $f \in \text{End}(\Pi_{\alpha \in A} P_\alpha)$ the set $\{R^0 \circ f, R^1 \circ f, \ldots, R^{n+1} \circ f = R^{n+1}\}$ contains a fence from f to R^{n+1}. Thus $\text{End}(\Pi_{\alpha \in A} P_\alpha)$ is connected. Chain-completeness is proved in Exercise 14, the result follows from Theorem 10.1.5.

"2⇒1": First note that the diameter of $\text{End}(\Pi_{\alpha \in A} P_\alpha)$ is bounded by twice the distance from $\text{id}_{\Pi_{\alpha \in A} P_\alpha}$ to any constant map. This means that since $\text{End}(\Pi_{\alpha \in A} P_\alpha)$ is connected, it automatically has finite diameter.

Now fix $(p_\alpha)_{\alpha \in A} \in \Pi_{\alpha \in A} P_\alpha$. Every set of endomorphisms $\text{End}(P_\gamma)$ can be embedded into $\text{End}(\Pi_{\alpha \in A} P_\alpha)$ via $f \mapsto F_f$, where

$$(F_f((q_\alpha)_{\alpha \in A}))_\beta := \begin{cases} f(q_\gamma); & \text{if } \beta = \gamma, \\ p_\beta; & \text{otherwise.} \end{cases}$$

Moreover for fixed $(p_\alpha)_{\alpha \in A} \in \Pi_{\alpha \in A} P_\alpha$ the map $m(F) := f_F$, where (with notation as in Lemma 10.2.2)

$$(f_F((q_\alpha)_{\alpha \in A}))_\beta := \begin{cases} \Pi_\gamma(F(((p_\alpha)_{\alpha \in A}, q_\gamma, \gamma))); & \text{if } \beta = \gamma, \\ p_\beta; & \text{otherwise,} \end{cases}$$

maps End $(\Pi_{\alpha \in A} P_\alpha)$ onto the isomorphic copy of End(P_γ). Hence the diameter of each End(P_α) is bounded by the diameter of End $(\Pi_{\alpha \in A} P_\alpha)$. Since every factor of a product is a retract of the product, all factors are chain-complete. ∎

Remark 10.3.6 As the following example[1] shows, the technical condition 1 in Proposition 10.3.5 can not be replaced with the condition "all P_α have uniformly bounded diameter". Let **2** be the two-element antichain and let F_{2n} be a fence of length $2n$. Let P_n be the lexicographic sum $\mathbf{2} \oplus F_{2n}$. Then diam$(P_n) = 2$ and diam(End$(P_n)) \geq n$. Hence $\Pi_{n \in \mathbb{N}} P_n$ is connected, but not dismantlable.

We have thus completely characterized the relationship between products and the conditions most easily linked to the fixed point property, connectivity and C-dismantlability. We have also in Remark 10.3.6 seen a natural candidate for an infinite product of finite ordered sets for which it is not trivially decidable if it has the fixed point property. The open question that beckons is

> **Open Question 10.3.7** *Let $\{P_\alpha\}_{\alpha \in A}$ be a family of ordered sets with the fixed point property such that there is an $n \in \mathbb{N}$ such that* diam$(P_\alpha) \leq n$ *for all $\alpha \in A$. Does $\Pi_{\alpha \in A} P_\alpha$ have the fixed point property?*

No progress on this problem has been made, even if all factors are the same ordered set. We conclude this section showing that a natural avenue to disprove the fixed point property, minimal finite automorphic retracts, is closed for infinite products. Indeed, by Proposition 10.3.8 and Theorem 10.2.11, every finite retract of an infinite product of finitely many nonisomorphic finite ordered sets with the fixed point property must also have the fixed point property.

Proposition 10.3.8 (Cf. [67], Lemma 3.2.) *Let $S_1, \ldots, S_N$ be pairwise nonisomorphic finite ordered sets and let $\{Q_\alpha\}_{\alpha \in A}$ be a family of ordered sets such that each Q_α is isomorphic to some set S_{k_α}. Let P be a finite retract of the product $\Pi_{\alpha \in A} Q_\alpha$. Then there is a finite set $\{\alpha_1, \ldots, \alpha_n\} \subseteq A$ such that P is isomorphic to a retract of $\Pi_{m=1}^n Q_{\alpha_m}$.*

Proof. Let $r : \Pi_{\alpha \in A} Q_\alpha \to P$ be a retraction on the product. For each α let $\phi_\alpha : Q_\alpha \to S_{k_\alpha}$ be an isomorphism. Let

$$\alpha \equiv \beta \quad \text{iff} \quad \phi_\alpha \circ \pi_\alpha|_P = \phi_\beta \circ \pi_\beta|_P.$$

---

[1] First shown to the author by A. Rutkowski in a different context.

Clearly $\equiv$ is an equivalence relation. Moreover, since P and the S_k are finite and since there are only finitely many sets S_k, there are only finitely many possibilities for maps from P to S_k with arbitrary k. Thus the equivalence relation $\equiv$ partitions the set A into finitely many equivalence classes. Assuming there are n equivalence classes, let $\alpha_1, \ldots, \alpha_n$ be a system of representatives for the equivalence classes. For $q \in \Pi_{m=1}^n Q_{\alpha_m}$ define $q' \in \Pi_{\alpha \in A} Q_\alpha$ componentwise as follows. For $\alpha \in \{\alpha_1, \ldots, \alpha_n\}$ let $q'_\alpha := q_\alpha$. For $\alpha \notin \{\alpha_1, \ldots, \alpha_n\}$ let α_m be the unique element of $\{\alpha_1, \ldots, \alpha_n\}$ such that $\alpha \equiv \alpha_m$ and let $q'_\alpha := \Phi_\alpha^{-1}(\Phi_{\alpha_m}(q_{\alpha_m}))$. Note that $q \mapsto q'$ is an embedding of $\Pi_{m=1}^n Q_{\alpha_m}$ into $\Pi_{\alpha \in A} Q_\alpha$.

Now define $r' : \Pi_{m=1}^n Q_{\alpha_m} \to P$ by $r'(q) := r(q')$. Since $q \mapsto q'$ is an embedding, r' is order-preserving. Define $c : P \to \Pi_{m=1}^n Q_{\alpha_m}$ to be the map that maps $(p_\alpha)_{\alpha \in A}$ to $(p_{\alpha_m})_{m=1}^n$. Clearly c is also order-preserving. Finally note that $r' \circ c = \mathrm{id}_P$. This means that P is a retract of $\Pi_{m=1}^n Q_{\alpha_m}$ in the sense of the alternative definition of retracts given in Exercise 1 in Chapter 4.   ■

Remark 10.3.9 The condition in Proposition 10.3.8 that only finitely many factors are nonisomorphic is needed. M. Roddy showed in [230] that $\Pi_{n \in \mathbb{N}} P_n$ as in Remark 10.3.6 can be retracted onto a four crown. (In particular, $\Pi_{n \in \mathbb{N}} P_n$ as in Remark 10.3.6 does not have the fixed point property.[2]) In regards to Proposition 10.3.8, in Exercise 23 of Chapter 4 it is shown that any retract of a finite $\mathcal{I}$-dismantlable ordered set is again $\mathcal{I}$-dismantlable, and Proposition 10.3.5 shows that any finite product of finite $\mathcal{I}$-dismantlable ordered sets is again $\mathcal{I}$-dismantlable. Since the four crown is not $\mathcal{I}$-dismantlable, the condition of finitely many nonisomorphic factors cannot be dropped.

Moreover, the set $\Pi_{n \in \mathbb{N}} P_n$ is a retract of an infinite power of linear sums of 2-antichains and the "spider" in Figure 1.1 f). Thus the assumption that the factors are all finite is also needed in Proposition 10.3.8.

10.4   Hashimoto's Theorem and Automorphisms of Products

Infinite products are also investigated in their own right to provide a representation theory for ordered sets. It is natural to ask what the relation between two product representations of an ordered set might be. Clearly their factors need not be isomorphic. Simply consider $P \times (P \times Q)$ and $(P \times P) \times Q$ as products of two factors each. Yet for this example it is also quite obvious that we get isomorphisms between the factors if we factor into three factors. An interesting fact about products is that by Hashimoto's Theorem 10.4.4 this refinement will always work for *connected* ordered sets, while for disconnected ordered sets it may not work (cf. Exercise 22).

---

[2]In preprints of this text this was posed as an open question.

Hashimoto's Theorem will also allow us to show that the property that all automorphisms of an ordered set have a fixed point is "infinitely productive". To this end we need to prove a stronger result than Hashimoto's original refinement theorem. This requires us to set up a more formal framework for products.

Definition 10.4.1 *A* (**product**) **decomposition** *of an ordered set P is an isomorphism $\Psi : P \to \Pi_{\alpha \in A} X_\alpha$ between P and a product $\Pi_{\alpha \in A} X_\alpha$ of ordered sets X_α.*

This definition of a product decomposition clearly separates different representations from each other. As we find common refinements of product decompositions (such as in the $P \times (P \times Q)$ versus $(P \times P) \times Q$ example) we will find product decompositions of the factors which we will need to turn into decompositions of the original ordered set. To do this, we need to turn decompositions of the factors into maps from the original product to the refined product. This is easily done using the idea of a product map.

Definition 10.4.2 *Let $\Pi_{\alpha \in A} X_\alpha$ and $\Pi_{\alpha \in A} Y_\alpha$ be two products with the same index set A. Let $\{f_\alpha : X_\alpha \to Y_\alpha\}_{\alpha \in A}$ be a family of order-preserving maps. Then the* **product map** *$f : \Pi_{\alpha \in A} X_\alpha \to \Pi_{\alpha \in A} Y_\alpha$ is defined componentwise as the unique map such that $\pi_\beta f((x_\alpha)_{\alpha \in A}) = f_\beta(x_\beta)$.*

The strict refinement property now says that any two product representations have refinements that differ from each other only in the way the factors are grouped. Hashimoto's refinement theorem then says that this property applies for all connected ordered sets.

Definition 10.4.3 *An ordered set P has the* **strict refinement property** *iff for any two decompositions $\Psi_A : X \to \Pi_{\alpha \in A} X_\alpha$ and $\Psi_B : X \to \Pi_{\beta \in B} Y_\beta$ there are decompositions $a_\alpha : X_\alpha \to \Pi_{\beta \in B} Z_{\alpha,\beta}$ and $b_\beta : Y_\beta \to \Pi_{\alpha \in A} Z_{\alpha,\beta}$ such that, with a and b being the product maps of the a_α and b_β respectively and γ being the natural map that maps $((z_{\alpha,\beta})_{\beta \in B})_{\alpha \in A}$ to $((z_{\alpha,\beta})_{\alpha \in A})_{\beta \in B}$, we have $\gamma \circ a \circ \Psi_A = b \circ \Psi_B$.*

For an illustration of the strict refinement property and its proof through an extension of Hashimoto's refinement theorem, cf. Figure 10.2.

Theorem 10.4.4 *(A strengthening of Hashimoto's refinement theorem, cf. [36] and [107].) Every connected ordered set has the strict refinement property.*

Note that in the definition of the strict refinement property we can eliminate the "middleman" X and simply talk about two infinite products that are isomorphic via $\Phi : \Pi_{\alpha \in A} X_\alpha \to \Pi_{\beta \in B} Y_\beta$. We will do so throughout the proof of Hashimoto's Theorem. We can (and do) also assume that all factors are connected. The proof starts with a series of lemmas. These lemmas show that product decompositions have a certain resilience under isomorphisms.

Consider the first lemma. It focuses on elements that have the same first component in one decomposition of an ordered set. The lemma says that the first component in this decomposition is invariant under finite interchanges of components of these elements in any other decomposition.

Lemma 10.4.5 (Cf. [107], Lemma 3.) *Let P, Q, U, V be connected ordered sets such that $\Phi : P \times Q \to U \times V$ is an isomorphism. If*

$$\Phi(p, q) = (u, v), \quad \Phi(p, q') = (u', v'), \quad \text{and} \quad \Phi^{-1}(u, v') = (p^*, q^*),$$

then $p^* = p$.

Proof. First assume q, q' have a common upper bound $\widehat{q}$. Let $(\widehat{u}, \widehat{v}) := \Phi(p, \widehat{q})$. Note that $(u, v) \leq (u, \widehat{v}) \leq (\widehat{u}, \widehat{v})$ and $(u', v') \leq (\widehat{u}, \widehat{v}) \leq (\widehat{u}, \widehat{v})$ implies that

$$p = \pi_P \Phi^{-1}(u, v) = \pi_P \Phi^{-1}(u', v')$$
$$\leq \{\pi_P \Phi^{-1}(u, \widehat{v}), \pi_P \Phi^{-1}(\widehat{u}, v')\} \leq \pi_P \Phi^{-1}(\widehat{u}, \widehat{v}) = p.$$

Thus there are $\widehat{q_1}$, $\widehat{q_2}$ such that

$$\Phi^{-1}(u, \widehat{v}) = (p, \widehat{q_1}) \quad \text{and} \quad \Phi^{-1}(\widehat{u}, v') = (p, \widehat{q_2}).$$

Now $(u, v') \leq \{(\widehat{u}, v'), (u, \widehat{v})\}$ implies $(p^*, q^*) \leq \{(p, \widehat{q_1}), (p, \widehat{q_2})\}$. In particular we have $p^* \leq p$.

Moreover, since $q^* \leq \{\widehat{q_1}, \widehat{q_2}\}$, we have $(p, q^*) \leq \{(p, \widehat{q_1}), (p, \widehat{q_2})\}$. This implies $\Phi(p, q^*) \leq \{(u, \widehat{v}), (\widehat{u}, v')\}$, which forces (by arguing componentwise) $\Phi(p, q^*) \leq (u, v')$. But then $(p, q^*) \leq (p^*, q^*)$ and we have $p \leq p^*$, which finishes the proof that $p = p^*$.

We now proceed by induction. Since Q is connected, there is a fence $q = q_0 \leq \widehat{q_1} \geq q_1 \leq \widehat{q_2} \geq q_2 \leq \cdots \leq \widehat{q_n} \geq q_n = q'$. The induction will be on the parameter n. For $n = 1$, the result is proved above. For the induction step let $n > 1$ and assume the result has been proved for all pairs (q, q') for which a fence as above but with fewer elements exists. With

$$\Phi(p, q_i) = (u_i, v_i) \quad \text{and} \quad \Phi(p, \widehat{q_i}) = (\widehat{u_i}, \widehat{v_i})$$

we obtain that there are $q_{0,n-1} \in Q$ and $q_{1,n} \in Q$ such that $\Phi^{-1}(u_0, v_{n-1}) = (p, q_{0,n-1})$ and $\Phi^{-1}(u_1, v_n) = (p, q_{1,n})$. By duality we also obtain $\Phi^{-1}(\widehat{u_1}, \widehat{v_n}) = (p, \widehat{q_{1,n}})$ for some $\widehat{q_{1,n}} \in Q$. Now $\{(u_0, v_{n-1}), (u_1, v_n)\} \leq (\widehat{u_1}, \widehat{v_n})$, which means that $\{q_{0,n-1}, q_{1,n}\} \leq \widehat{q_{1,n}}$.

Now apply what we proved first to $\{(p, q_{0,n-1}), (p, q_{1,n})\}$ to obtain that we have $\pi_P \Phi^{-1}(u, v') = \pi_P \Phi^{-1}(u_0, v_n) = p$ as was to be proved. ∎

Lemma 10.4.5 can now be turned into a result for arbitrary products.

Lemma 10.4.6 (Cf. [107], Lemma 2.) *Let $\Phi : \Pi_{\alpha \in A} X_\alpha \to \Pi_{\beta \in B} Y_\beta$ be an isomorphism between two connected products. For $i = 1, 2$ let $x^i = (x^i_\alpha)_{\alpha \in A}$ be elements of $\Pi_{\alpha \in A} X_\alpha$ such that $\Phi((x^i_\alpha)_{\alpha \in A}) = (y^i_\beta)_{\beta \in B}$. Fix $\lambda \in A$. If $x^1_\lambda = x^2_\lambda = x_\lambda$ and $y^3 \in \Pi_{\beta \in B} Y_\beta$ is such that for all $\beta \in B$ we have $y^3_\beta \in \{y^1_\beta, y^2_\beta\}$, then $\pi_\lambda \Phi^{-1}(y^3) = x_\lambda$.*

Proof. This follows directly from Lemma 10.4.5. Simply choose (with the obvious identifications made between sets)

$$P := X_\lambda, \quad Q := \Pi_{\alpha \in A \setminus \{\lambda\}} X_\alpha,$$

$$U := \Pi_{\beta \in \{\mu \in B : y_\mu^3 = y_\mu^1\}} Y_\beta, \quad V := \Pi_{\beta \in \{\mu \in B : y_\mu^3 \neq y_\mu^1\}} Y_\beta,$$

$$(p, q) := x^1, \quad (p, q') := x^2, \quad (u, v') := y^3.$$

∎

The key idea to obtain the common refinements now is to consider the embedded (via Lemma 10.2.2) factors in one decomposition and examine their images in the other decomposition. Interestingly (as we are about to show now), the image is itself a natural product and thus provides the desired decomposition of the factors.

Lemma 10.4.7 (Cf. [107], Lemma 4.) *Let* $\Phi : \Pi_{\alpha \in A} X_\alpha \to \Pi_{\beta \in B} Y_\beta$ *be an isomorphism between two connected products. Fix* $s^0 \in \Pi_{\alpha \in A} X_\alpha$ *and* $\lambda \in A$ *and let* $X_\lambda^{s^0}$ *be as in Lemma 10.2.2. For* $\beta \in B$ *let*

$$Y_\beta^\lambda := \pi_\beta \Phi[X_\lambda^{s^0}].$$

Then $\Phi|_{X_\lambda^{s^0}} : X_\lambda^{s^0} \to \Pi_{\beta \in B} Y_\beta^\lambda$ *is an isomorphism.*

Proof. Clearly the map $\Phi|_{X_\lambda^{s^0}}$ is an isomorphism onto its image and maps $X_\lambda^{s^0}$ into $\Pi_{\beta \in B} Y_\beta^\lambda$. What remains to be shown is that $\Phi|_{X_\lambda^{s^0}}$ is surjective. To do this note that $\Phi|_{X_\lambda^{s^0}}[X_\lambda^{s^0}]$ is a nonempty subset of the connected set $\Pi_{\beta \in B} Y_\beta^\lambda$. Thus by Exercise 19 we are done if we can show that $\Phi|_{X_\lambda^{s^0}}[X_\lambda^{s^0}]$ is an up-set and a down-set in $\Pi_{\beta \in B} Y_\beta^\lambda$. By duality it suffices to show it is an up-set.

For $i = 1, 2$ let $y^i = \Phi(x^i) = \Phi((x_\alpha^i)_{\alpha \in A}) = (y_\beta^i)_{\beta \in B}$ and assume that $y^1 \in \Phi|_{X_\lambda^{s^0}}[X_\lambda^{s^0}]$, $y^2 \in \Pi_{\beta \in B} Y_\beta^\lambda$ and $y^1 \leq y^2$. We will now first find an element $x^* \in X_\lambda^{s^0}$ such that $\Phi|_{X_\lambda^{s^0}}(x^*) \geq y^2$.

Let $\mu \in B$. By definition of $\Pi_{\beta \in B} Y_\beta^\lambda$, $y_\mu^2 = \pi_\mu \Phi((s^0, x_\lambda, \lambda))$ for some $x_\lambda \in X_\lambda$. Let $y^\mu := (y^1, y_\mu^2, \mu)$. Then by Lemma 10.4.6, $y^\mu \in \Phi|_{X_\lambda^{s^0}}[X_\lambda^{s^0}]$. Moreover, since $y^1 \leq y^\mu \leq y^2$, we have $\Phi^{-1}(y^1) \leq \Phi^{-1}(y^\mu) \leq \Phi^{-1}(y^2)$. This means $\Phi^{-1}(y^\mu) = (x^1, \tilde{x}, \lambda)$ for some $\tilde{x} \in X_\lambda$ with $x_\lambda^1 \leq \tilde{x} \leq x_\lambda^2$.

Let $x^* := (x^1, x_\lambda^2, \lambda)$. Then $y^\mu \leq \Phi(x^*)$. Since μ was arbitrary, this means that $\pi_\mu \Phi(x^*) \geq y_\mu^\mu = y_\mu^2$ for all μ. Therefore (after componentwise comparison) $\Phi(x^*) \geq y^2 \geq y^1$.

But then $(x^1, x^2_\lambda, \lambda) \geq \Phi^{-1}(y^2) \geq \Phi^{-1}(y^1)$, which means that all components of $\Phi^{-1}(y^2)$ except possibly the λth are equal to the corresponding component of s^0. Thus $\Phi^{-1}(y^2) \in X^{s^0}_\lambda$ and we are done. ∎

Proof of Theorem 10.4.4, first stage. Without loss of generality assume $A \cap B = \emptyset$. With s^0 and Y^λ_β chosen as in Lemma 10.4.7 there is an isomorphism $a_\lambda : X_\lambda \to \Pi_{\beta \in B} Y^\lambda_\beta$. Simply choose a_λ to be the composition of $\Phi|_{X^{s^0}_\lambda} : X^{s^0}_\lambda \to \Pi_{\beta \in B} Y^\lambda_\beta$ and the isomorphism between X_λ and $X^{s^0}_\lambda$. Similarly, using $\Phi(s^0)$ as the fixed element in $\Pi_{\beta \in B} Y_\beta$ we can define isomorphisms $b_\mu : Y_\mu \to \Pi_{\alpha \in A} X^\mu_\alpha$. Each b_μ is the composition of the inverse $\Phi^{-1}|_{Y^{\Phi(s^0)}_\mu} : Y^{\Phi(s^0)}_\mu \to \Pi_{\alpha \in A} X^\mu_\alpha$ and the isomorphism between Y_μ and $Y^{\Phi(s^0)}_\mu$.

We shall now show that X^μ_λ is isomorphic to Y^λ_μ. Let

$$
\begin{aligned}
S^\lambda_\mu &:= \{(\Phi(s^0), y_\mu, \mu) : y_\mu \in Y^\lambda_\mu\}, \\
S^\mu_\lambda &:= \{(s^0, x_\lambda, \lambda) : x_\lambda \in X^\mu_\lambda\}.
\end{aligned}
$$

Clearly S^λ_μ is isomorphic to Y^λ_μ and S^μ_λ is isomorphic to X^μ_λ. Let $x \in S^\mu_\lambda$. Then $x \in X^{s^0}_\lambda$ and thus by Lemma 10.4.7, $\Phi(x) \in \Pi_{\beta \in B} Y^\lambda_\beta$, especially $\pi_\mu \Phi(x) \in Y^\lambda_\mu$. Moreover $s^0 = \Phi^{-1}(\Phi(s^0))$ is in $\Pi_{\alpha \in A} X_\alpha$ and $x_\lambda \in X^\mu_\lambda$ by definition of S^μ_λ. Thus $x \in \Pi_{\alpha \in A} X^\mu_\alpha$, which means $x = \Phi^{-1}(y^*)$ for some $y^* \in Y^{\Phi(s_0)}_\mu$ or equivalently $\Phi(x) = (\Phi(s^0), y_\mu, \mu)$ for some $y_\mu \in Y_\mu$. Since $\pi_\mu \Phi(x) \in Y^\lambda_\mu$, this means that $\Phi(x) \in S^\lambda_\mu$. Since $x \in S^\mu_\lambda$ was arbitrary we conclude $\Phi[S^\mu_\lambda] \subseteq S^\lambda_\mu$. The symmetric argument can be applied to $\Phi^{-1}|_{S^\lambda_\mu}$, showing $\Phi^{-1}[S^\lambda_\mu] \subseteq S^\mu_\lambda$.

Thus $\Phi|_{S^\mu_\lambda}$ is an isomorphism between S^μ_λ and S^λ_μ. This in turn shows that X^μ_λ is isomorphic to Y^λ_μ.

The above is what is proved in [107]. Any two connected products have refinements with isomorphic factors. This property is also referred to as the **refinement property**. Note however, that the isomorphism between the refinements has not been put into relation with the isomorphism between the products. For a quick overview of the structures that are isomorphic to $\Pi_{\alpha \in A} X_\alpha$ and the isomorphisms between them, consider Figure 10.2. To finish the proof of Theorem 10.4.4, we will prove that the diagram in Figure 10.2 commutes. That is, independent of which path of arrows we follow, we obtain the same map. This fact was observed for example in [132].

Proof of Theorem 10.4.4, taking stock. We have so far that $\Pi_{\alpha \in A} \Pi_{\beta \in B} Y^\alpha_\beta$ is isomorphic to $\Pi_{\beta \in B} \Pi_{\alpha \in A} X^\beta_\alpha$. With all the other isomorphisms, this gives us two distinct isomorphisms between $\Pi_{\alpha \in A} X_\alpha$ and $\Pi_{\alpha \in A} \Pi_{\beta \in B} X^\beta_\alpha$.

For the first isomorphism we map $(x_\lambda)_{\lambda \in A} \in \Pi_{\lambda \in A} X_\lambda$ to

$$
\Phi((x_\lambda)_{\lambda \in A}) \in \Pi_{\beta \in B} Y_\beta.
$$

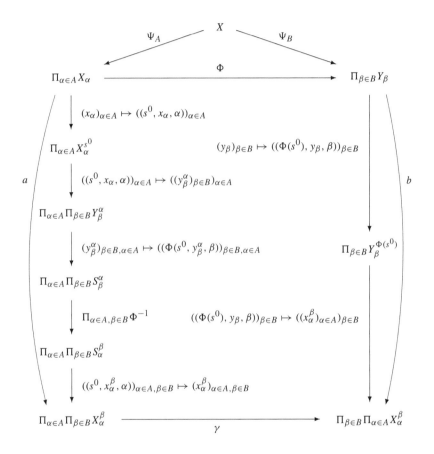

Figure 10.2: The strict refinement property and the isomorphisms constructed in the proof of Theorem 10.4.4.

This in turn is mapped to

$$((\Phi(s^0), \Phi((x_\lambda)_{\lambda \in A})_\beta, \beta))_{\beta \in B} \in \Pi_{\beta \in B} Y_\beta^{\Phi(s^0)}.$$

Via the B-fold product of Φ^{-1} we reach

$$\Psi_1((x_\lambda)_{\lambda \in A}) := (\Phi^{-1}(\Phi(s^0), \Phi((x_\lambda)_{\lambda \in A})_\beta, \beta))_{\beta \in B} \in \Pi_{\beta \in B} \Pi_{\alpha \in A} X_\alpha^\beta,$$

and $\Pi_{\beta \in B} \Pi_{\alpha \in A} X_\alpha^\beta$ is isomorphic to $\Pi_{\alpha \in A} \Pi_{\beta \in B} X_\beta^\alpha$ via the natural map γ^{-1}. On the other hand, we can also map $(x_\alpha)_{\alpha \in A} \in \Pi_{\alpha \in A} X_\alpha$ to

$$((s^0, x_\alpha, \alpha))_{\alpha \in A} \in \Pi_{\alpha \in A} X_\alpha^{s^0}.$$

The A-fold product of Φ maps this to

$$(\Phi(s^0, x_\alpha, \alpha))_{\alpha \in A} = ((\Phi(s^0, x_\alpha, \alpha)_\beta)_{\beta \in B})_{\alpha \in A} \in \Pi_{\alpha \in A} \Pi_{\beta \in B} Y_\beta^\alpha.$$

From here we can map to

$$((\Phi(s^0), \Phi(s^0, x_\alpha, \alpha)_\beta, \beta)_{\beta \in B})_{\alpha \in A} \in \Pi_{\alpha \in A} \Pi_{\beta \in B} S_\beta^\alpha.$$

The AB-fold product of Φ^{-1} maps this to

$$((\Phi^{-1}(\Phi(s^0), \Phi(s^0, x_\alpha, \alpha)_\beta, \beta))_{\beta \in B})_{\alpha \in A} \in \Pi_{\alpha \in A} \Pi_{\beta \in B} S_\alpha^\beta.$$

Finally each element of S_α^β is of the form (s^0, z, α) for some $z \in X_\alpha^\beta$. So to get to $\Pi_{\alpha \in A} \Pi_{\beta \in B} X_\alpha^\beta$ we need to take the αth projection inside and map to

$$\Psi_2((x_\lambda)_{\lambda \in A}) := ((\Phi^{-1}(\Phi(s^0), \Phi(s^0, x_\alpha, \alpha)_\beta, \beta)_\alpha)_{\beta \in B})_{\alpha \in A} \in \Pi_{\alpha \in A} \Pi_{\beta \in B} X_\alpha^\beta.$$

If we can now show that these two isomorphisms actually are equal, that is, $\gamma^{-1} \circ \Psi_1 = \Psi_2$, then we are done with the proof of this refinement of Hashimoto's theorem.

To do this we need to show that for all $(x_\lambda)_{\lambda \in A} \in \Pi_{\lambda \in A} X_\lambda$ and all $\alpha \in A$ and $\beta \in B$ we have that

$$\Phi^{-1}(\Phi(s^0), \Phi((x_\lambda)_{\lambda \in A})_\beta, \beta)_\alpha = \Phi^{-1}(\Phi(s^0), \Phi(s^0, x_\alpha, \alpha)_\beta, \beta)_\alpha.$$

First fix $\mu \in B$ and consider

$$(x_\lambda)_{\lambda \in A} \in \Pi_{\lambda \in A} X_\lambda^\mu = \{\Phi^{-1}(\Phi(s^0), y_\mu, \mu)) : y_\mu \in Y_\mu\}.$$

For $\beta \neq \mu$ both sides above return the αth component of s^0. For $\beta = \mu$ both sides above return x_α. Therefore we have $\gamma^{-1} \circ \Psi_1 = \Psi_2$ on $\Pi_{\lambda \in A} X_\lambda^\mu$ for arbitrary $\mu \in B$. To finish the proof we will now need to extend this equality to the whole product. To do this we require a lemma that is similar to Lemma 10.4.5. Lemma 10.4.8 shows that sometimes components in different product decompositions "stick together". That is, equality of components in one product decomposition can, under the right circumstances, force the equality of corresponding components in the other.

Lemma 10.4.8 *Let* P, Q, U, V *be connected ordered sets such that* $\Phi : P \times Q \to U \times V$ *is an isomorphism. If* $\Phi(p, q) = (u, v)$ *and* $\Phi(p, q') = (u, v')$, *then for each* $p' \in P$ *there is a* $w_{p'} \in U$ *such that* $\Phi(p', q) = (w_{p'}, z)$ *and* $\Phi(p', q') = (w_{p'}, z')$ *for some* $z, z' \in V$.

Proof. Since P is connected, it suffices to show the result for $p < p'$. The proof is an induction on $\text{dist}(q, q')$. The base step consists of the cases $q' < q$ and $q' > q$. Let $(w, z) := \Phi(p', q)$ and $(w', z') := \Phi(p', q')$.

First consider the case $q' < q$. In this case $w' \le w$. Moreover, the supremum of (p', q') and (p, q) is (p', q). This means the supremum of (u, v) and (w', z') is (w, z). Since $(p, q') \le (p', q')$ we have $(u, v') \le (w', z')$, which in particular means $u \le w'$. Therefore (w', z) is an upper bound of (u, v) and of (w', z'), which means $(w', z) \ge (w, z)$. Hence $w' \ge w$, which shows $w' = w$, proving our claim.

For the case $q' > q$, simply interchange the roles of q and q' above.

This leaves the induction step. Let $\text{dist}(q, q') = n$ and assume the statement is true for all pairs q, q' whose distance is less than n. Let $q = q_0, q_1, \ldots, q_n = q'$ be a fence that connects q and q'. For $i = 0, \ldots, n$ let $(u_i, v_i) := f(p, q_i)$. Note that $u_0 = u = u_n$. By Lemma 10.4.5 there is an $r \in Q$ such that $f(p, r) = (u_0, v_1)$. Now

$$\text{dist}_{U \times V}((u_0, v_1), (u_n = u_0, v_n)) = \text{dist}_V(v_1, v_n) < n,$$

which means that

$$n > \text{dist}_{P \times Q}((p, r), (p, q')) = \text{dist}_Q(r, q').$$

By induction hypothesis we have that $\pi_U(\Phi(p', r)) = \pi_U(\Phi(p', q'))$ and since $r \sim q$, the base case shows that $\pi_U(\Phi(p', r)) = \pi_U(\Phi(p', q))$. This proves the claim.   ∎

Proof of Theorem 10.4.4, final stage. Keep μ fixed and fix $\lambda_0 \in A$. Consider

$$(x_{\lambda_0}, (x_\lambda)_{\lambda \in A \setminus \{\lambda_0\}}) \in X_{\lambda_0} \times \Pi_{\lambda \in A \setminus \{\lambda_0\}} X_\lambda =: P \times Q.$$

We will denote elements of the subset $X_{\lambda_0} \times \Pi_{\lambda \in A \setminus \{\lambda_0\}} X_\lambda^\mu$ by $(x_{\lambda_0}, (x_\lambda^\mu)_{\lambda \in A \setminus \{\lambda_0\}})$. Decompose the range $\Pi_{\alpha \in A} \Pi_{\beta \in B} X_\beta^\alpha$ as

$$\Pi_{\beta \in B} X_\beta^{\lambda_0} \times \Pi_{\alpha \in A \setminus \{\lambda_0\}} \Pi_{\beta \in B} X_\beta^\alpha =: U \times V.$$

Let $\Psi : X_{\lambda_0} \times \Pi_{\lambda \in A \setminus \{\lambda_0\}} X_\lambda \to \Pi_{\beta \in B} X_\beta^{\lambda_0} \times \Pi_{\alpha \in A \setminus \{\lambda_0\}} \Pi_{\beta \in B} X_\beta^\alpha$ be any isomorphism such that for $(x_\lambda^\mu)_{\lambda \in A} \in \Pi_{\lambda \in A} X_\lambda^\mu$ we have that

$$(\Psi((x_\lambda^\mu)_{\lambda \in A})_\beta)_\alpha = s_\alpha^0 \text{ if } \beta \ne \mu \quad \text{and} \quad (\Psi((x_\lambda)_{\lambda \in A})_\mu)_\alpha = x_\alpha^\mu.$$

(Ψ will serve as a stand-in for $\gamma^{-1} \circ \Psi_1$ and Ψ_2.) By Lemma 10.4.8 we know that the $X_\beta^{\lambda_0}$-components of the Ψ-image of any point $(x_{\lambda_0}, (x_\lambda^\mu)_{\lambda \in A \setminus \{\lambda_0\}})$ with

$(x^\mu_\lambda)_{\lambda \in A \setminus \{\lambda_0\}} \in \Pi_{\lambda \in A \setminus \{\lambda_0\}} X^\mu_\lambda$ solely depend on x_{λ_0}. Thus the U-component of the Ψ-image of $(x_{\lambda_0}, (x^\mu_\lambda)_{\lambda \in A \setminus \{\lambda_0\}})$ is exactly that of $\Psi((s^0, x_{\lambda_0}, \lambda_0))$; call it $\Psi_{\lambda_0}(x_{\lambda_0})$.

Now consider the V-components of the Ψ-image of $(x_{\lambda_0}, (x^\mu_\lambda)_{\lambda \in A \setminus \{\lambda_0\}})$. We have $\Psi(x_{\lambda_0}, (x^\mu_\lambda)_{\lambda \in A \setminus \{\lambda_0\}}) = (u, v)$ and for a fixed $x^\mu_{\lambda_0} \in X^\mu_{\lambda_0}$ we have

$$\Psi(x^\mu_{\lambda_0}, (x^\mu_\lambda)_{\lambda \in A \setminus \{\lambda_0\}}) = (u', v').$$

By Lemma 10.4.5 this means that $\Psi^{-1}(u, v') = (\tilde{x}_{\lambda_0}, (x^\mu_\lambda)_{\lambda \in A \setminus \{\lambda_0\}})$. Since the U-component of a Ψ-image of a point in $X_{\lambda_0} \times \Pi_{\lambda \in A \setminus \{\lambda_0\}} X^\mu_\lambda$ only depends on the X_{λ_0}-component, we have that $\tilde{x}_{\lambda_0} = x_{\lambda_0}$. Thus $\Psi(x_{\lambda_0}, (x^\mu_\lambda)_{\lambda \in A \setminus \{\lambda_0\}}) = (\Psi_{\lambda_0}(x_{\lambda_0}), v)$ and for $\beta \neq \mu$ the αth component of v is s^0_α while for $\beta = \mu$ the αth component of v is x_α. Call $v =: \Psi_{\neq \lambda_0}((x^\mu_\lambda)_{\lambda \in A \setminus \{\lambda_0\}})$.

Now we essentially repeat the argument only with the factors reversed. We know that for $(x_{\lambda_0}, (x^\mu_\lambda)_{\lambda \in A \setminus \{\lambda_0\}}) \in X_{\lambda_0} \times \Pi_{\lambda \in A \setminus \{\lambda_0\}} X^\mu_\lambda$ we have

$$\Psi(x_{\lambda_0}, (x^\mu_\lambda)_{\lambda \in A \setminus \{\lambda_0\}}) = (\Psi_{\lambda_0}(x_{\lambda_0}), \Psi_{\neq \lambda_0}((x^\mu_\lambda)_{\lambda \in A \setminus \{\lambda_0\}})).$$

By Lemma 10.4.8 the V-component of any $\Psi(x_{\lambda_0}, (x_\lambda)_{\lambda \in A \setminus \{\lambda_0\}})$ only depends on the components $(x_\lambda)_{\lambda \in A \setminus \{\lambda_0\}}$. Call this V-component $\Psi_{\neq \lambda_0}((x_\lambda)_{\lambda \in A \setminus \{\lambda_0\}})$. We have

$$\Psi(x_{\lambda_0}, (x^\mu_\lambda)_{\lambda \in A \setminus \{\lambda_0\}}) = (\Psi_{\lambda_0}(x_{\lambda_0}), \Psi_{\neq \lambda_0}((x^\mu_\lambda)_{\lambda \in A \setminus \{\lambda_0\}}))$$

and

$$\Psi(x_{\lambda_0}, (x_\lambda)_{\lambda \in A \setminus \{\lambda_0\}}) = (u, \Psi_{\neq \lambda_0}((x_\lambda)_{\lambda \in A \setminus \{\lambda_0\}})).$$

Thus by Lemma 10.4.5 we have

$$\Psi^{-1}(\Psi_{\lambda_0}(x_{\lambda_0}), \Psi_{\neq \lambda_0}((x_\lambda)_{\lambda \in A \setminus \{\lambda_0\}})) = (x_{\lambda_0}, (\tilde{x}_\lambda)_{\lambda \in A \setminus \{\lambda_0\}}).$$

Since the V-component of any $\Psi(x_{\lambda_0}, (x_\lambda)_{\lambda \in A \setminus \{\lambda_0\}})$ only depends on the components $(x_\lambda)_{\lambda \in A \setminus \{\lambda_0\}}$ we have $(\tilde{x}_\lambda)_{\lambda \in A \setminus \{\lambda_0\}} = (x_\lambda)_{\lambda \in A \setminus \{\lambda_0\}}$. We have thus proved that for all $(x_{\lambda_0}, (x_\lambda)_{\lambda \in A \setminus \{\lambda_0\}}) \in X_{\lambda_0} \times \Pi_{\lambda \in A \setminus \{\lambda_0\}} X_\lambda$ we have

$$\Psi(x_{\lambda_0}, (x_\lambda)_{\lambda \in A \setminus \{\lambda_0\}}) = (\Psi_{\lambda_0}(x_{\lambda_0}), \Psi_{\neq \lambda_0}((x_\lambda)_{\lambda \in A \setminus \{\lambda_0\}})).$$

This in turn implies that $\Psi((x_\lambda)_{\lambda \in A}) = (\Psi_\lambda(x_\lambda))_{\lambda \in A}$. Note that the above argument applies to both $\gamma^{-1} \circ \Psi_1$ and Ψ_2. Since the λ_0-components of $\gamma^{-1} \circ \Psi_1((s^0, x_{\lambda_0}, \lambda_0))$ equal those of $\Psi_2((s^0, x_{\lambda_0}, \lambda_0))$, all the factor maps for $\gamma^{-1} \circ \Psi_1$ and Ψ_2 are equal. Thus we have proved that $\gamma^{-1} \circ \Psi_1 = \Psi_2$ and the proof of Theorem 10.4.4 is done.   ∎

10.4.1 Automorphisms of Products

With Hashimoto's Theorem proved we turn our attention to automorphisms. Unlike for the fixed point property, there are disconnected ordered sets for which every automorphism has a fixed point. The role of connectivity for the property that every automorphism has a fixed point is investigated in Exercise 20. We will concentrate here on connected sets exclusively.

The first step is to show that automorphisms of products are essentially made up of automorphisms of the factors. To prove this, we have to demand that our sets are decomposable into product indecomposable ordered sets defined as follows.

Definition 10.4.9 *Call an ordered set P* **product indecomposable** *iff for every isomorphism $\Phi : P \rightarrow A \times B$ we have that either A or B must be a singleton.*

This condition may look technical, yet it is needed. Unfortunately, while any two product decompositions of an ordered set have a common refinement, it is not guaranteed that (in infinite ordered sets) the refinement process ever stops. (Also cf. Remark 15 in this chapter.)

Corollary 10.4.10 (Cf. [61], Lemma 2.)

1. *Let $P = \Pi_{\alpha \in A} X_\alpha$ be a connected ordered set such that no two distinct X_α have a common factor. Then for any automorphism $\Phi : P \rightarrow P$ there are automorphisms $\Phi_\alpha : X_\alpha \rightarrow X_\alpha$ such that $\Phi((x_\alpha)_{\alpha \in A}) = (\Phi_\alpha(x_\alpha))_{\alpha \in A}$.*

2. *Let $P = \Pi_{\alpha \in A} X_\alpha$ be a connected ordered set such that all $X_\alpha = X$ and X is product indecomposable. Then for any automorphism $\Phi : P \rightarrow P$ there is a permutation σ of A and there are automorphisms $\Phi_\alpha : X \rightarrow X$ such that $\Phi((x_\alpha)_{\alpha \in A}) = (\Phi_\alpha(x_{\sigma(\alpha)}))_{\alpha \in A}$.*

Proof. In both cases we apply the strict refinement property to the decompositions $\Psi_A := \mathrm{id} : \Pi_{\alpha \in A} X_\alpha \rightarrow \Pi_{\alpha \in A} X_\alpha$ and $\Psi_B := \Phi : \Pi_{\alpha \in A} X_\alpha \rightarrow \Pi_{\beta \in A} X_\beta$. We infer there are decompositions $a_\alpha : X_\alpha \rightarrow \Pi_{\beta \in A} Z_{\alpha,\beta}$ and $b_\beta : X_\beta \rightarrow \Pi_{\alpha \in A} Z_{\alpha,\beta}$, such that $\gamma \circ a \circ \mathrm{id} = b \circ \Phi$.

To prove 1 note that X_α and X_β have no common factors unless $\alpha = \beta$. Thus we have that $Z_{\alpha,\beta}$ is trivial unless $\alpha = \beta$. Hence $\Pi_{\beta \in A} Z_{\alpha,\beta}$ is isomorphic to $Z_{\alpha,\alpha}$ which is isomorphic to $\Pi_{\delta \in A} Z_{\delta,\alpha}$. The isomorphism in each case is the projection on the αth component. Let $\eta_\alpha : \Pi_{\beta \in A} Z_{\alpha,\beta} \rightarrow \Pi_{\delta \in A} Z_{\delta,\alpha}$ be the isomorphism that fixes the αth component. Choose $\Phi_\alpha := b_\alpha^{-1} \circ \eta_\alpha \circ a_\alpha$. Then $\Phi_\alpha : X_\alpha \rightarrow X_\alpha$ is an isomorphism. Moreover $\Phi = b^{-1} \circ \gamma \circ a$, which means Φ is the product of the Φ_α.

To prove 2 note that X is product indecomposable. Thus there must be a permutation σ of A such that for each α only $Z_{\sigma(\alpha),\alpha}$ is nontrivial. Hence $\Pi_{\beta \in A} Z_{\sigma(\alpha),\beta}$ is isomorphic to $Z_{\sigma(\alpha),\alpha}$ which is in turn isomorphic to $\Pi_{\delta \in A} Z_{\delta,\alpha}$. The isomorphism in each case is the projection on the αth or $\sigma(\alpha)^{\mathrm{th}}$ component, respectively. Let $\eta'_\alpha : \Pi_{\beta \in A} Z_{\sigma(\alpha),\beta} \rightarrow \Pi_{\delta \in A} Z_{\delta,\alpha}$ be the isomorphism that maps the αth component to the $\sigma(\alpha)^{\mathrm{th}}$ component. Choose $\Phi_\alpha := b_\alpha^{-1} \circ \eta'_\alpha \circ a_{\sigma(\alpha)}$. ∎

Fixed points of automorphisms of sets as in part 1 of Corollary 10.4.10 are easy to construct. For sets as in part 2 the following lemma shows that fixed points also can be constructed.

Lemma 10.4.11 (Cf. [61], Lemma 3.) *Let $P = \Pi_{\alpha \in A} X_\alpha$ be a connected ordered set such that all $X_\alpha = X$ and X is product indecomposable. If each automorphism of X has a fixed point, then each automorphism of P has a fixed point.*

Proof. Let Φ be an automorphism of P. As in part 2 of Corollary 10.4.10 let σ be a permutation of A and let $\Phi_\alpha : X \to X$ be automorphisms of X such that $\Phi((x_\alpha)_{\alpha \in A}) = (\Phi_\alpha(x_{\sigma(\alpha)}))_{\alpha \in A}$. We construct a fixed point of Φ componentwise.

For $\alpha \in A$ call $\{\sigma^k(\alpha) : k \in \mathbb{Z}\}$ the *orbit* of α. A has two different types of elements, those with finite orbits and those with infinite orbits. Constructing a fixed point will have to take into account that with the representation as above Φ maps every component x_α of $(x_\alpha)_{\alpha \in A}$ to the $\sigma^{-1}(\alpha)^{\text{th}}$ component.

If the orbit of $\alpha \in A$ has size n, let $y_\alpha \in X_\alpha$ be a fixed point of $\Phi_\alpha \circ \Phi_{\sigma(\alpha)} \circ \Phi_{\sigma^2(\alpha)} \circ \cdots \circ \Phi_{\sigma^{n-1}(\alpha)}$. Let $y_{\sigma^k(\alpha)} := \Phi_{\sigma^k(\alpha)} \circ \cdots \circ \Phi_{\sigma^{n-1}(\alpha)}(y_\alpha)$. Then for all $p \in P$ with $p_{\sigma^k(\alpha)} = y_{\sigma^k(\alpha)}$ for all $k \in \{0, \ldots, n-1\}$ we have for all $k \in \{0, \ldots, n-1\}$,

$$
\begin{aligned}
\pi_{\sigma^k(\alpha)} \Phi(p) &= \Phi_{\sigma^k(\alpha)}(\pi_{\sigma^{k+1}(\alpha)}(p)) = \Phi_{\sigma^k(\alpha)}(y_{\sigma^{k+1}(\alpha)}) \\
&= \Phi_{\sigma^k(\alpha)}(\Phi_{\sigma^{k+1}(\alpha)} \circ \cdots \circ \Phi_{\sigma^{n-1}(\alpha)}(y_\alpha)) \\
&= y_{\sigma^k(\alpha)} = \pi_{\sigma^k(\alpha)}(p).
\end{aligned}
$$

If the orbit of α is infinite, we argue as follows. Fix $y_\alpha \in X_\alpha$. Inductively (in both directions) we define $y_{\sigma^k(\alpha)} := \begin{cases} \Phi^{-1}_{\sigma^{k-1}(\alpha)}(y_{\sigma^{k-1}(\alpha)}) & \text{if } k > 0, \\ \Phi_{\sigma^k(\alpha)}(y_{\sigma^{k+1}(\alpha)}) & \text{if } k < 0. \end{cases}$ Then for all $p \in P$ with $p_{\sigma^k(\alpha)} = y_{\sigma^k(\alpha)}$ for all $k \in \mathbb{Z}$ we have

$$
\begin{aligned}
\pi_{\sigma^k(\alpha)} \Phi(p) &= \Phi_{\sigma^k(\alpha)}(\pi_{\sigma^{k+1}(\alpha)}(p)) \\
&= \Phi_{\sigma^k(\alpha)}(y_{\sigma^{k+1}(\alpha)}) \\
&= \begin{cases} \Phi_{\sigma^k(\alpha)}(\Phi^{-1}_{\sigma^k(\alpha)}(y_{\sigma^k(\alpha)})) & \text{if } k \geq 0, \\ \Phi_{\sigma^k(\alpha)}(\Phi_{\sigma^{k+1}(\alpha)}(y_{\sigma^{k+2}(\alpha)})) & \text{if } k < 0, \end{cases} \\
&= \begin{cases} y_{\sigma^k(\alpha)} & \text{if } k \geq 0, \\ \Phi_{\sigma^k(\alpha)}(y_{\sigma^{k+1}(\alpha)}) & \text{if } k < 0, \end{cases} \\
&= y_{\sigma^k(\alpha)} \\
&= \pi_{\sigma^k(\alpha)}(p).
\end{aligned}
$$

From the above it is easy to construct a fixed point. ∎

The above now combines into an analogue of Theorem 10.2.11 for fixed points of automorphisms.

Theorem 10.4.12 (Cf. [61], Theorem 4.) *Let X and Y be connected ordered sets, each with a decomposition as product indecomposable sets. Then $X \times Y$ has a fixed point free automorphism iff at least one of X and Y has a fixed point free automorphism.*

Proof. The direction "$\Leftarrow$" is trivial. For "$\Rightarrow$" we prove the counterpositive. So assume all automorphisms of X and Y have a fixed point and let Φ be an automorphism of $X \times Y$. Since X and Y are decomposable into product indecomposable sets, there also is a decomposition $X \times Y$ that is isomorphic to $\prod_{\alpha \in A} Z_\alpha^{k_\alpha}$, with the Z_α being product indecomposable, pairwise non-isomorphic and the k_α being cardinals. By Corollary 10.4.10, part 1, Φ is the product of automorphisms Φ_α of $Z_\alpha^{k_\alpha}$.

Since all automorphisms of X and Y have a fixed point, all automorphisms of the Z_α have a fixed point. By Lemma 10.4.11 this means that all automorphisms of the powers $Z_\alpha^{k_\alpha}$ have a fixed point. In particular each Φ_α has a fixed point, which means that Φ has a fixed point. ∎

10.5   Arithmetic of Ordered Sets

So far, we have seen several binary operations that take two ordered sets and return a new set. Among them are the disjoint sum $P+Q$, the product $P \times Q$ and the exponentiation $P^Q = \mathrm{Hom}(Q, P)$. Interestingly enough, there are many analogies with regular algebraic operations. Consider that the n-fold sum of P with itself is isomorphic to the product of P with an antichain with n elements. This means that our notion of product really is an abbreviation of repeated summation. Similarly the exponentiation operation is an abbreviation of repeated multiplication. Indeed, the n-fold product of P with itself is isomorphic to the set of order-preserving maps from an antichain with n elements to P. Thus with isomorphism taking the place of equality we can define an arithmetic on ordered sets.

In this section we shall investigate how far the arithmetic properties of these operations on ordered sets can go. For the purposes of an arithmetic one does not distinguish between isomorphic sets here. Yet it seemed worth pointing out which laws are equalities and which are isomorphisms. Our starting point are the "usual laws of algebra" for the operations mentioned above.

Proposition 10.5.1 (Arithmetic of ordered sets, cf. [17], p. 55, ff.) *With $\approx$ denoting isomorphism of ordered sets the following are true.*

$$X + Y = Y + X \qquad (X + Y) + Z = X + (Y + Z)$$
$$X \times Y \approx Y \times X \qquad (X \times Y) \times Z \approx X \times (Y \times Z)$$
$$X \times (Y + Z) = (X \times Y) + (X \times Z) \qquad (X + Y) \times Z = (X \times Z) + (Y \times Z)$$
$$X^{Y+Z} \approx X^Y \times X^Z \qquad (X \times Y)^Z \approx X^Z \times Y^Z$$
$$(X^Y)^Z \approx X^{Y \times Z}$$

Proof. Commutativity and associativity of sum and product are trivial.

Because of the commutativity of the product, we only need to prove one distributive law. Clearly the underlying sets of $X \times (Y + Z)$ and $(X \times Y) + (X \times Z)$ are equal. Now consider (x_1, a_1) and (x_2, a_2) with $x_i \in X$ and $a_i \in Y \cup Z$. Then $(x_1, a_1) \leq_{X \times (Y+Z)} (x_2, a_2)$ iff $x_1 \leq_X x_2$ and $a_1 \leq_{Y+Z} a_2$ iff $x_1 \leq_X x_2$ and $a_1, a_2 \in A \in \{Y, Z\}$ and $a_1 \leq_A a_2$ iff $a_1, a_2 \in A \in \{Y, Z\}$ and $(x_1, a_1) \leq_{X \times A} (x_2, a_2)$ iff $(x_1, a_1) \leq_{X \times Y + X \times Z} (x_2, a_2)$. This shows that the two underlying orders are equal, too.

The remaining proofs are very similar to the above, so we will only give the isomorphisms, leaving the details to the reader.

For $X^{Y+Z} \approx X^Y \times X^Z$ the isomorphism is $f \mapsto (f|_X, f|_Y)$. Its inverse is $(f, g) \mapsto f \cup g$.

The identity $(X \times Y)^Z \approx X^Z \times Y^Z$ is left as Exercise 15.

Finally for $(X^Y)^Z \approx X^{Y \times Z}$ the isomorphism is $f \mapsto g(y, z) := f(z)(y)$ with inverse $g \mapsto f(z)(y) := g(y, z)$.   ∎

The search for neutral elements for our arithmetic operations is simple, too. The empty set is the neutral element for addition and the singleton ordered set is the neutral element for multiplication. The singleton ordered set also has all properties that the number 1 has for exponentiation.

Whenever we use finite sets with at least two elements, all our operations always return larger sets. (In Exercise 10 we show that this need not be the case for infinite ordered sets.) Thus it is impossible to associate a group structure with ordered sets in this fashion. With inverse elements out of the question, the next natural candidates for laws of algebra are cancelation laws. The reader will prove cancelation laws for sums (including linear lexicographic sums) in Exercise 17. For products we have the following result. Its main scope for our purposes are finite ordered sets, but it shows that some infinite ordered sets also follow a cancelation law. The reader will prove in Exercise 21 the result for finite ordered sets with the connectivity hypotheses dropped.

Theorem 10.5.2 *Let P, Q, R be connected ordered sets that can be factored into product indecomposable factors. Moreover, let R be factorable into finitely many indecomposable factors. If $P \times R$ is isomorphic to $Q \times R$, then P is isomorphic to Q.*

Proof. Factor both products into product indecomposable ordered sets. Say the factorization of $P \times R$ is $\Pi_{i \in I} X_i$ and the factorization of $Q \times R$ is $\Pi_{j \in J} Y_j$. By Hashimoto's theorem we can assume that $I = J$ and that X_i is isomorphic to Y_i.

Let $i_1, \ldots, i_n$ be the indices of the factors of R in $\Pi_{i \in I} X_i$. Then the product $\Pi_{i \in I \setminus \{i_1, \ldots, i_n\}} X_i$ is isomorphic to P. Let $i'_1, \ldots, i'_n$ be the indices of the factors of R in $\Pi_{i \in I} Y_i$. Then the set $\Pi_{i \in I \setminus \{i'_1, \ldots, i'_n\}} Y_i$ is isomorphic to Q. By Hashimoto's theorem again, we can assume without loss of generality that X_{i_k} is isomorphic to $Y_{i'_k}$ for $k \in \{1, \ldots, n\}$. Moreover, after further renumbering, we can assume that if $i_j = i'_{j'}$, then $j = j'$.

This means that the map $f : I \setminus \{i_1, \ldots, i_n\} \to I \setminus \{i'_1, \ldots, i'_n\}$ defined by

$$f(i) := \begin{cases} i; & \text{if } i \notin \{i_1, \ldots, i_n\} \cup \{i'_1, \ldots, i'_n\}, \\ i_k; & \text{if } i = i'_k \notin \{i_1, \ldots, i_n\}, \end{cases}$$ is a bijection between the sets

$I \setminus \{i_1, \ldots, i_n\}$ and $I \setminus \{i'_1, \ldots, i'_n\}$ such that X_i is isomorphic to $Y_{f(i)}$. This shows that P and Q are isomorphic.   ∎

The non-existence of an infinite factorization is essential as the following example shows. This means that while our ordered set arithmetic has no problems

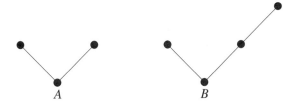

Figure 10.3: Ordered sets A and B that show that the cancelation problem has a negative solution for infinite exponents even if the bases are finite.

accommodating sums and products with arbitrarily many factors, it reaches its limitations in the cancelation laws.

Example 10.5.3 The condition that R is not factorable into infinitely many factors in Theorem 10.5.2 is needed. Let P, Q be two ordered sets and let A_ω be a countable antichain. Then $(Q \times P) \times P^{A_\omega}$ is isomorphic to $Q \times P^{A_\omega}$, but $Q \times P$ is not isomorphic to Q for any finite P, Q with at least two elements.

This leaves the question if our exponentiation operation is "injective", leading to a long-standing (in excess of 40 years) and recently solved problem by Birkhoff.

> **Formerly Open Question 10.5.4** (Birkhoff's cancelation problem. Recently solved by McKenzie in [179].) *Let P, Q, R be (finite) ordered sets. Does the fact that P^Q is isomorphic to R^Q imply that P is isomorphic to R?*

While the answer is negative if the sets are allowed to be infinite (cf. Examples 10.5.5 and 10.5.6) it was the finite case that baffled researchers for decades. We will be able to present some examples and a partial result here. The proof for all finite sets is left for the reader to consider in [179].

Example 10.5.5 (Cf. [131], Example 4, p.21.) *Let C be a four crown and let $A := L\{C_i | i \in \mathbb{N}\}$, where all C_i are pairwise disjoint four crowns and $\mathbb{N}$ carries the discrete order, that is, it is an antichain. Let $B := A + S$, where S is a singleton ordered set. Then A^C is isomorphic to B^C and A is not isomorphic to B.*

To see this, note that both A^C and B^C have countably many isolated points. Indeed, every map that maps C surjectively to any of the C_i is not comparable to any other maps. Let H be the set of these maps. Moreover, for B^C the map f that maps C into S is also not comparable to any other maps. The isomorphism is now made up of the identity between $A^C \setminus H$ and $B^C \setminus H \cup \{f\}$ and any bijection between H and $H \cup \{f\}$. ∎

Example 10.5.6 (Cf. [131], Example 5, p.21.) *Let A, B be the ordered sets in Figure 10.3. Let C be the direct product of infinitely many copies of the two element chain* **2**. *Then A^C is isomorphic to B^C, but A is not isomorphic to B.*

To see this note that every map from C to A (resp. B) must map C into one arm of A (resp. B), since the range must have a largest element. Thus A^C is obtained from two copies of 2^C with their bottom elements identified and B^C is obtained from a copy of 2^C and 3^C with their bottom elements identified (**3** naturally denotes the 3-element chain). But **3** is isomorphic to 2^2 and thus by Proposition 10.5.1 we have that 3^C is isomorphic to $2^{2 \times C}$ which naturally is isomorphic to 2^C. ∎

Thus the question that remained was if it was possible to cancel the exponent if all the involved sets were finite. Special cases were proved over the years, such as for example that arbitrary linearly ordered exponents could be canceled (cf. [16]). The general proof however remained elusive. All the partial results required that hypotheses had to be imposed on the exponent and/or the bases. Consider for example the following simple, yet fairly strong cancelation result.

Lemma 10.5.7 *Let D be an ordered set with an up-retraction r. Then for all ordered sets P we have that $r : P^D \to P^D$ defined by $R(f) := f \circ r$ is an up-retraction. Moreover $R[P^D]$ is isomorphic to $P^{r[D]}$.*

Proof. It is clear that R is an up-retraction. To see the "moreover" part, note that $f \mapsto f \circ r$ is a bijective map from $P^{r[D]}$ to $R[P^D]$ that preserves order both ways. Thus it is by Proposition 1.3.5 an isomorphism. ∎

Proposition 10.5.8 *Let P, D, R be finite ordered sets such that the set D is $\mathcal{I}$-dismantlable to a singleton and let P^D be isomorphic to R^D. Then $\mathcal{I} - \text{core}(P)$ is isomorphic to $\mathcal{I} - \text{core}(R)$.*

Proof. Recall that by Proposition 4.3.4 the $\mathcal{I}$-core and the $\mathcal{C}$-core of a finite ordered set are equal. By Lemma 10.5.7 we have that P^D can be $\mathcal{C}$-dismantled to a subset that is isomorphic to P. Thus P^D can be $\mathcal{C}$-dismantled to a subset that is isomorphic to $\mathcal{I} - \text{core}(P)$. By the uniqueness of the $\mathcal{I}$-core, we infer that $\mathcal{I} - \text{core}(P^D)$ is isomorphic to $\mathcal{I} - \text{core}(P)$. Similarly $\mathcal{I} - \text{core}(R^D)$ is isomorphic to $\mathcal{I} - \text{core}(R)$, and the isomorphic sets P^D and R^D naturally must have isomorphic $\mathcal{I}$-cores. ∎

Thus the above shows that if the bases have no irreducible points and the exponent is $\mathcal{I}$-dismantlable to a singleton, then the exponent can be canceled. This is a nice application of notions we introduced earlier, but not the solution of the cancelation problem. We will only state the solution here, as the proof is beyond the scope of this text.

Theorem 10.5.9 (For the proof, cf. [179].) *Let P, Q, R be finite ordered sets. If $P^R \approx Q^R$, then $P \approx Q$.*

Exercises

1. Prove that if L is a lattice (complete lattice, distributive lattice), then the ordered set $\mathrm{Hom}(Q, L)$ is a lattice (complete lattice, distributive lattice).

2. Let P be a finite ordered set. Prove that for every fixed point free order-preserving map $f : P \to P$ there is a fixed point free map $h : P \to P$ that maps minimal elements to minimal elements and maximal elements to maximal elements and such that $\mathrm{dist}_{\mathrm{End}(P)}(f, h) \leq 2$.

3. Show that for each $n \in \mathbb{N}$ there are connected ordered sets P and Q such that $h(P) > |Q| + n$ and yet P^Q is not connected.

4. Let P be a finite ordered set. Prove that $D(P)$ is isomorphic to $\mathrm{Hom}(P, \mathbf{2})$, where $\mathbf{2}$ is a 2-element chain.

5. Show that the Dedekind–MacNeille completion of a product can be a proper subset of the product of the Dedekind–MacNeille completions of the factors.

6. Let P, Q be ordered sets, let $\leq_{P \times Q}$ be the order of the product $P \times Q$ and let $\leq_{L(P,Q)}$ be the order of the lexicographic sum $L\{P_q | q \in Q\}$ with all P_q isomorphic to P. Prove that $\leq_{P \times Q} \subseteq \leq_{L(P,Q)}$.

7. Let P be a chain-complete ordered set and let Q be an ordered set (not necessarily chain-complete). Prove that if $f : P \times Q \to P \times Q$ is order-preserving and $f(p, q) \geq (p, q)$, then f has a fixed point.

8. (K. A. Baker, unpublished, cf. [102, 135].) Prove that forming the Dedekind–MacNeille completion does not change the dimension, that is, $\dim(P) = \dim(DM(P))$. (Hint. Combine Theorems 5.3.8 and 10.2.4.)

9. Let P be a chain-complete ordered set and let Q be an ordered set with the fixed point property. Prove that if P is $\mathcal{C}$-dismantlable to C and $C \times Q$ has the fixed point property, then $P \times Q$ has the fixed point property.

10. Prove that $\mathcal{P}(\mathbb{N}) \times \mathcal{P}(\mathbb{N})$ is isomorphic to $\mathcal{P}(\mathbb{N})$.

11. It would be nice if we were able to further extend Dilworth's theorem to (maybe) say that an ordered set without infinite antichains must be a union of countably many chains. Unfortunately this is not true.

    Let α be an infinite limit ordinal and consider the product $\alpha \times \alpha$. Prove that:

    (a) $\alpha \times \alpha$ has no infinite antichains.
    (b) $\alpha \times \alpha$ is not the union of fewer than α chains.

    (This particularly simple example was first observed in [193], also cf. [180].)

12. Prove that the product of two connected ordered sets is again connected.

13. Prove Lemma 10.2.2, which says that every factor X_λ of a product can be "spliced into any element" s^0 to form a retract.

14. Prove that any product of chain-complete ordered sets is chain-complete.

15. Prove that $(X \times Y)^Z \approx X^Z \times Y^Z$.

16. Show that in general $X \oplus Y \not\approx Y \oplus X$.

17. Let X, Y, Z be ordered sets. Prove that

    (a) If Z has finitely many components and $X + Z \approx Y + Z$, then $X \approx Y$.

    (b) If the complement of the comparability graph of Z has finitely many components and $X \oplus Z \approx Y \oplus Z$, then $X \approx Y$.

    (c) Give examples that show 17a and 17b are false if the respective finiteness condition is dropped.

    (d) Give an example to show that if $Z \oplus X \approx Y \oplus Z$, then it is not necessary that $X \approx Y$.

18. Prove that if $\text{Hom}(P, P) = \text{Hom}(Q, Q)$, then $P = Q$ or $P = Q^d$.

19. Let C be a connected ordered set and let $\emptyset \neq A \subseteq C$. Prove that if A is an up-set and a down-set, then $A = C$.

20. (Cf. [61], introduction.) To investigate Open Question 1.4.8 for products of disconnected ordered sets, let $n \cdot X$ be an ordered set with n components isomorphic to X.

    (a) Prove that $n_1 \cdot X_1 + n_2 \cdot X_2 + \cdots + n_k \cdot X_k$ with $X_1, \ldots, X_k$ pairwise nonisomorphic has a fixed point free automorphism iff for all i with $n_i = 1$ the set X_i has a fixed point free automorphism.

    (b) Find two ordered sets P, Q such that all automorphisms of P and Q have a fixed point, but $P \times Q$ has a fixed point free automorphism.

    (c) Find an ordered set P such that all automorphisms of P have a fixed point, but $P \times P$ has a fixed point free automorphism.

    (d) Prove that if P is finite with at most four components and all automorphisms of P have a fixed point, then all automorphisms of P^2 have a fixed point.

    (e) Can 20d be improved?

    (f) Prove that if P is finite, connected and all automorphisms of P have a fixed point, then for all finite Q such that each automorphism of Q has a fixed point, every automorphism of $P \times Q$ has a fixed point.

(g) Prove that if P is finite with two components, then there is a finite Q such that each automorphism of Q has a fixed point, and $P \times Q$ has a fixed point free automorphism.

(h) What is the smallest total number of components such that there are finite P, Q such that each automorphism of P and Q has a fixed point, and $P \times Q$ has a fixed point free automorphism?

21. (Cf. [131], Theorem 5.4.) Let P, Q, R be finite ordered sets. Prove that if $P \times R$ is isomorphic to $Q \times R$, then P is isomorphic to Q.

Hint: If no two connected components of R have the same number of elements, this can be proved via Theorem 10.5.2. For the general case, identify the indecomposable factors of the components of P, Q and R with variables. Identify P, Q and R with multivariable polynomials in these variables with integer coefficients and use the cancelation property in the ring of multivariable polynomials.

22. (There is no Hashimoto theorem for disconnected ordered sets, cf. [108, 132].)

(a) Find two factorizations of $P = 1 + 2 + 2^2 + 2^3 + 2^4 + 2^5$ such that no two factors are isomorphic and no factor can be represented nontrivially as a product.

(b) Find a characterization when a finite ordered set has a unique product decomposition. Hint. Identify each indecomposable factor of a component with a variable in a multivariate polynomial. Consider the factorizations of the polynomial.

23. Let $\Phi : P \times Q \to U \times V$, $\Phi_P : P \to U$ and $\Phi_Q : Q \to V$ be isomorphisms such that for a fixed $(x, y) \in P \times Q$ the following is true.

(a) For all $p \in P$ we have $\Phi(p, y) = (\Phi_P(p), \Phi_Q(y))$.

(b) For all $q \in Q$ we have $\Phi(x, q) = (\Phi_P(x), \Phi_Q(q))$.

Prove that $\Phi(p, q) = (\Phi_P(p), \Phi_Q(q))$ for all $(p, q) \in P \times Q$. (Hint. Lemmas 10.4.5 and 10.4.8.)

24. (More on f-chains. This problem is a precursor to Exercise 25.) Let P be a chain-complete ordered set and let $f : P \to P$ be order-preserving. For $x \in P$, if $f(x) \geq x$ $(f(x) \leq x)$ let C_x^f be the (dual) maximal f-chain that starts at x.

(a) Prove that for $x \in P$, $n \in \mathbb{N}$ if $f(x) \geq x$ we have $\bigvee C_x^f = \bigvee C_{f^n(x)}^f$. (Hint: Lemma 3.4.6.)

(b) Let $g : P \to P$ be order-preserving with $f \leq g$ and let $p \leq q$. Prove the following statements.

1) If $f(p) \geq p$ and $f(q) \geq q$, then $\bigvee C_p^f \leq \bigvee C_q^f$.

2) If $f(p) \geq p$ and $f(q) \leq q$, then $\bigvee C_p^f \leq \bigwedge C_q^f$.

3) If $f(p) \leq p$ and $f(q) \geq q$, then $\bigwedge C_p^f \leq \bigvee C_q^f$.

4) If $f(p) \geq p$ and $g(p) \geq p$, then $\bigvee C_p^f \leq \bigvee C_p^g$.

5) If $f(p) \leq p$ and $g(p) \geq p$, then $\bigwedge C_p^f \leq \bigvee C_p^g$.

(Hint: Lemma 3.4.6.)

(c) Prove that if $f^k(p) \geq p$ and $l \in \mathbb{N}$, then $f^{lk}(p) \geq p$ and $\bigvee C_p^{f^k} = \bigvee C_p^{f^{lk}}$.

25. Let P be an ordered set. We define $\mathrm{Retr}(P)$ to be the ordered subset of $\mathrm{End}(P)$ that consists of all the retractions on P.

   (a) Let P be a chain-complete ordered set that does not contain any infinite antichains. Prove that there is a retraction $R : \mathrm{End}(P) \to \mathrm{Retr}(P)$ such that for each $f \in \mathrm{End}(P)$ we have

$$R(f)[P] = \{p \in P : (\exists n \in \mathbb{N}) f^n(p) = p\}.$$

   Hint. For each p there is a $k \in \mathbb{N}$ with $p \sim f^k(p)$; also use Exercise 24.

   (b) Let L be a complete lattice. Prove that $\mathrm{Retr}(L)$ is a retract of $\mathrm{End}(L)$. Hint. Generalize Exercise 24 using completeness of the lattice.

26. Use the proof for Exercise 25 to show that for chain-complete ordered sets without infinite antichains we can replace order preserving factor maps with maps that satisfy the hypotheses of Theorem 10.2.9. This provides an alternative proof for Theorem 10.2.11 that does not necessitate the Li–Milner theorem.

27. (Motivated by a question in [46]. Also cf. [44], [45] or Exercise 25b for the proof that for complete lattices L the set $\mathrm{Retr}(L)$ is a complete lattice and [158] for an example that the retraction set of a lattice need not form a lattice.) Let L be a lattice. Prove that if $\mathrm{Retr}(L)$ is a lattice, then L must be a complete lattice.

   Hint. By Proposition 5.1.7 it is enough to prove L is chain-complete (why?). Suppose not. Find without loss of generality a well-ordered chain C without a supremum. Partition it into two cofinal subsets C_1, C_2 so that the C-successor of each element in C_i is in the respective other set. Retract onto $C_1 \cup \uparrow C_1$ via

$$r_1(x) := \begin{cases} x; & \text{if } x \geq C_1 \text{ or } x \in C_1, \\ \min_{C_1}\{c \in C_1 : c \not< x\}; & \text{if } x > c \text{ for some } c \in C_1, \\ \min(C_1); & \text{otherwise,} \end{cases}$$

   and similarly with r_2 onto $C_2 \cup \uparrow C_2$. These two retractions do not have a supremum, contradiction.

Remarks and Open Problems

1. For more structure theory for ordered sets, cf. [67]. For more on the arithmetic, cf. [131].

2. As opposed to previous chapters, treatment of the reconstruction problem was conspicuously absent in this chapter. This is because little is known about the reconstruction problems for products and homomorphism sets. It is not even known if products and homomorphism sets are recognizable from the deck.

3. For what finite *connected* ordered sets P, Q does P^Q have the fixed point property? The author is not aware of any progress on this question. The next open question emphasizes how little there is known.

4. Do the ordered sets in Example 10.1.6 have the fixed point property when P does?

5. Is there an analogue of Theorem 10.2.4 for other notions of dimension, e.g., interval dimension?

6. For refinements of Theorem 10.2.4, cf. Chapter 10 in [263].

7. Is it true that for a finite ordered set P with the fixed point property and A a countable antichain, the set $P^A = \Pi_{n \in \mathbb{N}} P_n$ (with all $P_n = P$) has the fixed point property?

   The author was unable to prove the above even for any specific set with the fixed point property that is not C-dismantlable to a singleton.

   Proposition 10.3.8 in conjunction with Theorem 10.2.11 shows that a natural avenue to *dis*prove the fixed point property is closed. All finite retracts will have the fixed point property. Moreover Lemma 10.4.11 shows that all automorphisms of this set have a fixed point. This does not seem to exclude the existence of an automorphic retract, though it would have to be infinite.

   Nonetheless, the author conjectures that sets as above have the fixed point property.

8. Is it true that every finite subset of an ordered set is actually contained in a finite retract? If this is true, then any finite retract of a product of two ordered sets with the fixed point property must also have the fixed point property.

9. Note that the topological fixed point property is not productive (cf. [33]). Is there a way to get two infinite ordered sets P, Q with the fixed point property such that $P \times Q$ does not have the fixed point property through topology? One possible way could be through infinite truncated lattices that arise via infinite triangulations of the manifolds involved in the topological counterexamples.

10. On the other hand, Roddy (cf. [229]) recently proved that if P, Q are ordered sets with the fixed point property and P is of width ≤ 3 (no further hypotheses), then $P \times Q$ has the fixed point property. Is this a start towards proving the fixed point property is "productive"?

11. One crucial challenge in trying to prove that the fixed point property is productive for infinite ordered sets is that we know very little about infinite ordered sets with the fixed point property. Indeed, the ordered set $\Pi_{n \in \mathbb{N}} P_n$ as in Remark 10.3.6 as well as Problem 7 were motivated by the desire to find nontrivial infinite ordered sets that have the fixed point property. It appears that all examples of infinite ordered sets with the fixed point property have the fixed point property for some "finitary reason". That is, the proof of the fixed point property focuses on some finitary structure or procedure.

Thus, finding infinite ordered sets that have the fixed point property for some "non-finitary" reason is an interesting challenge.

12. A property closely related to the fixed point property for products is Duffus and Sauer's **strong fixed point property** (cf. [71]). The idea for the strong fixed point property was first used at the end of [270] to prove that the product of an ordered set with the relational fixed point property with a set with the fixed point property again has the fixed point property.

An ordered set has the strong fixed point property iff there is an order-preserving function $\varphi : \mathrm{End}(P) \to P$ such that $f(\varphi(f)) = \varphi(f)$. That is, φ selects a fixed point for each order-preserving self map and it does so in an order-preserving fashion. In [71] it is proved that the product of two ordered sets with the strong fixed point property again has the strong fixed point property. Moreover the product of a set with the strong fixed point property and a set with the fixed point property again has the fixed point property.

It was hoped that the fixed point property and the strong fixed point property might be equivalent, thus settling the product problem. Unfortunately this is not the case as is shown in [194].

The above implies that if there is an infinite counterexample to the product conjecture, then one of the factors must have the fixed point property but not the strong fixed point property. The observation that such sets were hard to find in the finite case is an indication how difficult an infinite counterexample to the product conjecture (if it exists) may be.

13. Another approach towards the product problem was through projectivity as introduced in [43]. An ordered set is called **projective** iff the only order-preserving maps $f : P \times P \to P$ such that $f(x, x) = x$ for all $x \in P$ are the projections. In [43] it is shown that *if* all minimal automorphic finite ordered sets were projective (which is conjectured in [43]), then the product of two finite ordered sets with the fixed point property again has the fixed

point property. A conjecture stronger than projectivity of minimal automor-phic sets was refuted in [109]. The author does not know of a satisfactory generalization of the approach in [43] to fixed points in products of infinite ordered sets, and Roddy of course solved the product problem in the finite case. Nonetheless, as for example [110, 157] show, projectivity has become a property of independent interest.

In a way, questions related to projectivity stand representative of a large algebraic part of the theory of ordered sets that is beyond the scope of this text. The main ideas of representation theory go back to [67], where the rep-resentation theory is developed and some interesting questions are posed. The author cannot comment on which of the problems in [67] are solved and what the exact status of this part of the theory is. For some later devel-opments cf., for example, [280, 281].

14. According to [190], the **distortion** of an ordered set P is defined to be $D(P) := \max_{f \in \text{End}(P)} \min_{p \in P} \text{dist}(p, f(p))$. Nowakowski conjectures that $D(P \times Q) = \max\{D(P), D(Q)\}$. This conjecture has the product conjecture for the fixed point property as a special case. How much can be proved about the distortion parameter? The only work known to the author is in [112].

15. Every finite ordered set is a product of product-irreducible ordered sets. (Simply decompose until this is no longer possible.) However, not every ordered set is a product of product-irreducible ordered sets.

    Consider the following counterexample. An atom in a lattice is an upper cover of the smallest element. There is up to isomorphism exactly one countable Boolean algebra C that has no atoms. This is actually a pretty deep result. It depends on the Stone representation theorem for Boolean al-gebras and results leading up to it (for example, cf. [49], Theorem 10.8 and Propositions 10.6, 10.7), Urysohn's metrization theorem (cf. [275], Theo-rem 23.1) and on the fact that every perfect, compact, totally disconnected metric space is homeomorphic to the Cantor set (cf. [275], Corollary 30.4) together with Example 10.21(3) of [49].

    It is easy to show that any factor of C must also be atomless and thus in particular countable. Moreover, every factor of a Boolean algebra is again a Boolean algebra. Thus every factor of C is a countable atomless Boolean algebra and hence isomorphic to C. Finally C is not product-irreducible, since $C \times C$ is isomorphic to C (since $C \times C$, too is a countable atomless Boolean algebra).

16. In Exercise 25 we have considered two classes of ordered sets (chain-complete without infinite antichains and complete lattices) for which the fixed point property is productive. For both these classes we proved that the set $\text{Retr}(P)$ is a retract of $\text{End}(P)$. By Exercise 26 part of the proof can be used to prove Theorem 10.2.11 without using the Li–Milner theorem.

Is there a deeper connection between $\text{Retr}(P)$ being a retract of $\text{End}(P)$ and productivity of the fixed point property? Specifically, if P, Q are ordered sets with the fixed point property and $\text{Retr}(P)$ is a retract of $\text{End}(P)$, does it follow that $P \times Q$ has the fixed point property?

17. For what classes of ordered sets P is $\text{Retr}(P)$ a retract of $\text{End}(P)$?

18. Let P, Q, R be ordered sets. If $P \times R$ is embeddable into $Q \times R$, is P embeddable into Q?

19. There also is a cancelation law for bases. It says that if A, B, C are finite ordered sets and A is not an antichain, then $A^B \approx A^C$ implies $B \approx C$. For a proof cf. [66]. Clearly it is needed that A is not an antichain, since otherwise A^B is merely an $|A|c(B)$-element antichain, where $c(B)$ denotes the number of components of B.

For more on the arithmetic of ordered sets, cf. [131].

11

Enumeration

One of the most natural questions to ask in any discrete setting is "How many of these objects are there?" Dedekind's problem (Open Question 2.6.1) is such a question. A counting question can be motivated by pure curiosity or, as the Kelly Lemma (cf. Proposition 1.5.14) in reconstruction shows, it can be asked as a step in proving something else. The two most natural counting questions for ordered sets are still unanswered.

> **Open Question 11.0.1** *Given an n-element ground set, how many different orders can be imposed on the set?*

Closely related to the question how many orders there are for a finite n-element set is

> **Open Question 11.0.2** *Given an n-element ground set, how many non-isomorphic orders can be imposed on the set?*

In this chapter we will investigate some techniques that are motivated by the above questions. This will lead us eventually to another beautiful open problem in Section 11.5. Our approach towards enumeration will feature two natural steps that are taken when an enumeration problem appears too hard to solve exactly for the time being. These steps are the enumeration of subclasses (cf. Section 11.2) and the construction of expressions that are asymptotically equal to the unknown numbers (cf. Section 11.3). Tools that are needed will be introduced on the spot.

It should be said here that the selection of topics still only gives a glimpse of the vast topic of enumeration.

11.1   Graded Ordered Sets

A class of ordered sets that can be enumerated exactly and also asymptotically is the class of graded ordered sets. We will work extensively with this class in this chapter. This section shall serve as an introduction.

Definition 11.1.1 *A **graded ordered set** is an ordered set P for which there is an order-preserving function $g_P : P \to \mathbb{N}$ such that if p is a lower cover of q in P, then $g_P(q) - g_P(p) = 1$.*

*To have a common reference point we will always assume that every connected component of P contains a point b such that $g_P(b) = 1$ (normalization condition). We will refer to g_P as the **grading function** and to the sets $g_P^{-1}(k) \subseteq P$ as the **grade levels** of P.*

Note that the rank function also is order-preserving from P into $\mathbb{N}$. It is tempting to hope that every rank function is a grading function. Figure 2.1 shows that this is not the case. The set in part b) of Figure 2.1 is not graded. (The reader can prove this directly, or by first solving Exercise 2.) Knowing this, it is still tempting to assume that if an ordered set is graded, then the rank function plus 1 will also be a grading function. As the reader will show in Exercise 3c, even this is not the case. These examples should serve as a caution that it is easy to read too much or too little into an ordered set when it comes to grading functions. One nice result is that with the normalization as above, if a grading function exists, it must be unique.

Proposition 11.1.2 *Every ordered set P has at most one grading function $g_P : P \to \mathbb{N}$.*

Proof. Suppose P has two functions g_P and g'_P as in Definition 11.1.1. If $g_P \neq g'_P$, then there is a $p \in P$ such that $g_P(p) \neq g'_P(p)$. By Definition 11.1.1, p cannot be an isolated point. Let $b \in P$ be such that $g_P(b) = 1$ and such that p and b are in the same connected component of P. Then there is a sequence $b = p_1, \ldots, p_k = p$ such that p_{i+1} is a lower or an upper cover of p_i for all $i \in \{1, \ldots, k-1\}$. Let u be the number of times p_{i+1} is an upper cover of p_i and let l be the number of times p_{i+1} is a lower cover of p_i. By definition of g_P and g'_P we have that

$$u - l = \sum_{i=1}^{k-1} [g_P(p_{i+1}) - g_P(p_i)] = g_P(p_k) - g_P(p_1) = g_P(p) - g_P(b)$$

$$\neq \; g'_P(p) - g'_P(b) = g'_P(p_k) - g'_P(p_1) = \sum_{i=1}^{k-1} [g'_P(p_{i+1}) - g'_P(p_i)]$$

$$= u - l,$$

a contradiction. Thus we must have $g_P = g'_P$. ∎

11.2 The Number of Graded Ordered Sets

Throughout, we will use the set $\{1, \ldots, n\}$ as the ground set that carries all the orders involved. To abbreviate notation we will denote the set $\{1, \ldots, n\}$ as $[n]$.

Definition 11.2.1 We define $G_{n,k}$ as the set of graded ordered sets with ground set $[n]$ and up to k grade levels. Also let $g_{n,k} := |G_{n,k}|$.

Note that by the above definition $g_{n,1} = 1$, $g_{n,k} \le g_{n,k+1}$ for all k and $g_{n,k} = g_{n,n}$ for $k \ge n$. $g_{n,n}$ is the number of all graded ordered sets with ground set $[n]$. The natural way to build a graded ordered set is to assign each point a grade level and then to insert adjacencies only between points in adjacent grade levels. While this process can generate the same set in different ways, the number of sets generated this way will be important in the future.

Definition 11.2.2 Define $C_{n,k}$ as the set of $(k+1)$-tuples $(X_1, \ldots, X_k, \le)$ obtained as follows. $X_1, \ldots, X_k$ are k disjoint, not necessarily nonempty subsets of $[n]$ such that $\bigcup_{i=1}^{k} X_k = [n]$. $\le$ is an order on $[n]$ such that the only adjacencies for $\le$ are between elements of X_j and X_{j+1} for $j \in \{1, \ldots, k-1\}$. We let $c_{n,k} := |C_{n,k}|$ be the size of $C_{n,k}$.

Interestingly enough it is now possible to compute the $c_{n,k}$ directly and to obtain from them the numbers $g_{n,k}$. This is because every set in $C_{n,k}$ can be envisioned as built in two ways. On one hand the straightforward way described above and on the other hand as a certain type of union of graded sets as described in the proof below.

Lemma 11.2.3

$$c_{n,k} = \sum_{\substack{n_1 + n_2 + \cdots + n_k = n \\ n_i \ge 0}} \binom{n}{n_1, \ldots, n_k} 2^{n_1 n_2 + n_2 n_3 + \cdots + n_{k-1} n_k}$$

$$= \sum_{\substack{s_1 + s_2 + \cdots + s_k = n \\ s_i \ge 0}} \binom{n}{s_1, \ldots, s_k} g_{s_1,1} g_{s_2,2} \cdots g_{s_k,k},$$

where $\binom{n}{n_1, \ldots, n_k} = \dfrac{n!}{n_1! n_2! \cdots n_k!}$ is the multinomial coefficient.

Proof. The idea for this proof is to prove separately that either sum is equal to $c_{n,k}$. To prove the first sum is equal to $c_{n,k}$, we proceed as follows. Note that the sizes $n_1, \ldots, n_k$ of the sets $X_1, \ldots, X_k$ in the partition of $[n]$ are nonnegative numbers that must add up to n. Hence $C_{n,k}$ is the disjoint union

$$C_{n,k} = \bigcup_{\substack{n_1 + n_2 + \cdots + n_k = n \\ n_i \geq 0}} \{(X_1, \ldots, X_k, \leq) : (\forall 1 \leq j \leq k)|X_j| = n_j\}.$$

This means we are done if we can find the sizes of the sets in the above union. For any given sequence of sizes $n_1, \ldots, n_k$ the multinomial coefficient $\begin{pmatrix} n \\ n_1, \ldots, n_k \end{pmatrix}$ gives the number of ways to assign the elements of $[n]$ to disjoint sets $X_1, \ldots, X_k$ with $|X_j| = n_j$. For every such partition of $[n]$ there are $2^{n_{j-1}n_j}$ ways to insert comparabilities between X_{j-1} and X_j. Since all combinations are possible we have

$$|\{(X_1, \ldots, X_k, \leq) : (\forall 1 \leq j \leq k)|X_j| = n_j\}|$$

$$= \begin{pmatrix} n \\ n_1, \ldots, n_k \end{pmatrix} 2^{n_1 n_2 + n_2 n_3 + \cdots + n_{k-1} n_k},$$

which proves the first equality.

This leaves us to prove the second equality. For each $(X_1, \ldots, X_k, \leq)$ in $C_{n,k}$ partition $[n]$ as follows. Let S_k be the set of all numbers m that are connected in $([n], \leq)$ to an element of X_1. Let $S'_j := [n] \setminus \bigcup_{i=j+1}^{k} S_i$. Let S_j be the set of all numbers m that are connected to an element of X_{k-j+1} in $\left(S'_j, \leq |_{S'_j \times S'_j}\right)$. Then each $(S_j, \leq |_{S_j \times S_j})$ is a graded ordered set with at most j grade levels and ground set S_j. This means that $C_{n,k}$ is the disjoint union

$$C_{n,k} = \bigcup_{\substack{s_1 + s_2 + \cdots + s_k = n \\ s_i \geq 0}} \left\{(X_1, \ldots, X_k, \leq) : [n] = \bigcup_{j=1}^{k} S_j; \right.$$

$$|S_i| = s_i; \ S_i \cap S_j = \emptyset \text{ for } i \neq j; \ S_j \subseteq \bigcup_{i=k-j+1}^{k} X_i;$$

if $S_j \neq \emptyset$, then $S_j \cap X_{k-j+1} \neq \emptyset$;

$\leq |_{S_j \times S_j}$ is a graded order with at most j grade levels;

$$X_i = \bigcup_{j=1}^{i} \{m \in S_j : g_{S_j}(m) = i - j + 1\}\bigg\}.$$

So to prove the second equality we again need to compute the sizes of the sets within the disjoint union. For any given sequence of sizes $s_1, \ldots, s_k$ there are

$\left(\begin{array}{c} n \\ s_1, \ldots, s_k \end{array} \right)$ ways to assign the elements of $[n]$ to disjoint sets $S_1, \ldots, S_k$ with $|S_j| = s_j$. For each S_j there are $g_{s_j, j}$ ways to order S_j in the desired fashion. This proves the second equality. ∎

Lemma 11.2.3 essentially gives us an infinite set of nonlinear equations that allows us to compute the $g_{n,k}$ from the $c_{n,k}$. How does one solve such a system of equations? While there are no techniques that work in general, in our case the algebra of formal power series will allow us to solve the problem. The central definition for this type of argument is that of a generating function.

Definition 11.2.4 *Let* $\{a_n\}_{n=0}^{\infty}$ *be a sequence of numbers. Then the generating function of* $\{a_n\}_{n=0}^{\infty}$ *is the formal power series* $\displaystyle\sum_{n=0}^{\infty} \frac{a_n}{n!} x^n$.

Essentially the generating function is a formal power series (the radius of convergence can be and often is zero) such that the nth formal derivative at $x = 0$ is a_n. If the equations one tries to solve have the right type of pattern, the generating function can become a very powerful tool as we will see shortly. First another lemma.

Lemma 11.2.5 *If* $T(x) = \displaystyle\sum_{n=0}^{\infty} t_n x^n$ *and* $t_0 = 1$, *then*

$$\frac{1}{T(x)} = 1 + \sum_{n=1}^{\infty} \left[\sum_{\substack{n_1 + n_2 + \cdots + n_l = n \\ l > 0, \ n_i > 0}} (-1)^l t_{n_1} t_{n_2} \cdots t_{n_l} \right] x^n.$$

Proof. Clearly the zeroeth coefficient of the formal product of $T(x)$ and $\dfrac{1}{T(X)}$ is 1. For $n \geq 1$ we obtain that the nth coefficient is

$$t_n + \sum_{k=0}^{n-1} t_k \left[\sum_{\substack{n_1 + n_2 + \cdots + n_l = n - k \\ l > 0, \ n_i > 0}} (-1)^l t_{n_1} t_{n_2} \cdots t_{n_l} \right]$$

$$= \ t_n + \sum_{\substack{n_1 + n_2 + \cdots + n_l = n \\ l > 0, \ n_i > 0}} (-1)^l t_{n_1} t_{n_2} \cdots t_{n_l}$$

$$+ \sum_{\substack{n_1 + n_2 + \cdots + n_l + n_{l+1} = n \\ l > 0, \ n_i > 0}} (-1)^l t_{n_1} t_{n_2} \cdots t_{n_l} t_{n_{l+1}} = 0. \qquad \blacksquare$$

We now define the following generating functions.

Definition 11.2.6 *Let*

$$C_k(x) \quad := \quad \sum_{n=0}^{\infty} \frac{c_{n,k}}{n!} x^n,$$

$$G_k(x) \quad := \quad \sum_{n=0}^{\infty} \frac{g_{n,k}}{n!} x^n,$$

$$D_k(x) \quad := \quad \frac{1}{C_k(x)} =: \sum_{n=0}^{\infty} \frac{d_{n,k}}{n!} x^n.$$

With these generating functions we can finally prove enumeration formulas for graded sets. The number of graded ordered sets with n elements is obtained as $g_{n,n}$.

Theorem 11.2.7 *For $n, k \geq 2$,*

$$d_{n,k} \quad = \quad \sum_{\substack{n_1 + n_2 + \cdots + n_l = n \\ l > 0, \, n_i > 0}} (-1)^l \binom{n}{n_1, \ldots, n_l} c_{n_1,k} c_{n_2,k} \cdots c_{n_l,k},$$

$$g_{n,k} \quad = \quad \sum_{j=0}^{n} \binom{n}{j} c_{j,k} d_{n-j,k-1}.$$

Proof. By Lemma 11.2.3 we have $C_k(x) = G_1(x) G_2(x) \cdots G_k(x)$ (simply notice that in a product of formal power series the nth coefficient is the sum of the products of all coefficients whose powers add up to n). This means that $G_1(x) = C_1(x)$ and for $k > 1$ we have

$$G_k(x) = \frac{C_k(x)}{C_{k-1}(x)} = C_k(x) \frac{1}{C_{k-1}(x)}.$$

The formula for the $d_{n,k}$ is a direct consequence of Lemma 11.2.5. The formula for the $g_{n,k}$ follows from the formula for the nth coefficient of a product of formal power series.   ∎

11.3   The Asymptotic Number of Graded Ordered Sets

A natural question to ask if no exact formula for a certain quantity (such as the number of ordered sets with n elements) is available is how large that quantity is for large values of a given parameter (here the size n). This question can also be asked if a precise, but complex, formula (such as Theorem 11.2.7) is available.

We are thus entering the subject of asymptotic enumeration. In this subject one does not try to find an exact value for a given quantity, but a value to which it is asymptotically equal.

Definition 11.3.1 *Two sequences $\{a_n\}_{n\in\mathbb{N}}$ and $\{b_n\}_{n\in\mathbb{N}}$ of real numbers are called* **asymptotically equal** *(in symbols $a_n \sim_{\text{as}} b_n$) iff $\displaystyle\lim_{n\to\infty} \frac{a_n}{b_n} = 1$.*

The asymptotic value of the number of ordered sets with ground set $[n]$ is known (cf. Theorem 11.3.13). A standard approach to finding asymptotic values for the size of a class of objects is to first establish an asymptotic value for a large subclass and then show that all other subclasses are small in comparison. To quantify what large and small mean, we define

Definition 11.3.2 *Let $\{a_n\}_{n\in\mathbb{N}}$ and $\{b_n\}_{n\in\mathbb{N}}$ be sequences of real numbers. Then $a_n = O(b_n)$ (read "a_n is big-oh of b_n") iff $\displaystyle\limsup_{n\to\infty} \frac{a_n}{b_n} < \infty$.*

The above means that if $a_n = O(b_n)$, then a_n grows at most at a rate comparable to b_n. For example, any sequence that is $O(\log_2(n))$ is eventually much smaller than n. This means that in some scenarios an object of size $O(\log_2(n))$ can be ignored. This approach is taken to prove Theorem 11.3.13 in [142, 143]. However this proof is very technical[1]. Thus we will merely exemplify the described approach by proving an asymptotic formula for the number of graded ordered sets in Theorem 11.3.3. Asymptotically, the graded ordered sets are the largest class of ordered sets. This means that the gap we leave is that we will not show that the class of nongraded ordered sets is small in comparison. The reader is referred to [50, 142, 143] for the proof of Theorem 11.3.13 and also to the Open Questions 6 and 7 at the end of this chapter.

Theorem 11.3.3 *The number of graded ordered sets with ground set $[n]$ is asymptotically equal to $2^{\frac{n^2}{4}+\frac{3}{2}n+O(\sqrt{n})}$.*

As the reader can see, the estimate that we will prove is not exact. The reason is that the author did not believe the technical details needed for the exact bound would illuminate the subject any more. We will spend the rest of this section proving Theorem 11.3.3 in a series of lemmas. All of these lemmas, except for the first one, are of an analytical nature. This is not surprising, as we are dealing with an analytical notion.

Convention 11.3.4 *For the remainder of this section, inequalities that do not hold for all n are inequalities that hold for large n (i.e., for all n beyond a certain N_0). Also, divisions will be performed regardless of whether the result is an integer or not. The small perturbations effected by rounding to integer values have no influence on the asymptotic results. Thus to keep the presentation uncluttered, these parts of the computation are suppressed.*

---

[1] Indeed one of the authors of [142, 143] considers it "horrible".

In this fashion notation will be a little more economical. The following lemma is the main tool to achieve the estimate. The upper bound derived here will generically be called "the upper bound" in what follows.

Lemma 11.3.5 *The number of graded ordered sets with exactly k nonempty grade levels is bounded below by*

$$\sum_{\substack{n_1+n_2+\cdots+n_k=n \\ n_i>0}} \binom{n}{n_1,\ldots,n_k} (2^{n_2}-n)^{n_1}(2^{n_3}-n)^{n_2}\cdots(2^{n_k}-n)^{n_{k-1}}$$

and it is bounded above by

$$\sum_{\substack{n_1+n_2+\cdots+n_k=n \\ n_i>0}} \binom{n}{n_1,\ldots,n_k} 2^{n_1 n_2+n_2 n_3+\cdots n_{k-1} n_k}.$$

Proof. To construct all graded ordered sets with exactly k grade levels and ground set $[n]$ we proceed as follows. Partition the ground set into k nonempty levels and then insert covering relations between adjacent levels. This procedure will produce all graded ordered sets with k levels. We will find upper and lower bounds on the number of structures that can be produced in this fashion.

First note that $\binom{n}{n_1,\ldots,n_k} = \dfrac{n!}{n_1!\cdots n_k!}$ is the number of possible distributions of n elements into k "levels" each having n_i elements. With the sums running over all possible choices $n_1,\ldots,n_k$ we cover all possible distributions of n elements into k bins so that no bin is empty. To verify the upper bound, notice that there can be at most $2^{n_i n_{i+1}}$ connections between the ith and the $(i+1)$st "level". (The upper bound counts a few orders more than once. For example, some disconnected graded ordered sets can be generated in several ways here.)

The lower bound is less than the number of graded orders such that in the ith level $(i < k)$ each element has at least one upper cover in the $(i+1)^{\text{st}}$ level and such that no two elements have the same upper covers. Indeed, when building such a set, for the jth element of level i for which we choose a set of upper bounds, we have $2^{n_{i+1}} - 1 - (j-1) \geq 2^{n_{i+1}} - n$ choices for a set of upper covers. Thus there are at least $(2^{n_{i+1}}-n)^{n_i}$ possible ways to connect the ith and the $(i+1)^{\text{st}}$ level. Since no two elements on the same level have the same upper covers and all elements are below an element on level k, no order is doubly counted. This establishes the lower bound.

Note that some summands will even turn out negative for the lower bound. Since we are looking for a lower bound this is of no consequence.   ∎

Lemma 11.3.6 *The number of graded ordered sets with three grade levels is asymptotically bounded above by the quantity $2^{\frac{n^2}{4}+\frac{3}{2}n+O(\sqrt{n})}$ and it is bounded below by the quantity $2^{\frac{n^2}{4}+\frac{3}{2}n-O(\log_2(n))}$.*

Proof. For a lower bound notice that in the sum from Lemma 11.3.5, the term for $n_1 = \dfrac{n}{4}$, $n_2 = \dfrac{n}{2}$, and $n_3 = \dfrac{n}{4}$ is very large.

$$
\begin{pmatrix} n \\ \frac{n}{4}, \frac{n}{2}, \frac{n}{4} \end{pmatrix} (2^{\frac{n}{4}} - n)^{\frac{n}{2}} (2^{\frac{n}{2}} - n)^{\frac{n}{4}} \underset{\text{as}}{\sim} \begin{pmatrix} n \\ \frac{n}{2} \end{pmatrix} \begin{pmatrix} \frac{n}{2} \\ \frac{n}{4} \end{pmatrix} 2^{\frac{n}{4}\frac{n}{2} + \frac{n}{2}\frac{n}{4}}
$$

$$
\geq \; 2^{n - O(\log_2(n))} 2^{\frac{n}{2} - O(\log_2(n))} 2^{\frac{n^2}{4}} = 2^{\frac{n^2}{4} + \frac{3}{2}n - O(\log_2(n))}.
$$

The first asymptotic equality is because $\left(1 - \dfrac{n}{2^{\frac{n}{4}}}\right)^{\frac{n}{2}} \to 1$ as $n \to \infty$ and

$\left(1 - \dfrac{n}{2^{\frac{n}{2}}}\right)^{\frac{n}{4}} \to 1$ as $n \to \infty$. The inequality holds because $n\begin{pmatrix} n \\ \frac{n}{2} \end{pmatrix} \geq 2^n$.

To obtain the upper bound, we argue as follows. For $n_1 + n_2 + n_3 = n$ we have that $n_1 n_2 + n_2 n_3 = n_2(n - n_2)$. For the part of the upper bound in Lemma 11.3.5 in which $\left|\dfrac{n}{2} - n_2\right| \geq \sqrt{n}$ this is maximized for $n_2 = \dfrac{n}{2} \pm \sqrt{n}$. Thus, in this part of the sum we can bound the power of 2 with a term that is independent of the summation. Now notice (cf. Exercise 7) the sum of multinomial coefficients is

$$
\sum_{n_1 + \cdots + n_k = n, n_i \geq 0} \begin{pmatrix} n \\ n_1, n_2, \ldots, n_k \end{pmatrix} = k^n.
$$

$$
\sum_{\substack{n_2 \leq \frac{n}{2} - \sqrt{n} \text{ or} \\ n_2 \geq \frac{n}{2} + \sqrt{n} \\ n_1 + n_2 + n_3 = n}} \begin{pmatrix} n \\ n_1, n_2, n_3 \end{pmatrix} 2^{n_2(n - n_2)}
$$

$$
\leq \; 3^n 2^{\frac{n^2}{4} - n} = 2^{\frac{n^2}{4} + n(\log_2 3 - 1)} \leq 2^{\frac{n^2}{4} + n}.
$$

The main part of the sum is estimated as follows.

$$
\sum_{\substack{\frac{n}{2} - \sqrt{n} \leq n_2 \leq \frac{n}{2} + \sqrt{n} \\ n_1 + n_2 + n_3 = n}} \begin{pmatrix} n \\ n_1, n_2, n_3 \end{pmatrix} 2^{n_1 n_2 + n_2 n_3}
$$

$$
= \sum_{\substack{\frac{n}{2} - \sqrt{n} \leq n_2 \leq \frac{n}{2} + \sqrt{n} \\ n_1 + n_2 + n_3 = n}} \begin{pmatrix} n \\ n_2 \end{pmatrix} \begin{pmatrix} n - n_2 \\ n_1 \end{pmatrix} 2^{n_2(n_1 + n_3)}
$$

$$
\leq \; n 2^n 2^{\frac{n}{2} + \sqrt{n}} 2^{\frac{n^2}{4}} = 2^{\frac{n^2}{4} + \frac{3}{2}n + \sqrt{n} + \log_2(n)}.
$$

∎

We now know (almost) the asymptotic size of what turns out to be the largest class of graded ordered sets. The remainder of the proof of Theorem 11.3.3 consists of a series of arguments that show that all other classes of graded ordered sets are of negligible size compared to this class. Thus in the following "negligible" will mean "negligible in comparison to the class of graded sets with three grade levels".

Lemma 11.3.7 *The numbers of all graded ordered sets with one or two grade levels are negligible for large n.*

Proof. There is only one graded ordered set with one grade level. For the graded ordered sets with two levels notice that

$$\sum_{n_1=1}^{n-1} \binom{n}{n_1} 2^{n_1(n-n_1)} \leq 2^{\frac{n^2}{4}+n}.$$

■

Lemma 11.3.8 *The number of graded ordered sets with four grade levels is negligible for large n.*

Proof. First we estimate the parts of the upper bound with $n_1 \geq n^{\frac{9}{10}}$ or (by symmetry) $n_4 \geq n^{\frac{9}{10}}$.

$$\sum_{\substack{n_1+\cdots+n_4=n \\ n_i > 0; \\ n_1 \geq n^{\frac{9}{10}}}} \binom{n}{n_1, n_2, n_3, n_4} 2^{n_1 n_2 + n_2 n_3 + n_3 n_4}$$

$$= \sum_{\substack{n_1+\cdots+n_4=n \\ n_i > 0; \\ n_1 \geq n^{\frac{9}{10}}}} \binom{n}{n_1, n_2 + n_4, n_3} \binom{n_2 + n_4}{n_4}$$

$$\times 2^{n_1(n_2+n_4)+(n_2+n_4)n_3} 2^{-n_1 n_4}$$

$$\leq \sum_{\substack{n_1+n_2+n_3=n \\ n_i > 0; \\ n_1 \geq n^{\frac{9}{10}}}} \binom{n}{n_1, n_2, n_3} 2^{n_1 n_2 + n_2 n_3} \sum_{n_4=1}^{n} \binom{n}{n_4} (2^{-n_1})^{n_4}$$

$$\leq 2^{\frac{n^2}{4}+\frac{3}{2}n+O(\sqrt{n})} \left(\left(1 + 2^{-n^{\frac{9}{10}}}\right)^n - 1 \right)$$

$$= 2^{\frac{n^2}{4}+\frac{3}{2}n+O(\sqrt{n})} \left(\left(\left(1 + \frac{1}{2^{n^{\frac{9}{10}}}}\right)^{2^{n^{\frac{9}{10}}}} \right)^{\frac{n}{2^{n^{\frac{9}{10}}}}} - 1 \right)$$

$$\leq \quad 2^{\frac{n^2}{4}+\frac{3}{2}n+O(\sqrt{n})} \left(e^{\left(\frac{n}{2n^{\frac{9}{10}}}\right)} - 1 \right)$$

$$\leq \quad 2^{\frac{n^2}{4}+\frac{3}{2}n+O(\sqrt{n})} \frac{2n}{2n^{\frac{9}{10}}},$$

where the last estimate shows that the sum is negligible compared to the number of graded orders with three levels. For $n_1, n_4 \leq n^{\frac{9}{10}}$ we have

$$\binom{n}{n_1, n_2, n_3, n_4} \leq \binom{n}{n_1}\binom{n}{n_4}\binom{n}{n_2} \leq n^{n^{\frac{9}{10}}} n^{n^{\frac{9}{10}}} 2^n = 2^{n+2n^{\frac{9}{10}}\log_2(n)}.$$

Therefore we are done because the remaining sum in the upper bound is at most $2^{\frac{n^2}{4}+n+2n^{\frac{9}{10}}\log_2(n)}$. ∎

The above arguments clearly show one of the great benefits of asymptotic enumeration. Since the estimates only have to hold for large enough n, one can sometimes use extremely crude inequalities and still achieve the result.[2] As the reader also can see in the previous proofs, the term $2^{n_1 n_2 + n_2 n_3 + \cdots + n_{k-1} n_k}$ in the estimate of Lemma 11.3.5 plays a dominating role. We shall thus investigate this term in more detail. The maximization of the exponent subject to the constraints $\sum_{i=1}^{k} n_i = n$ and $n_i \geq 1$ can be an epic exercise in the use of Lagrange multipliers. Yet there is also a quicker combinatorial way to obtain estimates. The following lemma shows that the more clustered the elements in the levels are, the larger an exponent we have in our upper bound estimate.

Lemma 11.3.9 *Let* $k \geq 4$, $F(x_1, \ldots, x_k) := \sum_{i=1}^{k-1} x_i x_{i+1}$ *and* $n_1, \ldots, n_k \in \mathbb{N}$. *Then there is a permutation* $\sigma : [k] \to [k]$ *and a* $j \in \{2, \ldots, k-1\}$ *such that:*

1. *For* $1 \leq i \leq j-1$ *we have* $n_{\sigma(i)} \leq n_{\sigma(i+1)}$.

2. *For* $j \leq i \leq k-1$ *we have* $n_{\sigma(i)} \geq n_{\sigma(i+1)}$.

3. $F(n_1, \ldots, n_k) \leq F(n_{\sigma(1)}, \ldots, n_{\sigma(k)})$.

4. *For* $k \geq 4$ *we have*

$$F(n_{\sigma(1)}, \ldots, n_{\sigma(k)})$$
$$\leq \quad F\Big(1, \ldots, 1, n_{\sigma(j-1)} + n_{\sigma(j+1)} - 1,$$
$$n_{\sigma(j)} + \sum_{i \neq j-1, j, j+1} (n_{\sigma(i)} - 1), 1, \ldots, 1\Big),$$

---

[2] This benefit can come back to haunt the researcher however, if someone is interested in the rate of convergence. We shall not address this issue in this text.

where the term with $n_{\sigma(j)}$ occurs at the earliest in the third position and at the latest in the $(k-1)^{\text{st}}$ position.

Proof. Let σ be a permutation such that part 3 holds and such that the value $F(n_{\sigma(1)}, \ldots, n_{\sigma(k)}) \geq F(n_1, \ldots, n_k)$ is maximal. To prove parts 1 and 2 we shall make a sequence of assumptions that each lead to a contradiction.

First assume there is a $j \in \{2, \ldots, k-1\}$ such that $n_{\sigma(j)} < n_{\sigma(i)}$ for all i. Then with

$$\tau = (\tau(1), \ldots, \tau(k)) := (\sigma(j), \sigma(1), \ldots, \sigma(j-1), \sigma(j+1), \ldots, \sigma(k))$$

we have

$$
\begin{aligned}
F(n_{\sigma(1)}, & \ldots, n_{\sigma(k)}) \\
&= \sum_{i=1}^{k-1} n_{\sigma(i)} n_{\sigma(i+1)} \\
&= n_{\sigma(j-1)} n_{\sigma(j)} + n_{\sigma(j)} n_{\sigma(j+1)} - n_{\sigma(1)} n_{\sigma(j)} - n_{\sigma(j+1)} n_{\sigma(j-1)} \\
&\quad + \sum_{i \neq j-1, j} n_{\sigma(i)} n_{\sigma(i+1)} + n_{\sigma(1)} n_{\sigma(j)} + n_{\sigma(j+1)} n_{\sigma(j-1)} \\
&= n_{\sigma(j-1)} (n_{\sigma(j)} - n_{\sigma(j+1)}) + n_{\sigma(j)} (n_{\sigma(j+1)} - n_{\sigma(1)}) \\
&\quad + F(n_{\tau(1)}, \ldots, n_{\tau(k)}) \\
&< n_{\sigma(j)} (n_{\sigma(j)} - n_{\sigma(j+1)}) + n_{\sigma(j)} (n_{\sigma(j+1)} - n_{\sigma(j)}) \\
&\quad + F(n_{\tau(1)}, \ldots, n_{\tau(k)}) \\
&= F(n_{\tau(1)}, \ldots, n_{\tau(k)}),
\end{aligned}
$$

contradicting the choice of σ. Thus by maximality of $F(n_{\sigma(1)}, \ldots, n_{\sigma(k)})$ we must have (without loss of generality) $n_{\sigma(1)} \leq n_{\sigma(i)}$ for all i.

Now assume there is a $j \in \{2, \ldots, k-1\}$ such that $n_{\sigma(j-1)} > n_{\sigma(j)} < n_{\sigma(j+1)}$. Find $l \in \{2, \ldots, j-1\}$ such that $n_{\sigma(l-1)} \leq n_{\sigma(j)} \leq n_{\sigma(l)}$. Then with

$$\tau = (\sigma(1), \ldots, \sigma(l-1), \sigma(j), \sigma(l), \ldots, \sigma(j-1), \sigma(j+1), \ldots, \sigma(k))$$

we have

$$
\begin{aligned}
F(n_{\sigma(1)}, & \ldots, n_{\sigma(k)}) \\
&= n_{\sigma(j-1)} n_{\sigma(j)} + n_{\sigma(j)} n_{\sigma(j+1)} + n_{\sigma(l-1)} n_{\sigma(l)} \\
&\quad - n_{\sigma(l-1)} n_{\sigma(j)} - n_{\sigma(l)} n_{\sigma(j)} - n_{\sigma(j+1)} n_{\sigma(j-1)} + F(n_{\tau(1)}, \ldots, n_{\tau(k)}) \\
&= n_{\sigma(j+1)} (n_{\sigma(j)} - n_{\sigma(j-1)}) + n_{\sigma(l)} (n_{\sigma(l-1)} - n_{\sigma(j)}) \\
&\quad + n_{\sigma(j)} (n_{\sigma(j-1)} - n_{\sigma(l-1)}) + F(n_{\tau(1)}, \ldots, n_{\tau(k)}) \\
&< n_{\sigma(j)} (n_{\sigma(j)} - n_{\sigma(j-1)}) + n_{\sigma(j)} (n_{\sigma(l-1)} - n_{\sigma(j)}) \\
&\quad + n_{\sigma(j)} (n_{\sigma(j-1)} - n_{\sigma(l-1)}) + F(n_{\tau(1)}, \ldots, n_{\tau(k)}) \\
&= F(n_{\tau(1)}, \ldots, n_{\tau(k)}),
\end{aligned}
$$

contradicting the choice of σ. Thus by maximality of $F(n_{\sigma(1)}, \ldots, n_{\sigma(k)})$ we must have that $\sigma(1), \ldots, \sigma(j)$ is increasing and $\sigma(j), \ldots, \sigma(k)$ is decreasing for some $j \in [k]$. To finish the proof of 1 and 2 we have to show $j \notin \{1, k\}$.

In case $j = 1$, $n_{\sigma(i)}$ would be decreasing, in case $j = k$, $n_{\sigma(i)}$ would be increasing. Suppose the $n_{\sigma(i)}$ are increasing with $\sigma(2) < \sigma(k)$. Then we have the strict inequality $F(n_{\sigma(2)}, \ldots, n_{\sigma(k)}, n_{\sigma(1)}) > F(n_{\sigma(1)}, \ldots, n_{\sigma(k)})$, contradicting the choice of σ. Similarly we exclude the case in which the $n_{\sigma(i)}$ are decreasing with $\sigma(1) > \sigma(k-1)$. However this means in case $j \in \{1, k\}$ we can simply pick a different $j \in \{2, \ldots, k-1\}$ and 1 and 2 both hold.

For part 4 note that for $i < j - 1$ and $i > j + 1$ we have

$$n_{\sigma(i)}n_{\sigma(i+1)} \le (n_{\sigma(i)} - 1)(n_{\sigma(j+1)} + n_{\sigma(j-1)} - 1) + n_{\sigma(i+1)}.$$

Thus all of these terms that are lost by going from the left side of the inequality to the right side are replaced by corresponding larger terms. The terms $n_{\sigma(j-1)}n_{\sigma(j)} + n_{\sigma(j)}n_{\sigma(j+1)}$ on the left occur as $(n_{\sigma(j-1)} + n_{\sigma(j+1)} - 1)n_{\sigma(j)} + n_{\sigma(j)}$ on the right. The term $n_{\sigma(j+1)}n_{\sigma(j+2)}$ on the left (if it exists) has the upper bound $(n_{\sigma(j-1)} + n_{\sigma(j+1)} - 1)n_{\sigma(j+2)} + n_{\sigma(j+2)}$ on the right. The $(k-2)$ "−1"s that were neglected above are picked up by the $(k-4)$ "+1"s that are generated by multiplying the "+1"s on the sides and by the term $(n_{\sigma(j-1)} + n_{\sigma(j+1)} - 1)$ (if this term is ≥ 2) that is multiplied by the 1 to its left. If $(n_{\sigma(j-1)} + n_{\sigma(j+1)} - 1) < 2$, then all n_i except for $n_{\sigma(j)}$ are 1 and the statement is trivial. ∎

Lemma 11.3.10 *Let $k \ge 4$, let $F(x_1, \ldots, x_k) := \displaystyle\sum_{i=1}^{k-1} x_i x_{i+1}$ and let the numbers $n_1, \ldots, n_k \in \mathbb{N}$ be such that $\displaystyle\sum_{i=1}^{k} n_i = n$. Then:*

1. $F(n_1, \ldots, n_k) \le \dfrac{n^2}{4} + \left(2 - \dfrac{k}{2}\right)n + \left(1 - \dfrac{k}{2}\right)^2 - 2.$

2. *For $k = 6$ and all $n_i \ge 2$ we have $F(n_1, \ldots, n_6) \le \dfrac{n^2}{4} - 2n + 8.$*

3. *For $k = 5$ and all $n_i \ge 2$ we have $F(n_1, \ldots, n_5) \le \dfrac{n^2}{4} - n + 1.$*

Proof. To prove 1 note that

$$
\begin{aligned}
F(n_1, \ldots, n_k) &\le F\left(1, \ldots, 1, \frac{n}{2} - \frac{k}{2} + 1, \frac{n}{2} - \frac{k}{2} + 1, 1, \ldots, 1\right) \\
&= \left(\frac{n}{2} - \frac{k}{2} + 1\right)^2 + n - k + 2 + k - 4 \\
&= \frac{n^2}{4} + \left(2 - \frac{k}{2}\right)n + \left(1 - \frac{k}{2}\right)^2 - 2.
\end{aligned}
$$

The proofs of 2 and 3 are similar, so we only prove 3. Part 2 and some details of part 3 are left to Exercise 9. Note that an argument similar to part 4 of Lemma 11.3.9 gives the first estimate below.

$$
\begin{aligned}
F(n_1, \ldots, n_5) &\leq F\left(2, 2, \frac{n}{2} - 3, \frac{n}{2} - 3, 2\right) \\
&= \left(\frac{n}{2} - 3\right)^2 + 2n - 12 + 4 \\
&= \frac{n^2}{4} - n + 1.
\end{aligned}
$$

∎

Lemma 11.3.11 *The number of graded ordered sets with five or six grade levels is negligible for large n.*

Proof. We only prove the estimate for the number of five leveled graded ordered sets. By part 3 of Lemma 11.3.10 we have that

$$
\sum_{i=1}^{4} n_i n_{i+1} \leq \frac{n^2}{4} - n + 1
$$

if all n_i are at least 2. This implies via the multinomial theorem

$$
\sum_{\substack{n_1 + \cdots + n_5 = n \\ n_i > 1}} \binom{n}{n_1, \ldots, n_5} 2^{n_1 n_2 + \cdots + n_4 n_5}
$$

$$
\leq 5^n 2^{\frac{n^2}{4} - n + 1} \leq 2^{-\frac{n}{10}} 2^{\frac{n^2}{4} + \frac{3}{2}n},
$$

which is negligible. This leaves us to estimate the part of the sum in which one of the n_i is equal to 1. By part 3 of Lemma 11.3.9, this part of the sum is bounded by the sum for which $n_1 = 1$, which is in turn estimated by

$$
\sum_{\substack{n_1 + \cdots + n_5 = n \\ n_i > 0;\, n_1 = 1}} \binom{n}{n_1, n_2, n_3, n_4, n_5} 2^{n_1 n_2 + n_2 n_3 + n_3 n_4 + n_4 n_5}
$$

$$
\leq \sum_{\substack{1 + n_2 + \cdots + n_5 = n \\ n_i > 0}} \binom{n}{1, n_2, n_3, n_4, n_5} 2^{n_2 + n_2 n_3 + n_3 n_4 + n_4 n_5}
$$

$$
\leq n \sum_{\substack{1 + n_2 + \cdots + n_5 = n \\ n_i > 0}} \binom{n}{n_2, n_3 + 1, n_4, n_5} 2^{n_2(n_3 + 1) + (n_3 + 1)n_4 + n_4 n_5}
$$

$$\leq \quad n \sum_{\substack{n_1 + \cdots + n_4 = n \\ n_i > 0}} \binom{n}{n_1, n_2, n_3, n_4} 2^{n_1 n_2 + n_2 n_3 + n_3 n_4}$$

$$\leq \quad n 2^{\frac{n^2}{4} + \frac{3}{2}n + O(\sqrt{n})} \frac{2n}{2^{n^{\frac{9}{10}}}},$$

by the estimates in the proof of Lemma 11.3.8. Again, this is negligible. The estimate for the number of graded ordered sets with six grade levels is obtained in similar fashion. ∎

Lemma 11.3.12 *The number of graded ordered sets with more than six grade levels is negligible for large n.*

Proof. Use Lemma 11.3.10, part 1 to estimate the power of 2 with a term that is independent of the summation. The summation of the multinomial coefficients yields k^n. Thus we estimate the upper bound for the number of graded ordered sets with k levels by $2^{\log_2(k)n + \frac{n^2}{4} + \left(2 - \frac{k}{2}\right)n + \left(1 - \frac{k}{2}\right)^2 - 2}$. For all $k \geq 7$ this quantity is bounded by $2^{-\frac{n}{10}} 2^{\frac{n^2}{4} + \frac{3}{2}n}$ for large enough n. ∎

This concludes the proof of Theorem 11.3.3. We finally state the exact asymptotic values for the number of ordered sets with ground set $[n]$.

Theorem 11.3.13 (For a proof cf. [50, 142, 143].) *Let P_n be the number of ordered sets with ground set $[n]$. For $n \to \infty$ we have for odd n:*

$$P_n^{(\text{odd})} \sim_{\text{as}} \left(\frac{2}{\pi}\right)^{\frac{1}{2}} \left(\sum_{x=-\infty}^{\infty} 2^{-x^2}\right) 2^{\frac{n^2}{4} + \frac{3}{2}n - \frac{1}{2}\log n}$$

and for even n:

$$P_n^{(\text{even})} \sim_{\text{as}} \left(\frac{2}{\pi}\right)^{\frac{1}{2}} \left(\sum_{x=-\infty}^{\infty} 2^{-\left(x^2 - \frac{1}{2}\right)^2}\right) 2^{\frac{n^2}{4} + \frac{3}{2}n - \frac{1}{2}\log n}.$$

11.4   The Number of Nonisomorphic Ordered Sets

In Section 11.3 we found the asymptotic number of all possible graded ordered sets with ground set $[n]$ and stated the asymptotic number for all ordered sets with ground set $[n]$. What we did not take into account was that while two ordered sets might not be equal, they could still be isomorphic. For most purposes, isomorphic structures need not be considered different from each other. Thus it also makes sense to ask for the number of "truly different" (that is, nonisomorphic) ordered sets with n elements.

Unfortunately, counting nonisomorphic structures is often harder than counting all structures. This has led to the approach of first counting all structures and then determining which fraction of them is large enough to contain one representative from each isomorphism class. This fraction is at least $\frac{1}{n!}$ of the number of all structures, since there are $n!$ permutations for an n-element set. The following theorem shows that for ordered sets this fraction is (asymptotically) indeed $\frac{1}{n!}$. We will make extensive use of the set of automorphisms of an ordered set, so we first define

Definition 11.4.1 *Let P be an ordered set. Then* $\mathrm{Aut}(P)$ *denotes the set of all automorphisms of P.*

Theorem 11.4.2 (Cf. [201], Corollary 2.3.) *Let $\mathcal{O}_n$ be the number of ordered sets with ground set $[n]$ and let $[\mathcal{O}]_n$ be the number of nonisomorphic ordered sets with n elements. Then*

$$\frac{\mathcal{O}_n}{n![\mathcal{O}]_n} \to 1 \qquad \text{as } n \to \infty.$$

Proof. This proof is an adaptation of the proof of the Main Lemma in [201] for the case of ordered sets.

First note that $\mathcal{O}_n \leq n![\mathcal{O}]_n$. If we can now find an upper bound for $n![\mathcal{O}]_n$ that grows at the same speed as $\mathcal{O}_n$, then we are done. To do this we first derive another expression for $n![\mathcal{O}]_n$.

Let $\mathcal{P}_n$ be the set of ordered sets with ground set $[n]$ and let $[\mathcal{P}]_n$ be the set of isomorphism classes of ordered sets with ground set $[n]$. Then $|\mathcal{P}_n| = \mathcal{O}_n$ and $|[\mathcal{P}]_n| = [\mathcal{O}]_n$. Let π be a permutation of $[n]$ and let $F(\pi)$ be the number of ordered sets in $\mathcal{P}_n$ for which π is an automorphism. Note that for each ordered set $([n], \leq) \in \mathcal{P}_n$ there are exactly $|\mathrm{Aut}([n], \leq)|$ permutations π such that π is an automorphism for $([n], \leq)$. With σ_n denoting the set of permutations of $[n]$, we compute

$$\sum_{\pi \in \sigma_n} F(\pi) = \sum_{([n], \leq) \in \mathcal{P}_n} |\mathrm{Aut}([n], \leq)|$$

$$= \sum_{[([n], \leq)] \in [\mathcal{P}]_n} \sum_{([n], \leq') \in [([n], \leq)]} |\mathrm{Aut}([n], \leq')|.$$

Now consider $\displaystyle\sum_{([n], \leq') \in [([n], \leq)]} |\mathrm{Aut}([n], \leq')|$. Fix a representative $([n], \leq)$ of $[([n], \leq)]$. Any ordered set $([n], \leq')$ in $[([n], \leq)]$ is obtained from $([n], \leq)$ by permuting the elements of $[n]$ and defining the order $\leq'$ as follows. If σ is a permutation of n elements, then $p \leq' q$ iff $\sigma^{-1}(p) \leq \sigma^{-1}(q)$. There are $n!$ ways to permute the elements of $([n], \leq)$. Sometimes distinct permutations do not lead to distinct orders. Let $([n], \leq')$ be an ordered set obtained from $([n], \leq)$ as indicated above and let $\sigma_1, \ldots, \sigma_m$ be all distinct permutations that can be used to construct $([n], \leq')$ from $([n], \leq)$ as indicated. Then $|\mathrm{Aut}([n], \leq')| \geq m$, since the identity and $\sigma_2 \circ \sigma_1^{-1}, \ldots, \sigma_m \circ \sigma_1^{-1}$ are distinct automorphisms of $([n], \leq')$. Moreover, $|\mathrm{Aut}([n], \leq')| \leq m$, since for each automorphism τ of $([n], \leq')$ we must have that $\tau = \tau \circ \sigma_1 \circ \sigma_1^{-1} = \sigma_k \circ \sigma_1^{-1}$ for some k. Therefore

$$\sum_{([n], \leq') \in [([n], \leq)]} |\mathrm{Aut}([n], \leq')| = n!$$ and we conclude

$$\sum_{\pi \in \sigma_n} F(\pi) = \sum_{[([n], \leq)] \in [\mathcal{P}]_n} \sum_{([n], \leq') \in [([n], \leq)]} |\mathrm{Aut}([n], \leq')|$$

$$= \sum_{[([n], \leq)] \in [\mathcal{P}]_n} n! = n! [\mathcal{O}]_n.$$

To finish our proof we will show that $\displaystyle\sum_{\pi \in \sigma_n} F(\pi) \leq \mathcal{O}_n \left(1 + \frac{1}{2^{\frac{n}{8}}}\right)$ for large enough n. As in Section 11.3, inequalities that do not hold for all n are to be seen as holding for large enough n. Since $F(\mathrm{id}_n) = \mathcal{O}_n$, we need to show that the contribution from the nontrivial permutations is negligible. To do this, we partition the remaining permutations into two sets.

$$X := \left\{ \pi \in \sigma_n \setminus \{\mathrm{id}_n\} : \pi \text{ has more than } \frac{9n}{10} \text{ fixed points} \right\},$$

$$Y := \left\{ \pi \in \sigma_n \setminus \{\mathrm{id}_n\} : \pi \text{ has at most } \frac{9n}{10} \text{ fixed points} \right\}.$$

We start with the permutations in X. Let $\pi \in X$. We find an overestimate for the number of ordered sets on $[n]$ for which π is an automorphism as follows. Let A be a set that contains exactly one element from every nontrivial cycle of π. Let B contain the remaining elements of the nontrivial cycles of π. Let $a := |A|$, $b := |B|$ and note that $a \leq b$.

Let $\leq$ be an order relation on $[n]$ such that π is an automorphism of $([n], \leq)$. Then B and $[n] \setminus B$ are ordered subsets with the induced order relations. Conversely, given order relations on B and on $[n] \setminus B$, we claim there are at most $2^{2|A| \cdot |B|}$ order relations on $[n]$ that induce these orders and that are such that π is an automorphism. Indeed, with restrictions to B and $[n] \setminus B$ as given, there are $2|A||B|$ possible comparabilities between elements of A and B. Thus there are $2^{2|A||B|}$ possible sets of comparabilities between A and B. The comparabilities between elements of $A \cup B$ and the fixed points of π are dictated by π (cf. Exercise 12). Thus $F(\pi)$ is bounded by the number of orders on $[n] \setminus B$ times the number of orders on B times $2^{2|A||B|}$. With the result of Exercise 13 and $f \geq \frac{9n}{10}$ denoting the number of fixed points of π we obtain

$$F(\pi) \leq 2^{\frac{(f+a)^2}{4} + \frac{3}{2}(f+a) + O(\log_2(f+a))} 2^{\frac{b^2}{4} + \frac{3}{2}b + O(\log_2(b))} 2^{2ab}$$

$$\leq 2^{\frac{1}{4}\left(f + \left(\frac{a+b}{2}\right)\right)^2 + \frac{3}{2}n + \frac{1}{4}\left(\frac{a+b}{2}\right)^2 + 2\left(\frac{a+b}{2}\right)^2 + O(\log_2(b)) + O(\log_2(f+a))}.$$

Now if π has at least $\dfrac{9n}{10}$ fixed points, then π is such that $a + b \le \dfrac{1}{10}n$. Moreover $a + b \ge 2$, since the smallest possible cycle has two elements. Finally there are $\dbinom{n}{a+b}$ ways to choose the $a + b$ points that are not fixed and there are $(a + b)!$ ways to permute them. Thus with $i := a + b$ we obtain

$$
\begin{aligned}
\sum_{\pi \in X} F(\pi) \quad &\le \quad \sum_{i=2}^{\lfloor \frac{n}{10} \rfloor} \binom{n}{i} i! 2^{\frac{1}{4}\left(n-\frac{i}{2}\right)^2 + \frac{3}{2}n + \frac{9}{4}\left(\frac{i}{2}\right)^2 + O(\log_2(n))} \\
&\le \quad \sum_{i=2}^{\lfloor \frac{n}{10} \rfloor} 2^{i \log_2(n) + \frac{1}{4}n^2 + \frac{3}{2}n - \frac{1}{2}n\frac{i}{2} + \frac{5}{2}\left(\frac{i}{2}\right)^2 + O(\log_2(n))} \\
&\le \quad \sum_{i=2}^{\lfloor \frac{n}{10} \rfloor} 2^{\frac{1}{4}n^2 + \frac{3}{2}n + \frac{i}{2}\left(2\log_2(n) + \frac{5}{4}i - \frac{1}{2}n\right) + O(\log_2(n))} \\
&\le \quad 2^{\log_2\left(\frac{n}{10}\right)} 2^{\frac{1}{4}n^2 + \frac{3}{2}n + \frac{n}{20}\left(2\log_2(n) + \frac{1}{8}n - \frac{1}{2}n\right) + O(\log_2(n))} \\
&\le \quad \mathcal{O}_n 2^{-\frac{1}{80}n^2} \le \mathcal{O}_n 2^{-\frac{1}{4}n}.
\end{aligned}
$$

This leaves us with the estimation for Y. Let $\pi \in Y$. Choose A, B to be sets of points that are not fixed by π such that $|B| = \left\lfloor \dfrac{n}{20} \right\rfloor \le \dfrac{n}{20}$ and such that for each $x \in B$ there is exactly one element $y \in A$ such that y is in the same cycle of π as x. Then $|A| \le |B|$ and $[n] \setminus (A \cup B)$ also contains points that are not fixed by π.

Nonetheless the number of orders such that π is an automorphism can be bounded similar to what was done above. Again let $a := |A|$ and $b := |B|$. There are $2^{\frac{1}{4}(n-b)^2 + \frac{3}{2}(n-b) + O(\log_2(n-b))}$ ways to order $[n] \setminus B$, $2^{\frac{1}{4}b^2 + \frac{3}{2}b + O(\log_2(b))}$ ways to order B, and at most 2^{2ab} possible ways to impose orders between points of A and points of B. By choice of A and B the remaining orders are dictated by the fact that π is supposed to be an automorphism (cf. Exercise 12). Thus

$$
\begin{aligned}
F(\pi) \quad &\le \quad 2^{\frac{(n-b)^2}{4} + \frac{3}{2}(n-b) + O(\log_2(n-b))} 2^{\frac{b^2}{4} + \frac{3}{2}b + O(\log_2(b))} 2^{2ab} \\
&\le \quad 2^{\frac{n^2}{4} + \frac{3}{2}n - \frac{1}{2}nb + \frac{b^2}{4} - \frac{3}{2}b + \frac{b^2}{4} + \frac{3}{2}b + 2b^2 + O(\log_2(n-b)) + O(\log_2(b))} \\
&= \quad 2^{\frac{n^2}{4} + \frac{3}{2}n - \frac{1}{2}nb + \frac{5}{2}b^2 + O(\log_2(n))} \\
&= \quad 2^{\frac{n^2}{4} + \frac{3}{2}n + \frac{1}{2}b(5b-n) + O(\log_2(n))} \\
&\le \quad 2^{\frac{n^2}{4} + \frac{3}{2}n + \frac{1}{2}\frac{n}{20}\left(\frac{n}{4} - n\right) + O(\log_2(n))} \\
&= \quad 2^{\frac{n^2}{4} + \frac{3}{2}n - \frac{3}{160}n^2 + O(\log_2(n))}.
\end{aligned}
$$

With i denoting the number of points not fixed by π, there are again $\binom{n}{i}$ ways to choose the points not fixed by π and $i!$ ways to permute them. Thus

$$
\begin{aligned}
\sum_{\pi \in Y} F(\pi) &\leq \sum_{i=\lfloor \frac{n}{10} \rfloor}^{n} \binom{n}{i} i! 2^{\frac{n^2}{4}+\frac{3}{2}n-\frac{3}{160}n^2+O(\log_2(n))} \\
&\leq \sum_{i=\lfloor \frac{n}{10} \rfloor}^{n} 2^{i \log_2(n)} 2^{\frac{n^2}{4}+\frac{3}{2}n-\frac{3}{160}n^2+O(\log_2(n))} \\
&\leq 2^{\log_2(n)+n\log_2(n)+\frac{n^2}{4}+\frac{3}{2}n-\frac{3}{160}n^2+O(\log_2(n))} \\
&\leq \mathcal{O}_n 2^{-\frac{1}{4}n}.
\end{aligned}
$$

Adding our estimates now proves the result.   ∎

The ratio of $\dfrac{1}{n!}$ between isomorphism classes and all structures can only be realized if most structures have no symmetry. This is formalized in the next result.

Corollary 11.4.3 *Let $\mathcal{R}_n$ be the number of rigid orders on $[n]$ and let $[\mathcal{R}]_n$ be the number of nonisomorphic rigid orders on an n-element set. Then*

$$
\frac{\mathcal{R}_n}{\mathcal{O}_n} \to 1 \quad \text{and} \quad \frac{[\mathcal{R}]_n}{[\mathcal{O}]_n} \to 1 \quad \text{as } n \to \infty.
$$

Proof. First note that $\mathcal{R}_n = n![\mathcal{R}]_n$. Assume that $\liminf_{n \to \infty} \dfrac{[\mathcal{R}]_n}{[\mathcal{O}]_n} = c < 1$. Denote the number of nonrigid orders on $[n]$ by $\mathcal{F}_n$ and let $[\mathcal{F}]_n$ be the number of nonisomorphic nonrigid orders on $[n]$. Every nonrigid ordered set has at least one nontrivial automorphism. This means $\mathcal{F}_n \leq \dfrac{n!}{2}[\mathcal{F}]_n$. This in turn implies that for a sequence of numbers n that goes to infinity, there are $d_n \leq \dfrac{c+1}{2} < 1$ such that

$$
\begin{aligned}
\mathcal{O}_n &= \mathcal{R}_n + \mathcal{F}_n \leq n![\mathcal{R}]_n + \frac{1}{2}n![\mathcal{F}]_n = d_n n![\mathcal{O}]_n + (1-d_n)\frac{1}{2}n![\mathcal{O}]_n \\
&= \frac{1+d_n}{2}n![\mathcal{O}]_n \leq \frac{3+c}{4}n![\mathcal{O}]_n.
\end{aligned}
$$

This is a contradiction to Theorem 11.4.2. Therefore we must have that the limit is $\lim_{n \to \infty} \dfrac{[\mathcal{R}]_n}{[\mathcal{O}]_n} = 1$. Finally note that

$$
\lim_{n \to \infty} \frac{\mathcal{R}_n}{\mathcal{O}_n} = \lim_{n \to \infty} \frac{\mathcal{R}_n}{\mathcal{O}_n} \frac{n![\mathcal{O}]_n}{n![\mathcal{R}]_n} = \lim_{n \to \infty} \frac{n![\mathcal{O}]_n}{\mathcal{O}_n} = 1.
$$

■

11.5   The Number of Automorphisms

Corollary 11.4.3 says that most finite ordered sets are rigid. This indicates that nontrivial automorphisms are rare indeed. It is now natural to ask how rare automorphisms are in general. Is there any family of ordered sets in which they (asymptotically) make up an appreciable fraction of the total number of endomorphisms? The automorphism conjecture states that asymptotically for any ordered set most endomorphisms are not automorphisms.

Open Question 11.5.1 The automorphism problem. (Cf. [226].) *Is it true that*

$$\lim_{|P| \to \infty} \frac{|\text{Aut}(P)|}{|\text{End}(P)|} := \lim_{n \to \infty} \max_{|P|=n} \frac{|\text{Aut}(P)|}{|\text{End}(P)|} = 0 ?$$

The **automorphism conjecture** *is that the above limit is indeed zero.*

Natural stronger versions would include precise upper bounds on the above quotient, even for restricted classes of ordered sets. Given the information we have, the automorphism conjecture appears reasonable. Experiences with another problem, namely how many maps of an ordered set are fixed point free, should however caution us against jumping to conclusions.

Indeed, it seems that most endomorphisms of an ordered set should have a fixed point. It was even conjectured (very briefly) that the ratio of fixed point free endomorphisms to all endomorphisms should converge to zero. In [226] it is observed that for an antichain the limit of the quotient of the number of fixed point free order-preserving maps and the number of all order-preserving maps is $\frac{1}{e}$. It is not known if this is the largest possible quotient or not. Incidentally, the consideration of fixed point free endomorphisms versus all endomorphisms was the context in which the automorphism conjecture first arose. Please consider Open Problem 10 in this chapter for the statement and references on this problem.

In the following we will present some results on the number of endomorphisms and automorphisms that will allow us to at least settle the automorphism conjecture for lattices. The reader is to settle it for interval orders in Exercise 18.

We start with a lower bound on the total number of endomorphisms. In [69] the better lower bound $2^{\frac{2}{3}n}$ is proved. To abbreviate the presentation we give a slightly weaker result here. We prove the better lower bound in Exercise 14. The key for a better lower bound is to get a better lower bound on the number of endomorphisms of height 1 sets.

Theorem 11.5.2 (Compare with [69].) *Let P be a finite ordered set with n elements. Then P has at least $2^{\frac{n}{2}}$ order-preserving maps that are not automorphisms.*

Proof. Our presentation follows that of [69], p. 20, which is a nice application of notation and results from Chapter 10. The reader is invited to find a direct proof in Exercise 15.

If P is a chain or an antichain, there is nothing to prove. (Also cf. Exercise 26 for a recursive formula for the number of order-preserving maps between chains.) Thus we can assume in the following that P is neither a chain, nor an antichain. By Exercise 7 in Chapter 4, if P has height 1, then P has at least $2^{\frac{n}{2}}$ order-preserving self-maps that are not automorphisms. This leaves the case in which P has at least height 2. We will prove that an ordered set P of height $r \geq 2$ has at least $2^{\frac{r}{r+1}n}$ order-preserving maps that are not automorphisms.

Recall (cf. Proposition 2.6.7) that for an ordered set X, the set $\mathbf{2}^X$ of order-preserving maps from X into a 2-chain has as many elements as X has antichains. Let **a** denote a generic chain with a elements. Note that $\mathbf{2^r}$ is a chain with $r+1$ elements. Let $C \subsetneq P$ be a chain with $r+1$ elements. We obtain that P^P contains C^P, which is isomorphic to $(\mathbf{2^r})^P$, which is in turn (by Proposition 10.5.1) isomorphic to $\mathbf{2}^{\mathbf{r} \times P}$. Thus to finish the proof we have to show that $\mathbf{r} \times P$ contains at least $2^{\frac{r}{r+1}n}$ antichains.

For every subset $S \subseteq P$, let S_k be the set of elements of S that have rank k in P. Moreover let j be chosen such that $|P_j| \leq |P_i|$ for all $i \in \{0, \dots, r\}$. Then $Y := P \setminus P_j$ has at least $\dfrac{r}{r+1}n$ elements. Let $i_1 < i_2 < \cdots < i_r$ be such that $\{0, \dots, r\} \setminus \{j\} = \{i_1, \dots, i_r\}$ and assume that the chain **r** is $\{1 > 2 > \cdots > r\}$. Then for every subset $Z \subseteq Y$ the set

$$A(Z) := \bigcup_{k=1}^{r} \{k\} \times Z_{i_k}$$

is an antichain in $\mathbf{r} \times P$. Moreover $Z \mapsto A(Z)$ is injective, which means that $|\mathbf{2}^{\mathbf{r} \times P}|$ is at least $2^{\frac{r}{r+1}n}$. ∎

As is often the case in enumeration, we need to establish an inequality before we state the next result on automorphisms in sets with irreducible points.

Lemma 11.5.3 *For all $a \in \mathbb{N}$ we have $(a+1)^a \geq 2^a a!$.*

Proof. This is a proof by induction on a. For $a \in \{1, 2, 3, 4\}$ the assertion is easily verified directly.

For the induction step $a \to (a+1)$ assume the assertion is true for an $a \geq 4$. We will need to prove it for $a+1$. Note that

$$
\begin{aligned}
((a+1)+1)^{(a+1)} &= \left(\frac{a+2}{a+1}\right)^a (a+2)(a+1)^a \\
&\geq \left(\frac{a+2}{a+1}\right)^a (a+2)2^a a! \\
&\geq \left(\left(1+\frac{1}{a+1}\right)^{a+1}\right)^{\frac{a}{a+1}} 2^a(a+1)!
\end{aligned}
$$

From analysis we know the terms $\left(1 + \dfrac{1}{a+1}\right)^{a+1}$ form an *increasing* sequence with limit e. Thus $\left(\left(1 + \dfrac{1}{a+1}\right)^{a+1}\right)^{\frac{a}{a+1}}$ also is increasing. For $a = 4$ we have

$$\left(\left(1 + \frac{1}{a+1}\right)^{a+1}\right)^{\frac{a}{a+1}} = \left(1 + \frac{1}{a+1}\right)^a = \left(\frac{6}{5}\right)^4 = \frac{1296}{625} > 2.$$

Thus this term exceeds 2 for all values $a \geq 4$. This finishes the proof of the inequality $((a+1)+1)^{(a+1)} \geq 2^{a+1}(a+1)!$.   ■

Theorem 11.5.4 (Cf. [169], Theorem 1.) *For every finite ordered set P we have*

$$\frac{|\mathrm{Aut}(P)|}{|\mathrm{End}(P)|} \leq 2^{-\sqrt{\max\{|J(P)|,|M(P)|\}}}.$$

Proof. Without loss of generality let $\max\{|J(P)|, |M(P)|\} = |J(P)| =: \alpha$. Then there are either $m \geq \sqrt{\alpha}$ join-irreducible elements with the same lower cover, or there are $m \geq \sqrt{\alpha}$ join-irreducible elements such that no two have the same lower cover. In either case, denote these m join-irreducible elements by $v_1, \ldots, v_m$. For every automorphism f we define the sets

$$A(f) := \{g \in \mathrm{Aut}(P) : f|_{P \setminus \{v_1, \ldots, v_m\}} = g|_{P \setminus \{v_1, \ldots, v_m\}}\}, \quad \text{and}$$
$$H(f) := \{h \in \mathrm{End}(P) : f|_{P \setminus \{v_1, \ldots, v_m\}} = h|_{P \setminus \{v_1, \ldots, v_m\}}\}.$$

The sets $A(f)$ form a partition of $\mathrm{Aut}(P)$. Moreover $A(f) \cap A(g) = \emptyset$ is equivalent to $H(f) \cap H(g) = \emptyset$. Thus if we can prove $\dfrac{|A(f)|}{|H(f)|} \leq 2^{-m}$, then we have

$$|\mathrm{Aut}(P)| = \left|\bigcup_{f \in \mathrm{Aut}(P)} A(f)\right| \leq 2^{-m} \left|\bigcup_{f \in \mathrm{Aut}(P)} H(f)\right| \leq 2^{-m}|\mathrm{End}(P)|,$$

which was to be proved.

In case all v_i have distinct lower covers, we have $|A(f)| = 1$ and $|H(f)| = 2^m$ for all $f \in \mathrm{Aut}(P)$. This leaves the case in which all v_i have the same lower cover v.

Note that $h \mapsto f^{-1}h$ is a bijection between $A(f)$ and $A(\mathrm{id}_P)$ as well as between $H(f)$ and $H(\mathrm{id}_P)$. Thus we can concentrate on $A(\mathrm{id}_P)$. Let $S_1, \ldots, S_k$ be the partition of $\{v_1, \ldots, v_m\}$ that is obtained from the equivalence relation $v_i \equiv v_j$ iff $\uparrow v_i =\uparrow v_j$. Then every map in $A(\mathrm{id}_P)$ must map the S_i to themselves and we obtain $|A(\mathrm{id}_P)| = \Pi_{i=1}^k |S_i|!$. On the other hand every map in $H(\mathrm{id}_P)$ can map every element of S_i to any element of $S_i \cup \{v\}$, thus giving us by Lemma 11.5.3 that

$$|H(\mathrm{id}_P)| = \Pi_{i=1}^k (|S_i| + 1)^{|S_i|} \geq \Pi_{i=1}^k 2^{|S_i|}(|S_i|!) = 2^m \Pi_{i=1}^k |S_i|! = 2^m |A(\mathrm{id}_P)|.$$

This finishes the proof.   ∎

We are now in position to prove the automorphism conjecture for lattices. Remember that automorphisms of lattices not only must map irreducible points to irreducible points. Lattice automorphisms are uniquely determined by their values on the set of (join-)irreducible points (cf. Proposition 5.4.4). This is the key to the next result.

Theorem 11.5.5 (Cf. [169], Theorem 2.) *Let L be an n-element lattice. Then*

$$\frac{|\mathrm{Aut}(L)|}{|\mathrm{End}(L)|} \le C2^{-\sqrt{\log_2(n)}},$$

for some C > 0.

Proof. If $|J(L)| \ge \log_2(n)$, we are done by Theorem 11.5.4. If $|J(L)| \le \log_2(n)$ we argue as follows. By Proposition 5.4.4 two automorphisms of L are equal iff they are equal on the join-irreducible elements of L. Thus L has at most $|J(L)|!$ automorphisms. By Theorem 11.5.2 we have

$$\frac{|\mathrm{Aut}(L)|}{|\mathrm{End}(L)|} \le \frac{\log_2(n)!}{2^{\frac{n}{2}}} \le 2^{-\frac{n}{2}+\log_2(n)\log_2(\log_2(n))} \le C2^{-\sqrt{\log_2(n)}}.$$

    ∎

This concludes our excursion into enumeration. Naturally, whenever counting is involved one can also write a program to count the objects in question. In this fashion one can obtain exact numbers for those parts of a counting problem that are small enough to allow the program to terminate within the investigator's lifespan. Algorithmic approaches to counting problems are part of our discussion in Chapter 12. Also cf. Remarks 3 and 4.

Exercises

1. Recall the definition of a topology from Exercise 2 in Chapter 5. Let X be a finite set.

   (a) Prove that the number of topologies on X is equal to the number of preorders on X.

   (b) A topology τ is called T_0 iff for all distinct $x, y \in X$ there is an $A \in \tau$ such that $x \in A$, $y \notin A$ or $y \in A$, $x \notin A$. Prove that the number of T_0-topologies on X is equal to the number of orders on X.

2. Prove that a finite ordered set is graded iff for any sequence of points $x = x_0, x_1, \ldots, x_n = x$ such that for all $i \in [n]$ we have that x_{i-1} is adjacent to x_i we have $|\{i : x_{i-1} \prec x_i\}| = |\{i : x_i \prec x_{i-1}\}|$.

   (Notice that this is a discrete analogue of the analytical result that a vector field is a gradient field iff line integrals over closed curves vanish.)

3. Chain conditions related to graded ordered sets.

   (a) Prove that a finite ordered set that has a largest and a smallest element is graded iff any two maximal chains have the same length.

   (b) Prove that every ordered set that satisfies the above condition is graded by the function that maps each element to its rank plus 1.

   (c) Give an example of a graded ordered set for which the function that maps each element to its rank plus 1 is *not* a grading function.

   (d) Give an example of an ordered set that is graded by the function that maps each element to its rank plus 1 and which does not satisfy the chain condition in part 3a.

4. The ordered set P satisfies the **Jordan–Dedekind chain condition** iff for all $a < b$ in P any two maximal chains between a and b have the same length.

   (a) Prove that in finite ordered sets with a largest and a smallest element, the Jordan–Dedekind chain condition is equivalent to the condition in Exercise 3a.

   (b) Construct a nongraded ordered set that satisfies the Jordan–Dedekind chain condition.

5. Recognizability of conditions related to graded sets.

   (a) Prove that ordered sets that satisfy the Jordan–Dedekind chain condition are recognizable.

   (b) Prove that ordered sets for which the function that maps each element to its rank plus 1 is a grading function are recognizable.

   (c) Prove that graded ordered sets are recognizable.

6. Let P, Q be ordered sets and let P be graded by g_P. Prove that $f : P \to Q$ is order-preserving iff for all $k \in g_P[P]$ the restriction $f|_{g_P^{-1}[\{k,k+1\}]}$ of f to "grade levels k and $k+1$" is order-preserving.

7. The multinomial theorem.

   (a) Prove the multinomial theorem, that is, prove that if $a_1, \ldots, a_k \in \mathbb{R}$ and $n \in \mathbb{N}$, then

$$(a_1 + \cdots + a_k)^n = \sum_{n_1 + \cdots + n_k = n, n_i \geq 0} \binom{n}{n_1, \ldots, n_k} a_1^{n_1} \cdots a_k^{n_k},$$

where the $\binom{n}{n_1, \ldots, n_k} := \dfrac{n!}{n_1! \cdots n_k!}$ are multinomial coefficients.

(b) Prove that any multinomial coefficient $\begin{pmatrix} n \\ n_1, n_2, \ldots, n_k \end{pmatrix}$ is bounded by k^n.

8. Improve the estimate of $2^{\frac{n^2}{4}+\frac{3}{2}n+O(\sqrt{n})}$ for the number of graded ordered sets in Theorem 11.3.3 to $2^{\frac{n^2}{4}+\frac{3}{2}n+O(\log_2(n))}$. The easy part is to estimate the number of graded ordered sets with three levels and $\left| \frac{n}{2} - n_2 \right| \leq \log_2(n)$. The hard part is the estimate of that number for $\sqrt{n} \geq \left| \frac{n}{2} - n_2 \right| \geq \log_2(n)$.

9. Finishing Lemma 11.3.10.

   (a) Fill in the remaining detail in the proof of part 3 of Lemma 11.3.10 by producing an argument similar to part 4 of Lemma 11.3.9.

   (b) Prove part 2 of Lemma 11.3.10.

10. Finish the proof of Lemma 11.3.11 by proving an upper bound on the number of graded ordered sets with six levels.

11. Prove that the sizes of the following classes of ordered sets (with ground set $[n]$) are asymptotically negligible.

   (a) Decomposable ordered sets,

   (b) Ordered sets that have an irreducible point (hint: consider the sets treated in Lemma 11.3.6),

   (c) Ordered sets that have a retractable point (hint: consider the sets treated in Lemma 11.3.6),

   (d) Ordered sets that have the fixed point property (hint: consider the sets treated in Lemma 11.3.6 and show most contain a four-crown-tower).

12. Let $\leq$ be an order on $[n]$ such that the permutation $\pi : [n] \to [n]$ is an automorphism. Let $b \in [n]$ be an element of a nontrivial cycle of π. Let $a \in [n]$ be an element of the same cycle, say, $a = \pi^k(b)$. Let $m > k$ be the smallest natural number so that $b = \pi^m(b)$. Let $x \in [n]$ be arbitrary. Prove that $b \leq x$ iff $a \leq \pi^k(x)$.

13. Let f, a, b be positive numbers such that $a \leq b$ and $f \geq \frac{1}{2}(b - a)$. Prove that
$$(f + a)^2 + b^2 \leq \left(f + \left(\frac{a+b}{2} \right) \right)^2 + \left(\frac{a+b}{2} \right)^2.$$

14. Prove that every n-element ordered set of height 1 has at least $3^{\frac{n}{2}}$ order-preserving self maps. (Hint: Retract onto suitable 3-fences.) Conclude that every n-element ordered set has at least $2^{\frac{2}{3}n}$ order-preserving self maps.

15. Prove that for every ordered set of height r, there are at least $2^{\frac{r}{r+1}(n-r-1)}$ retractions onto an $(r+1)$-chain. Then show that there are at least $2^{\frac{r}{r+1}n}$ maps that are not automorphisms.

16. Prove that the number of self maps of fences with n elements is bounded by $n3^{n-1}$.

17. Prove that for any finite ordered set P we have $\dfrac{\text{Aut}(P)}{\text{End}(P)} \leq 2^{-\frac{\sqrt{|S(P)|}}{2}}$, where $S(P)$ is the set of all elements a that are retractable to an element $b \not> a$.

18. Let P be an n-element interval ordered set. Prove that there is a $C > 0$ such that $\dfrac{|\text{Aut}(P)|}{|\text{End}(P)|} \leq C2^{-\sqrt{\log_2(n)}}$. (Hint. Use Exercise 8 in Chapter 8 and Exercise 17 above.)

19. Call a finite ordered set P k-**thin** iff every point $p \in P$ is incomparable with at most k other points. For fixed k prove the automorphism conjecture for k-thin ordered sets.

20. Let $P = L\{P_t | t \in T\}$ be a lexicographic sum as in Exercise 5 in Chapter 9. Prove that $\dfrac{|\text{Aut}(P)|}{|\text{End}(P)|} \leq \Pi_{t\in T} \dfrac{|\text{Aut}(P_t)|}{|\text{End}(P_t)|}$.

21. Prove that for an ordered set of width 3 we have that $|\text{Aut}(P)| \leq 6^{\frac{n}{3}}$. Use the result for the number of endomorphisms for sets of height r in the proof of Theorem 11.5.2 to show that for ordered sets of width 3 we have $\dfrac{|\text{Aut}(P)|}{|\text{End}(P)|} \to 0$ as $|P| \to \infty$.

22. Let k be a fixed integer. Prove the automorphism conjecture for truncated lattices of width $\leq k$.

23. Prove that the standard example of an n-dimensional set (which has $2n$ elements) has $n!$ automorphisms and at least n^n endomorphisms. Conclude that $\lim\limits_{n\to\infty} \dfrac{|\text{Aut}(St_n)|}{|\text{End}(St_n)|} = 0$.

24. (a) Let P be a finite ordered set and let $a \in P$ be retractable. Prove that $|\text{End}(P \setminus \{a\})| < |\text{End}(P)|$.

    (b) Give an example of a finite ordered set P and an element $p \in P$ such that $|\text{End}(P \setminus \{a\})| > |\text{End}(P)|$ and $|\text{Aut}(P \setminus \{a\})| < |\text{Aut}(P)|$.

    (c) Show that for any $n \in \mathbb{N}$ there is a P as in 24b such that $|P| > n$.

25. Prove that for *graphs* that contain a clique of size $\dfrac{n}{2}$ there is a fixed $c > 0$ such that $\dfrac{|\text{Aut}(G)|}{|\text{End}(G)|} \leq c2^{-\frac{7}{80}n}$.

26. Let $MC_{n,m}$ be the number of order-preserving maps from an n-chain to an m-chain. Prove that $MC_{n,m} = MC_{n-1,m} + MC_{n,m-1}$.

27. Prove that Theorem 11.5.5 also is valid for lattices from which top *or* bottom element have been removed. Why does this proof not work for truncated lattices in general?

Remarks and Open Problems

1. For more on enumeration the reader should consult [74]. For more on generating functions, cf. [274].

2. For precise formulas for the number of nonisomorphic graded ordered sets with n elements, cf. [141].

3. Until recently the numbers of different orders for an n-element ordered set were known up to $n = 14$ (cf. [76]) and the numbers of nonisomorphic orders for an n-element ordered set were known up to $n = 13$ (cf. [38]). In recent published work, in [118] a new approach was utilized to obtain these numbers up to 16 (orders) and 14 (nonisomorphic orders). In [32] another approach was used to obtain the numbers of nonisomorphic orders with 15 and 16 elements respectively and the total number of orders with 17 elements. They are

n	nonisomorphic orders	total number of orders on $[n]$
1	1	1
2	2	3
3	5	19
4	16	219
5	63	4,231
6	318	130,023
7	2,045	6,129,859
8	16,999	431,723,379
9	183,231	44,511,042,511
10	2,567,284	6,611,065,248,783
11	46,749,427	1,396,281,677,105,899
12	1,104,891,746	414,864,951,055,853,499
13	33,823,827,452	171,850,728,381,587,059,351
14	1,338,193,159,771	98,484,324,257,128,207,032,183
15	68,275,077,901,156	77,567,171,020,440,688,353,049,939
16	4,483,130,665,195,087	83,480,529,785,490,157,813,844,256,579
17		122,152,541,250,295,322,862,941,281,269,151

These numbers were found by computation. Algorithmic approaches to theoretical questions are the subject of Chapter 12.

4. Similar to Remark 3, the first eight Dedekind numbers have been found via computation. They are (cf. [273]) 2; 3; 6; 20; 168; 7,581; 7,828,354; 2,414,682,040,998 and 56,130,437,228,687,557,907,788.

Proofs of asymptotic formulas for Dedekind numbers can be found in [144, 149]. Since the enumeration of ordered sets is the underlying thread (and since these proofs are rather involved), Dedekind numbers were not explicitly treated in this chapter.

5. Aside from graded ordered sets, there are other classes of ordered sets for which exact formulas for the number of n-element orders *in this class* are available. Among these classes are series-parallel posets, interval ordered sets and tiered posets. An overview of these classes is given in [74].

6. Find a proof for Theorem 11.3.13 that is easier than the proof in the literature. The proof of Theorem 11.3.3 together with [50] allows us to establish the bounds of Theorem 11.3.13 for the number of graded ordered sets. All that "remains" would be to show that the number of nongraded ordered sets is asymptotically negligible. A step towards this goal might be to first solve Problem 7 and then use sharp bounds on the number of graded sets to establish bounds on the number of nongraded sets.

7. Prove or disprove that the number of graded ordered sets with four grade levels is asymptotically $2^{\frac{n^2}{4}+\frac{5}{4}n+O(\log_2(n))}$. Prove or disprove that the number of graded ordered sets with $k \geq 5$ grade levels is asymptotically equal to $2^{\frac{n^2}{4}+\left(1-\frac{k-5}{2}\right)n+O(\log_2(n))}$. This conjecture is motivated by the fact that most graded ordered sets with three grade levels have structure $\frac{n}{4} - \frac{n}{2} - \frac{n}{4}$. To the author this suggests that most graded ordered sets with $k > 3$ levels will have three large levels that look like the typical graded set of size $n - k + 3$ and all other levels contain exactly one element. The author also is not sure of the degree of difficulty of this conjecture. While an approach as given in Section 11.3 appears promising, it also appears tedious.

Asymptotic formulas for the number of ordered sets of bounded *width* are given in [31].

8. Theorem 11.4.2 is an example of a "zero-one-law". Many combinatorial properties (to be precise, those that can be stated using a sentence in first-order logic) are of such a nature that as the size of the underlying structure gets large, the property either almost always holds (that is, the probability that a randomly chosen structure has the property approaches 1) or it almost always fails (that is, the probability that a randomly chosen structure has the property approaches 0). This result can be found in [78] or also in Section II.2, Theorem 6 in [21]. In a sense this is not surprising. For ordered sets we have seen that for large n almost all ordered sets are graded with three levels. The largest class of ordered sets is thus very homogeneous.

9. Knowing the number of ordered sets with a given ground set can help answer certain questions without actually constructing the answer.

For example, there are at most $n!$ linear extensions of an ordered set. Thus the number of linear extensions that an ordered set has is a number between 1 and $n!$. Since the number of nonisomorphic ordered sets grows faster than $n!$, there must be nonisomorphic ordered sets with the same number of linear extensions. (This application of what is also called the "probabilistic method" was shown to the author by S. Felsner. It answered the author's simple-minded question if reconstruction of the number of linear extensions would solve the reconstruction problem.) In particular (unsurprisingly), the number of linear extensions alone is not enough to reconstruct an ordered set.

As another example, recall (cf. Exercise 18 in Chapter 10) that no two distinct non-dual ordered sets can have the same set of endomorphisms. It is also easy to see that an ordered set and its dual have the same number of endomorphisms. Via an argument similar to the above there are two non-isomorphic and non-dually-isomorphic ordered sets P and Q of the same size such that $|\text{End}(P)| = |\text{End}(Q)|$.

For more work of a probabilistic nature in ordered sets, cf. for example Chapter 7 in [263].

10. (The original questions that motivated the automorphism problem.) Let $FPF(P)$ be the number of fixed point free maps of the ordered set P.

   (a) Find an overall bound for $\dfrac{|FPF(P)|}{|\text{End}(P)|}$.

   (b) Find $\limsup\limits_{|P|\to\infty} \dfrac{|FPF(P)|}{|\text{End}(P)|}$.

   (c) Find the above quantities when P is restricted to a special class of ordered sets.

The author is not aware of any progress in this direction beyond [226]. Natural candidates for first partial results might be ordered sets of small width, as the analogous automorphism problem also seems to be solvable there (cf. Exercises 21 and 22).

11. Other interesting problems that have a flavor similar to the automorphism problem are.

   (a) The automorphism problem for graphs. For a simple first result in this direction consider Exercise 25.

   (b) For graphs the notion of a strict endomorphism (adjacent vertices can only be mapped to the same vertex if there is a loop at that vertex) is widely used. Let the set of strict endomorphisms of G be $\text{End}_s(G)$. The quotient $\dfrac{|\text{Aut}(G)|}{|\text{End}_s(G)|}$ can remain stationary at 1 (complete graphs without loops) or converge to zero (discrete graphs). What is the limiting set of the quotients $\dfrac{|\mathrm{Aut}(G)|}{|\mathrm{End}_s(G)|}$ as $|G| \to \infty$? What is the class of graphs for which the limiting set is $\{0\}$? How about $\{1\}$?

(c) For lattices there is the notion of a lattice endomorphism (an endomorphism that preserves suprema and infima). Let the set of lattice endomorphisms of L be $\mathrm{End}_l(L)$. The quotient $\dfrac{|\mathrm{Aut}(L)|}{|\mathrm{End}_l(L)|}$ can remain stationary at 1 (antichain with a top and a bottom attached) or converge to zero (chains). What is the limiting set of the quotients $\dfrac{|\mathrm{Aut}(L)|}{|\mathrm{End}_s(L)|}$ as $|L| \to \infty$? What is the class of lattices for which the limiting set is $\{0\}$? How about $\{1\}$?

12. Let $m, n \in \mathbb{N}$ be given and let the union $T := \bigcup_{a=1}^{n}\{1, \ldots, m\}^a$ be ordered by $(x_1, \ldots, x_k) \le (y_1, \ldots, y_l)$ iff $k \le l$ and for all $i \in \{1, \ldots, k\}$ we have $x_i = y_i$. Let $c_1, \ldots, c_b$ be elements of T.

Find estimates for the number of elements of

$$S := T \setminus \bigcup_{i=1}^{b} \uparrow c_i.$$

The above can be used to describe the tree that is searched by a search algorithm such as backtracking or forward checking. These algorithms are discussed in Section 12.5.1. Insights into the above question may allow us to estimate run times of search algorithms.

The shown description of the search space is from [276].

13. Can the methods of Theorem 11.5.4 be extended to confirm the automorphism conjecture (Open Question 11.5.1) for dismantlable ordered sets?

14. The automorphism conjecture can be settled trivially for classes of ordered sets for which $|\mathrm{End}(P)|$ grows faster than $n!$. Unfortunately the hope that this is true in general is false, since by [69] fences and crowns, for example, have $< O(3^n) \ll n!$ (for large n) endomorphisms.

15. What general results are there to compute the number of homomorphisms from one ordered set to another? The author is not aware of any results beyond [80] and Exercise 26. A good list of earlier references is in [80]. Related to this, for each n, what is the ordered set P of size n with the fewest endomorphisms?

16. The **endomorphism spectrum** (cf. [98]) of an ordered set P is the set

$$S := \{s \in \{1, \ldots, |P|\} : (\exists f \in \mathrm{End}(P))|f[P]| = s\}.$$

That is, it is the set of possible range sizes of order-preserving self maps. What numbers (aside from trivial choices such as $1, 2, |P|$) are guaranteed to be always in the spectrum? What are examples of ordered sets with "small" spectra? In [64], Theorem 1.1, it is shown that every ordered set has an endomorphism with $|P|^{\frac{1}{7}}$ elements in its image. It is also shown in [64], Theorem 1.2, that there is a $c > 0$ such that for each n there is an ordered set of size n such that every endomorphism that is not the identity has at most $c(n \log(n))^{\frac{1}{3}}$ elements in its image.

17. For every k in the spectrum of P let e_k^P be the number of endomorphisms with image of size k. The spectrum analyzes the zeroes of this sequence. What more can be said about the properties of the sequence e_k^P?

18. Define the **retract spectrum** analogous to the endomorphism spectrum, only for retractions. What can we say about the analogous questions for the retract spectrum?

19. What is the asymptotic number of $(n + 2)$-element lattices? According to [74] there are asymptotic upper and lower bounds, but there is no asymptotic expression.

20. Another entity related enumeration in ordered sets is the Möbius function. For more on this function, cf. [154, 232].

12
Algorithmic Aspects

So far, we have freely used a luxury in mathematics that is so fundamental, we often take it for granted. In any proof, if one needs to work with an object that is known or assumed to exist, one says "let x be (the object in question)" and one moves on with the proof. No work is "wasted" on thinking about how to find the object. Indeed, especially in the finite setting it is obvious that given enough patience we should be able to find the object, provided it exists. Simply try out all possibilities and one of them will work. As long as one is not interested in the object itself, this approach is very efficient, especially in ruling out blind alleys.

However there are situations in which knowing that an answer exists may not be enough. For example, in our setting asking for the existence or non-existence of a fixed point free order-preserving map on a finite ordered set could be replaced with the question of what such a map would look like. The search for such a map (if it exists) can require great effort. Proposition 1.4.6 shows this effort for a comparatively small and well-behaved set. Similarly the counting tasks in Chapter 11 can "in principle" require (for a fixed ordered set or a fixed size) only enough patience. In practice the effort may be insurmountable.

At least since the late 1970s, as they became more widely available, computers have become the tool of choice for any kind of repetitive task that one may wish to perform. Clearly, with computers being widely available and continually becoming more powerful and easier to use, mathematicians can now answer questions that previously were hard to answer or unanswered. For example, while it is not impossible to show by hand that the ordered set in Figure 1.1 b) has the fixed point property (cf. Exercise 18b in Chapter 1), it is quite tedious. Extending this example, it is even more tedious, though not impossible, to show that this set has

30,126 order-preserving self maps. With a sufficiently fast computer, either task takes less than a second.

In this chapter we will investigate the use of various algorithms to perform computations that can support mathematical research. The main practical question we will consider is that of complexity of the algorithm. Briefly put, we want to know or estimate before we start it how long an algorithm may run. This knowledge or estimate will give an indication of how feasible a computation may be, without doing the computation. A more basic question is that of algorithm correctness. This is a reliability question. If we want to be sure an algorithm produces the desired results, we better prove it.

The philosophical issues that arise from the use of computers in mathematical work are complex and numerous. The main question is whether a proof by computer really is a proof or not. How does one know a proof is correct if one did not personally check the millions of cases a computer checked? (Do we really know the number above is 30,126?) Is it enough to know that a program (which is assumed to do the right thing) was (hopefully) correctly executed on a computer? These questions the reader must answer for him/herself and the answer will affect the reader's future. Excellent mathematical work can be done with and without computers.[1]

There are many ways to approach algorithmic issues. The central notion for this chapter will be that of a constraint satisfaction problem (cf. Definition 12.3.4). The early sections of the chapter give the framework in which constraint satisfaction problems are to be seen. Later sections describe solution algorithms.

12.1   Algorithms

The central point of any theoretical work related to computer applications is the notion of an algorithm. We will use the definition given in a mainstream computer science text.

Definition 12.1.1 (Cf. [5], p.2.) *An* **algorithm** *is a finite sequence of instructions each of which has a clear meaning and can be performed with a finite amount of effort in a finite length of time.*

Note that the definition of an algorithm does not include any statements as to what the algorithm is going to output. It simply states that all must be intelligible and doable in finite time. It is left up to whoever designs the algorithm to make sure that it does what it is supposed to do.

---

[1] The author's personal viewpoint is that the computer should be used as a tool when appropriate. Even people who refuse to use computers could benefit from their use. They might be able to test conjectures and generate new ideas. On the other hand very "computer dependent" researchers can benefit from a more theoretical approach. Some problems that one tends to solve with computing power have simpler theoretical solutions.

Let us first consider a few examples. In Proposition 1.2.7 we have seen that finite ordered sets can be represented through their cover relation or through the order relation. In many situations one representation is preferable to the other. As examples, just consider the idea of irreducible points, which depends on knowledge of the diagram, and contrast it with the idea of results such as Proposition 3.3.3, for which knowledge of the whole order relation is crucial. Consequently we need to be able to translate between both representations. This process has likely become automatized in the reader's mind by now. How would we "explain" it to a computer?

Example 12.1.2 (Simple algorithm to compute the cover relation from the order relation.)

Given. An order relation $\leq$ on a finite set P.
Task. Find the corresponding cover relation $\prec$.

The real-life idea is of course that in going from the order relation to the diagram we erase superfluous connections. This is exactly what is spelled out below.

We will describe our algorithms in English, thereby circumventing a long (and tedious) process of defining a language in which algorithms will be formulated. The "For $\cdots$ do" structures we use are to be understood as follows. We assume the set in question has been linearly ordered in some fashion. The operation between "For $\cdots$ do" and "end for $\cdots$" are executed exactly once for each element of the set. The operations are executed on the elements in the order that we imposed.

Start with $\prec^*$ being equal to $\leq$,

Erase all pairs (x, x) from $\prec^*$,

For every pair $(x, y) \in \leq$ do

For every $z \in P \setminus \{x, y\}$ do

If $x < z$ and $z < y$, then erase (x, y) from $\prec^*$ and continue to the next pair in $\leq$,

If not, continue with the next z,

end for z,

end for (x, y),

The covering relation will be stored in $\prec^{*,\text{end}}$, which is the relation $\prec^*$ after the above loops terminate.

It is a small but important issue to verify that the above algorithm really will produce the covering relation for $\leq$ in its output $\prec^{*,\text{end}}$. To see this we have to show $\prec = \prec^{*,\text{end}}$.

For $\prec \subseteq \prec^{*,\text{end}}$, note that if $(x, y) \in \prec$, then (x, y) is in $\prec^*$ at the start. Moreover (x, y) will never be erased, since $x \neq y$ and since there are no elements z of P strictly between x and y.

For $\prec \supseteq \prec^{*,end}$, assume $(x, y) \notin \prec$. Then either $x = y$ and (x, y) will be erased before the (x, y)-loop starts, or there is a z with $x < z < y$ and (x, y) will be erased during the loop. Either way $(x, y) \notin \prec^{*,end}$.

This shows that $\prec = \prec^{*,end}$.   ∎

Now let us see if going from the diagram to the order is equally easy. The algorithm we give here is in fact applicable to compute the transitive closure of any finite relation.

Example 12.1.3 (Warshall's algorithm for computing the transitive closure of a relation.)

Given. A relation $\prec$ on a set P.
Task. Find the transitive closure of the relation.

Here the definition of the transitive closure (cf. Definition 1.2.5) is not very efficient in the design of an algorithm. It essentially says to try out all possible chains of adjacencies and add the pairs of endpoints for each chain. Warshall's algorithm avoids this straightforward, but inefficient, approach. Note that to compute the order relation that belongs to a given cover relation we would precede Warshall's algorithm by adding all relations (x, x) to $\prec$. In the following we use notation that suggests order and cover relations. Yet, the algorithm is, in fact, applicable to the computation of *any* transitive closure.

Set $\leq^*$ equal to $\prec$,

For every $z \in P$ do

    For every $x \in P$ do

        For every $y \in P$ do

            If $x \leq^* z$ and $z \leq^* y$, add (x, y) to $\leq^*$,

        end for y,

    end for x,

end for z,

The transitive closure $\leq$ is stored in the relation $\leq^{*,end}$, which is the relation $\leq^*$ at the end of all the loops.

At every stage of the algorithm the relation $\leq^*$ is a subset of the transitive closure $\leq$ of $\prec$. This is because the relations that are added are all relations that arise by transitivity.

The proof of the inclusion $\leq \subseteq \leq^{*,end}$ is harder. Let

$$x = z_0 \prec z_1 \prec \cdots \prec z_n \prec z_{n+1} = y.$$

Now let σ be a permutation of $\{1, \ldots, n\}$ such that the order in which the z_k for $k = 1, \ldots, n$ occur in the outer (z-) loop is $z_{\sigma(1)}, z_{\sigma(2)}, \ldots, z_{\sigma(n)}$. After the execution of the z-loop for $z = z_{\sigma(1)}$ we have

$$x = z_0 \leq^* \cdots \leq^* z_{\sigma(1)-1} \leq^* z_{\sigma(1)+1} \leq^* \cdots \leq^* z_{n+1} = y.$$

That is $\leq^*$ has been extended in such a way that $z_{\sigma(1)}$ can be eliminated from the original progression of $\leq^*$-relations that go from x to y. Similarly for every $k \in \{1, \ldots, n\}$ the execution of the z-loop for $z_{\sigma(k)}$ will extend $\leq^*$ in such a way that $z_{\sigma(k)}$ can be erased from the connection between x and y. That means after the execution of the z-loop for $z_{\sigma(n)}$ we must have $x \leq^* y$ and we are done.   ∎

Example 12.1.4 Though we shall not encode them for a computer here, the reader should note that many proofs we have given so far are "algorithmic". For example, consider the constructive proofs of Theorem 7.4.1 and Lemma 7.4.2. Either proof could be re-written into a procedure that inputs an ordered set and outputs a realizer with the desired properties. At this stage it is a good exercise to review existence proofs (in this text or elsewhere) and decide if the object proven to exist is constructed in the proof or not. If not, one could ask if a non-constructive proof can be replaced with a constructive one.

12.2   Polynomial Efficiency

The sample algorithms in Section 12.1 are both "polynomial algorithms". From a theoretical point-of-view, polynomial algorithms are considered nice, since they terminate in comparatively short time. To make the above more precise, we shall define polynomial efficiency as follows.

Definition 12.2.1 *An algorithm is said to be of* **polynomial efficiency** *iff there is a polynomial p such that for input of size n the algorithm terminates after $\leq p(n)$ steps.*

It is important to also specify what is counted as a step in an algorithm. Ultimately one could define a "step" as one clock cycle on a computer. This however would make the definition very unwieldy. Any analysis would depend on what language, what compiler and what computer one uses. For theoretical results of lasting value such a dependency is unacceptable. Thus the most elementary operations that occur in an algorithm are normally used as steps to be counted. These operations usually can be performed in a polynomial number of machine cycles. Moreover, to keep the theoretical picture uncluttered, one does not explicitly count the overhead the computer encounters in controlling a loop or for the branching associated with an if-then statement. Given this underlying idea, an algorithm once proven to be of polynomial efficiency is (essentially) polynomially efficient on any machine and in any language. (We shall not delve into more subtle details such as Turing machines and quantum computers here.)

We shall now prove that the algorithms in Section 12.1 are of polynomial efficiency.

Proposition 12.2.2 *Let P be an ordered set with n elements, let $\leq$ be its order relation and let $\prec$ be its covering relation. Let a step be the erasure of a pair in a relation or the checking if two elements are related. Then the simple algorithm to compute $\prec$ from $\leq$ in Example 12.1.2 terminates in at most $|P| + 3|P|^3$ steps.*

Proof. Erasing all pairs (x, x) takes exactly $|P|$ steps. There are at most $|P| \cdot |P|$ pairs in any relation on P and there are $|P| - 2$ elements in any set $P \setminus \{x, y\}$. Thus the instructions inside the nested loops are executed at most $|P|^2(|P|-2) < |P|^3$ times. These instructions include two comparisons and possibly one erasure, meaning a total of up to three steps. Thus the total number of steps is at most $|P| + 3|P|^3$.   ∎

Note that the estimates in Proposition 12.2.2 are not terribly sophisticated. As long as one is only interested in establishing that a problem can be solved in polynomial time, they need not be. An interesting follow-up question that we will not address in this text is to establish the smallest polynomial such that a given task can be solved in $p(\langle\text{input size}\rangle)$ steps. To answer it, one generally needs very specific knowledge of the problem. Often a general algorithm might already be polynomial, but a specialized algorithm that uses the specific structure of the problem can be made to run faster. The significance of such improvements is normally measured in economic terms. For example, the faster your VLSI layout algorithms are, the faster chips can be designed.

Also note that we chose $|P|$ as our underlying variable, even though one can argue that the input size is in fact $|P|^2$, since we work with the order relation $\leq$. As long as the relation between the sizes of the objects in question is polynomial this is not a problem. However one can see that keeping track of all the different terms can be tedious. If one is only interested in the degree of the bounding polynomial, one uses terminology analogous to that of asymptotic enumeration (cf. Definition 11.3.2).

Definition 12.2.3 *Given an algorithm that takes input of size n, we will say the algorithm is $O(n^k)$ iff there is a polynomial p of degree k such that for all input sizes n, the algorithm terminates in at most $p(n)$ steps. If we do not specify what a step is, it is assumed that a step can be executed in an amount of time independent of the input size.[2]*

For example, Proposition 12.2.2 says that the algorithm in Example 12.1.2 is $O(|P|^3)$.

---

[2]This is a possible trap when implementing algorithms or analyzing their complexity. In implementation, simple-looking steps can become quite complex if the right data types are not available. In the analysis, orders of magnitude of the actual complexity can be hidden in sub-steps whose length does depend on the input size.

Proposition 12.2.4 *Warshall's algorithm is $O(|P|^3)$, with a step being the check if two elements are related or the adding of an ordered pair to a relation.*

Proof. Since P has $|P|$ elements, the nested loops in Warshall's algorithm are such that the instructions inside are executed exactly $|P|^3$ times. Thus Warshall's algorithm is $O(|P|^3)$. ∎

Other parameters of an ordered set that we often take for granted are the height and the width. In Exercise 2 the reader will show the height of an ordered set is computable in polynomial time. What about the width? One might be tempted to assume that the width of an ordered set is the size of the largest set of points of the same rank. The reader will show in Exercise 3 that this is not the case. For references to an algorithm that computes the width of an arbitrary ordered set, cf. Remark 2.

We shall consider here the special case of interval orders. Recall that if the interval order represents the time frames for a given set of tasks, the width will tell us how many tasks are at most executed at the same time. This has applications, for example, in the register allocation on a computer CPU. The graph-theoretically versed reader will recognize greedy coloring in the following proof.

Proposition 12.2.5 *Let P be a finite interval ordered set and assume we know an interval representation of P. Assume comparability between endpoints of intervals in the representation can be checked in a constant number of steps. Then $w(P)$ can be computed in $O(|P|^2)$ steps.*

Proof. (Adapted from [272], Proposition 5.1.11.) We will use the notation of the proof of Theorem 8.1.5 for the interval representation. That is, x is represented by $[L(x), R(x)]$ and the endpoints are ordered by inclusion.

Linearly order the vertices of P by $x \sqsubseteq y$ iff $L(x) \subseteq L(y)$ breaking ties arbitrarily. Recursively define a map $f : P \to \mathbb{N}$ as follows. For the $\sqsubseteq$-smallest element s of P set $f(s) := 1$. Now suppose $x \in P$ and $f(y)$ has been defined for all $y \in P$ with $y \sqsubseteq x$ and $y \neq x$. In this case we let $f(x)$ be the smallest number k such that for all $y \sqsubseteq x$, $y \neq x$, if $L(x) \subsetneq R(y)$ then $f(y) \neq k$.

We claim the function f can be computed in at most $O(|P|^2)$ steps. Indeed, for each $x \in P$ only the f-values of the elements of P before x in the order $\sqsubseteq$ have to be checked. For each (of the less than $|P|$) elements y that is $\sqsubseteq$-before x and satisfies $L(x) \subsetneq R(y)$ (note that we assume comparability of endpoints can be checked quickly) we enter $f(y)$ in a list. Once all y's are checked, we find the desired number for $f(x)$ in at most $|P|$ steps.

To finish the proof we claim that $\max\{f(p) : p \in P\} = w(P)$. The inequality "$\geq$" is clear, since if $x_1, \ldots, x_j$ is an antichain of P in its original order, then f must assign distinct values to all x_i. For the inequality "$\leq$" assume that $x \in P$ is such that $m := f(x) = \max\{f(p) : p \in P\}$. Then for each $i \in \{1, \ldots, m-1\}$ there is an $x_i \in P$ with $x_i \sqsubseteq x$ and $f(x_i) = i$. Since $L(x_i) \subseteq L(x) \subsetneq R(x_i)$, x and x_i are not comparable in P. Now let $i \neq j$ be distinct elements of $\{1, \ldots, m-1\}$. Assume without loss of generality $L(x_i) \subseteq L(x_j)$. Then $L(x_i) \subseteq L(x_j) \subseteq$

$L(x) \subsetneq R(x_i)$, so x_i and x_j are not comparable in P. This means $m \le w(P)$ and we are done.   ∎

The above result naturally leads to the question how hard it is to obtain the interval representation of the ordered set. We can at least get an idea by looking at the proof of Theorem 8.1.5 in which we constructed an interval representation. In this spirit, we conclude this section by looking at some constructive proofs and results given in the text (and elsewhere) which lead to polynomial algorithms. We will merely establish if a proof gives a polynomial algorithm or not. We will not consider if there is a more efficient algorithm and we will also not try to prove best possible bounds. The improvement of algorithms and the proving of their exact worst-case efficiency is a large and important occupation beyond the scope of this text (except for one algorithm improvement in the proof of Theorem 12.6.2).

- In the proof of Theorem 8.1.5 we constructed an interval representation for an ordered set that did not contain a copy of a set $\mathbf{2} + \mathbf{2}$. We claim this construction takes $O(|P|^3)$ steps if the order relation is known. Indeed, for each $x \in P$ it takes $O(|P|)$ steps to construct $L(x) = (\downarrow x) \setminus \{x\}$ and it takes $O(|P|^2)$ to construct $R(x) := \bigcap \{(\downarrow y) \setminus \{y\} : y \in P, y > x\}$. Thus in $O(|P|^3)$ steps the intervals $[L(x), R(x)]$ have been constructed.

  Therefore it takes $O(|P|^4)$ steps to check if an ordered set has an interval order (check for subsets $\mathbf{2}+\mathbf{2}$) and, if so, it takes $O(|P|^3)$ steps to construct an interval representation. Note that computing the explicit order relation of the $L(x)$ and $R(x)$ (as needed in Proposition 12.2.5) takes another $O(|P|^3)$ steps. Indeed, there are up to $2|P|$ endpoints, so there are $O(|P|^2)$ pairs of endpoints and the containment of the sets is checked in $O(|P|)$ steps if the sets are encoded appropriately.

- Polynomial verifiability of isomorphism for interval orders will be treated in Proposition 12.3.3. An $O(|P|^2)$ algorithm is to be constructed in Exercise 9.

- The above shows that certain tasks become easy once we know if a given order belongs to a certain class. This brings up the question how hard it is to check if an order belongs to a certain class. We give some examples of classes that are polynomially recognizable from the order relation.

  1. It can be checked in $O(|P|^4)$ steps if a given ordered set is a lattice. There are $O(|P|^2)$ pairs of points, $O(|P|)$ candidates for their supremum and their infimum and it takes $O(|P|)$ steps to check if a given point is the supremum or infimum of two others. This process not only checks if an ordered set is a lattice. It also computes all suprema and infima.

  2. The same result as in 1 also holds for truncated lattices. Simply add a check for boundedness.

3. By Theorem 5.6.4 it can be checked in $O(|P|^3)$ steps if a given lattice is distributive, provided suprema and infima can be computed in a constant number of steps.

- There is a polynomial algorithm to decide if the dimension of a given ordered set is ≤ 2. Consider [263], Corollary 2.6 (originally [97], Chapter 5) or [254] for the best algorithm.

- In [156] a polynomial algorithm to decide if the interval dimension of a given ordered set is ≤ 2 is given.

12.3   NP problems

Having seen examples of problems that are solvable in polynomial time, it is natural to ask for examples of problems that need more than polynomial time to be solved. At first glance, it is trivial to come up with such examples. Any problem that requires us to produce exponentially many answers cannot be solved in polynomial time. For example, if one wants to have a list of all order-preserving maps on a given ordered set, then any algorithm will need to take at least as many steps as there are order-preserving maps on this set. By Theorem 11.5.2 this number is at least $2^{\frac{n}{2}}$.

Interestingly enough, beyond examples similar to the above it is hard to prove that a certain problem cannot be solved in polynomial time. We will now turn to a class of problems called nondeterministically polynomial or NP problems. For many NP problems the existence of a polynomial algorithm can be established for the problem itself or for important special cases. However for large classes of NP problems it is unknown if there is a polynomial algorithm for the problem.

Definition 12.3.1 *A* **decision problem** *is a problem for which the answer will be "yes" or "no".*

The problems "Does a given ordered set have the fixed point property?" and "Are two given ordered sets isomorphic?" definitely are decision problems. The given formulation however (just as for many decision problems) does not suggest any avenue towards a solution or towards building an algorithm. The definition of NP problems provides a more formal framework in which to capture many decision problems.

Definition 12.3.2 *A decision problem for input of size n is said to be* **nondeterministically polynomial** *or* **NP** *iff there is an algorithm A whose run time is polynomial in the size of its input and a certain structure C (a "certificate") of size polynomial in n such that running A on C "proves" that the answer is "yes".*

To cast our two examples above into this light we first need to re-phrase the problems. Instead of asking "Does a given ordered set have the fixed point property?", the NP version of this question is as follows.

Given. A finite ordered set P.

Question. Is there a fixed point free order-preserving map $f : P \to P$?

So instead of asking if the set has the fixed point property we ask if the set does not have the fixed point property.[3] Proving that a given ordered set does not have the fixed point property is easy if some "oracle" provides us with the right conjecture for a fixed point free order-preserving map. Indeed it takes $|P|$ steps to check if a given self map of P has a fixed point and (cf. Exercise 6) another $O(|P|^2)$ steps to check if the map is order-preserving.

Similarly we can translate the isomorphism problem into this setting. It becomes

Given. Two finite ordered sets P and Q.

Question. Is there an order-isomorphism $\Phi : P \to Q$?

Again for any given map it can be verified in polynomially many steps if the map is an order-isomorphism.

The practical weaknesses of the definition of NP problems are obvious. The polynomial verifiability of the answer being "yes" only helps if we guess right. Moreover there is no provision made for verification of a negative answer. Thus it may not come as a surprise to the reader that there are no polynomial algorithms known which solve the above problems in general. Yet the framework of NP problems so far is the best framework in which to analyze the level of difficulty of many decision problems. The crucial advantage of working with NP problems is that there are results that allow us to formally distinguish those problems that are "hard". We will discuss this idea in Section 12.4.

As mentioned earlier, in special cases there can be a polynomial algorithm for a problem even if there is no general polynomial algorithm. We give an example for isomorphism here. Another example is given in Exercise 8.

Proposition 12.3.3 *Isomorphism of interval orders can be checked in polynomial time.*

Proof. (A more efficient algorithm is to be given in Exercise 9.) Let P and Q be ordered sets that carry interval orders. We shall describe the algorithm that verifies isomorphism in polynomial time. Since sets with different numbers of elements are trivially not isomorphic, we shall only consider the situation $|P| = |Q|$. (In a program the following would thus only execute if the check $|P| \overset{?}{=} |Q|$ succeeds.)

Linearly order the elements of both sets according to the following rules.

1. If $\mathrm{rank}(x) < \mathrm{rank}(y)$, then x comes before y.

2. If $\mathrm{rank}(x) = \mathrm{rank}(y)$ and $|\uparrow x| > |\uparrow y|$, then x comes before y.

---

[3]Formally (cf. [94], p.156) this means that the decision if an ordered set has the fixed point property is a co-NP problem (the complement of an NP problem). It would only be in NP if there was a way to design a "certificate" that proves in polynomial time that the set has the fixed point property. By Theorem 7.2 in [94] this is a genuinely "hard" task.

3. If $\text{rank}(x) = \text{rank}(y)$, $|\uparrow x| = |\uparrow y|$, and $|\downarrow x| > |\downarrow y|$, then x comes before y.

4. Break the remaining ties arbitrarily.

Since all quantities in question can be computed in polynomial time, the above orderings for the elements of P and Q can be achieved in polynomial time.

Let $p_1, \ldots, p_n$ and $q_1, \ldots, q_n$ be the elements of P and Q respectively, each listed in one of the orders as above. Isomorphisms preserve the rank and the numbers of upper and lower bounds of an element. Thus if there is an i such that $\text{rank}_P(p_i) \neq \text{rank}_Q(q_i)$ or $|\uparrow_P p_i| \neq |\uparrow_Q q_i|$ or $|\downarrow_P p_i| \neq |\downarrow_Q q_i|$, then P cannot be isomorphic to Q. Consider the remaining case $\text{rank}_P(p_i) = \text{rank}_Q(q_i)$ and $|\uparrow_P p_i| = |\uparrow_Q q_i|$ and $|\downarrow_P p_i| = |\downarrow_Q q_i|$ for all $i \in \{1, \ldots, n\}$. Note that by Scholium 8.1.7 and its dual we have that for $x \not\sim y$ in an interval order $|\uparrow x| = |\uparrow y|$ implies $(\uparrow x) \setminus \{x\} = (\uparrow y) \setminus \{y\}$ and $|\downarrow x| = |\downarrow y|$ implies $(\downarrow x) \setminus \{x\} = (\downarrow y) \setminus \{y\}$. Thus for any p_i, p_j that were tied as in 4 we have $(\uparrow_P p_i) \setminus \{p_i\} = (\uparrow_P p_j) \setminus \{p_j\}$ and $(\downarrow_P p_i) \setminus \{p_i\} = (\downarrow_P p_j) \setminus \{p_j\}$. A similar result holds in Q. Thus if P is isomorphic to Q, then there must be an isomorphism that for all $i \in \{1, \ldots, n\}$ maps p_i to q_i.

For our algorithm this means that we consider the map $\Phi : P \to Q$ that is defined by $\Phi(p_i) := q_i$ and check if it is an isomorphism. If Φ is an isomorphism, then P and Q are isomorphic. If not, then by the above P and Q are not isomorphic. Since this last check again only requires a polynomial number of steps, the proposed algorithm is polynomial.   ∎

We shall spend the rest of this section giving a formal definition of constraint satisfaction problems. This framework will allow us to introduce and analyze general algorithms that solve NP problems.

12.3.1 Constraint Satisfaction Problems (CSPs)

Constraint satisfaction problems (also referred to sometimes as constraint networks) provide a framework for problems in which values are assigned to variables subject to certain constraints. For an idea on the wide applicability of this concept, cf. [52, 153, 174, 185, 265]. In our examples we will concentrate on the translation of the fixed point problem into this realm. As is mentioned in the introduction to [265] the wide applicability of constraint satisfaction problems has led to many rediscoveries of the setup and of solution algorithms as well as to a multitude of terminologies. This presentation can only be an introduction using one terminology. In particular we will strive to give graphical interpretations of the constructions made. The author believes this approach to be a fruitful one that is not very explicit in the literature the author is aware of. Our presentation should give an overview of this area and how it interfaces with ordered sets and other branches of mathematics.

Definition 12.3.4 *A* **binary constraint satisfaction problem (CSP)** *(compare [265], Section 1.2; or* **binary constraint network***, compare [52], p. 276) consists of the following.*

1. *A set of variables $x_1, \ldots, x_r$.*

2. *A set of domains[4] $D_1, \ldots, D_r$, one for each variable.*

3. *A set C of unary and binary constraints.*

   - *Each unary constraint consists of a variable x_i and a set $C_i \subseteq D_i$,*

   - *Each binary constraint consists of a set of two variables $\{x_i, x_j\}$ and a binary relation $C_{ij} \subseteq D_i \times D_j$,[5]*

   - *For each set of variables we have at most one constraint.*

One can also define higher order CSPs using k-ary constraints. For our purposes binary CSPs will be sufficient. Moreover there is a translation process that turns higher order constraint satisfaction problems into binary ones (cf. [55], p.355), though we will not elaborate on it here.

Note that the above definition does not say anything about which assignments are allowed and which are not. Allowed assignments for some or all variables are defined as follows.

Definition 12.3.5 *(Compare [52], p.276.) For a given CSP, let $Y \subseteq \{1, \ldots, r\}$. Any set $\{(x_i, a_i) : a_i \in D_i, i \in Y\}$ is an* **instantiation** *of the variables $\{x_j : j \in Y\}$. An instantiation of the variables $\{x_j : j \in Y\}$ is called* **consistent** *iff for all $i, j \in Y$ we have that $a_i \in C_i$ and $(a_i, a_j) \in C_{ij}$.[6]*
A consistent instantiation for all variables $\{x_i : i \in \{1, \ldots, r\}\}$ is called a **solution***.*

The above means that the unary constraints encode the consistent instantiations of single variables and the binary constraints encode the consistent instantiations of sets of two variables. (These ideas are easily generalized to ternary and higher order constraints.) The question one is interested in is of course if there is a solution for a given binary CSP. That is, the central problem in binary constraint satisfaction is

Given. A binary CSP.
Question. Is there a solution for the given binary CSP?

---

[4]Careful here. These are the values that will *be assigned to the variables*. In mathematics, one would be tempted to call these sets ranges.

[5]This specification of constraints for *sets* $\{x_i, x_j\}$ of variables has some ambiguity. After all, the order of the variables matters in the specification of C_{ij}. The alternative, which makes things unnecessarily technical, would be to specify constraints for ordered pairs of variables and demand the appropriate symmetry for the constraints C_{ij} and C_{ji}.

[6]Careful with notation here. In some work an instantiation is called consistent if $(a_i, a_j) \notin C_{ij}$, which is in keeping with a constraint being something that forbids configurations.

Clearly the above is another NP problem, since for a given instantiation of all variables it can be verified in $O(n^2)$ time if it is a solution. Other questions that can be asked or tasks that can be given if the context is appropriate are

- How many solutions are there?

- List all solutions.

- In case a weight function for the instantiations is given, find a solution with lowest possible weight.

Let us now see how to cast the fixed point property into the framework of CSPs. This was first done (though apparently without any connection to the main body of literature on constraint satisfaction) in [278], where the following CSP is presented in the setting of formal concept analysis.

Example 12.3.6 For determining if the ordered set P has a fixed point free self map, we let $\{x_1, \ldots, x_r\} := P$, that is, we let our variable set be equal to P. For each x_i the domain D_i is $D_i := \{p \in P : p \not\geq x_i\}$. The constraints C_{ij} are

- If $x_i \not\leq x_j$, there is no constraint between x_i and x_j.

- If $x_i \leq x_j$, let $C_{ij} := \{(u_i, u_j) \in D_i \times D_j : u_i \leq u_j\}$.

Note that the above gives exactly one constraint for each set of two variables. Any consistent instantiation $\{(x_{i_1}, u_{i_1}), \ldots, (x_{i_k}, u_{i_k})\}$ of k variables $x_{i_1}, \ldots, x_{i_k}$ corresponds to an order-preserving map from $\{x_{i_1}, \ldots, x_{i_k}\}$ to P, namely the map that maps each x_{i_j} to u_{i_j}.

Other problems that can be translated into this venue are given in Exercise 10. Another way to translate the problem if a given ordered set has a fixed point free order-preserving self map is outlined in Exercise 28. The above translation embeds the specific problem of finding a fixed point free map of an ordered set into the framework of a more general problem. This allows us to apply the tools that exist for CSPs. We will present some of these tools in Sections 12.5.1 and 12.5.2. The reader who would like to investigate CSPs from an order-theoretical point-of-view may ask if there is a way to express CSPs as problems of certain ordered sets having the fixed point property. There is such a way. Essentially one could combine Cook's theorem (cf. Theorem 12.4.2) and Theorem 12.4.5 to embed CSPs into fixed point problems. Since this approach is very cumbersome, it is not used.

12.3.2 Expanded Constraint Networks/Cliques in r-colored Graphs

A more visual, equivalent presentation of constraint satisfaction problems is the representation through colored graphs. This is also called the **expanded constraint network** in [185], p. 195.[7]

For the translation in this section we will need to recall the definition of a coloring of a graph and the definition of a clique.

Definition 12.3.7 *Let $G = (V, E)$ be a graph and let C be a set (of "colors"). Then any function $\gamma : V \to C$ is a $|C|$-(**vertex-**)**coloring** of G. An r-**colored graph** is a triple $G = (V, E, \gamma)$, where (V, E) is a graph and γ is a coloring with r colors.*

Since there is no risk of confusion with edge colorings (which we will not discuss at all), we will drop the "vertex" and talk about r-colored graphs. Also recall (cf. Definition 6.3.1) that a clique was a set of vertices such that any two distinct vertices have an edge between them. We will call cliques with r vertices r-**cliques.** The NP problem that is the graph theoretical version of the constraint satisfaction problem in the previous section is as follows.

Given. A vertex-r-colored graph G.
Question. Does G contain an r-clique?

How are these problems equivalent? The following theorem states that they are. More importantly, the proof gives an exact construction to go from one problem to the other. The lack of unary constraints is no loss of generality whatsoever. Unary constraints only restrict domains, which can be done before formulating a CSP.

Theorem 12.3.8 *For every binary CSP with r variables and without unary constraints there is an r-colored graph $G = (V, E, \gamma)$ such that the consistent instantiations of the CSP are in one-to-one correspondence with the cliques of the graph. Moreover for every r-colored graph H there is a CSP with r variables and without unary constraints such that the graph as above is isomorphic to H.*

Proof. Given a binary CSP with r variables $x_1, \ldots, x_r$, let

$$C := \{x_1, \ldots, x_r\}$$

be the set of colors. The set of vertices will be

$$V := \{(x_i, u) : u \in D_i, i \in \{1, \ldots, n\}\},$$

that is, it can be viewed as the set of consistent instantiations of single variables. For unconstrained sets of variables $\{x_i, x_j\}$ we shall assume the dummy constraint

---

[7]The reason why it is not called the constraint graph is simply that term is already used for another entity, cf. Definition 12.5.9.

$C_{ij} := D_i \times D_j$ has been added. To be on formally completely solid ground we can also assume that for constrained sets of variables $\{x_i, x_j\}$ we have constraints C_{ij} and C_{ji} that are duals of each other. The set of edges will be

$$E := \{\{(x_i, u), (x_j, v)\} : (x_i, u), (x_j, v) \in V, (u, v) \in C_{ij}, i \neq j\},$$

that is, the edges are the consistent instantiations of pairs of variables. Finally we define $\gamma(x, u) := x$. Then $G = (V, E, \gamma)$ is an r-colored graph. Now suppose $\{(x_{i_1}, u_{i_1}), \dots, (x_{i_k}, u_{i_k})\}$ is a consistent instantiation of the variables $\{x_{i_1}, \dots, x_{i_k}\}$. Then for all $a \neq b$ we have $\{(x_{i_a}, u_{i_a}), (x_{i_b}, u_{i_b})\} \in E$, which means that the set $\{(x_{i_1}, u_{i_1}), \dots, (x_{i_k}, u_{i_k})\}$ is a clique. Conversely, if $\{(x_{i_1}, u_{i_1}), \dots, (x_{i_k}, u_{i_k})\}$ is a clique, then for all $a \neq b$ we have that the instantiation $\{(x_{i_a}, u_{i_a}), (x_{i_b}, u_{i_b})\}$ is a consistent instantiation of $\{x_{i_a}, x_{i_b}\}$. Therefore $\{(x_{i_1}, u_{i_1}), \dots, (x_{i_k}, u_{i_k})\}$ is a consistent instantiation of the variables $\{x_{i_1}, \dots, x_{i_k}\}$. Thus the consistent instantiations of the CSP are the cliques of the graph we defined.

To see that every r-colored graph $H = (V_H, E_H, \gamma_H)$ can be obtained in this fashion, let $H = (V_H, E_H, \gamma_H)$ be an r-colored graph. Let $C = \{x_1, \dots, x_r\} := \gamma[V]$ be our set of variables. The domain for every variable x_i is $D_i := \gamma^{-1}(x_i)$. We will not have any unary constraints and the binary constraints are $C_{ij} := \{(v, w) : v \in D_i, w \in D_j, \{v, w\} \in E_H\}$. The translation of this CSP to an r-colored graph yields a graph isomorphic to H.   ∎

In the graph-theoretical framework notation is very compact. Moreover, graph-theoretical and geometrical intuition are available to help with the solution of the problem at hand. In the future, we will use graph-theoretical visualizations of CSPs whenever appropriate, assuming that the reader can easily make the above translation.

12.4   NP-completeness

NP-completeness is a formalized way of saying/proving that a problem is "hard". This section will define what exactly this statement means. To discuss the concept of NP-completeness we will need to define a problem in formal logic which can be viewed as another prototype of NP problems. Recall that a logical variable (or literal) is a variable that can be assigned the values TRUE or FALSE and that "∨" denotes the logical OR operation.

Example 12.4.1   The problem 3SAT. (Cf. [94].) Given n logical variables ("literals") and m clauses $c_j = q_j \vee r_j \vee s_j$, where each q_j, r_j, s_j can be a literal or its negation, decide if there is a truth value assignment to the literals such that all clauses are true.

3SAT is a problem in NP. For any truth value assignment to the n literals it will take polynomial time $p(n)$ to check if all clauses are satisfied. The importance of 3SAT stems from a result by Cook, which shows that 3SAT is of a very universal nature.

Theorem 12.4.2 (Cook's theorem, cf. [42] or [94], Theorem 2.1.) *Suppose there is an algorithm that solves 3SAT in polynomial time. Then for every NP problem there is a polynomial time solution algorithm.*

The proof of Theorem 12.4.2 does not fit our scope and the interested reader is directed to [42] and [94], Theorem 2.1, where the result is proved for SAT. Theorem 12.4.2 shows a fact that the translations shown previously may have suggested to the reader. There are certain "universal problems" into which all NP problems can be translated. So if one can solve these problems efficiently, then all NP problems can be solved efficiently. These universal problems are called NP-complete.

Definition 12.4.3 *An NP problem X is said to be **NP-complete** iff an algorithm that solves X in polynomial time would induce an algorithm that solves 3SAT with n variables in polynomial time $p(n)$.* [8]

One very common practice in analyzing problems for which one has not found a polynomial algorithm yet is to attempt to prove that the problem is NP-complete. This proof normally involves an embedding of 3SAT (or another NP-complete problem) into the problem in question. In the following we will exhibit Duffus and Goddard's proof that checking for the existence of a fixed point free order-preserving map is an NP-complete problem. We start by showing that a related problem is NP-complete.

Lemma 12.4.4 (Cf. [62], Theorem 4.1.) *The following problem is NP-complete.*

Given. *Two ordered sets P and Q and $p_1, \ldots, p_t \in P$, $q_1, \ldots, q_t \in Q$.*
Question. *Is there an order-preserving map $f : P \to Q$ with $f(p_i) \geq q_i$?*

Proof. We shall first embed 3SAT into a similar problem for graphs. Let an instance of 3SAT with clauses $C_1, \ldots, C_m$ and literals $x_1, \ldots, x_n$ be given. Order the literals in each clause in a fixed fashion (for example sort them such that the indices increase). Define graphs $G = (V_G, E_G)$ and $H = (V_H, E_H)$ as follows.

The vertices of G are called $C_1^*, \ldots, C_m^*$ and $x_1^*, \ldots, x_n^*$. There is an edge in G between C_i^* and x_j^* iff the literal x_j or its negation occurs in C_i and there are no other edges.

In H we have vertices $C_1^1, C_1^2, C_1^3, C_2^1, C_2^2, C_2^3, \ldots, C_m^1, C_m^2, C_m^3$ and vertices $x_1^T, x_1^F, x_2^T, x_2^F, \ldots, x_n^T, x_n^F$. Again edges only exist between C's and x's. If x_j or its negation occurs in C_i, but not in the l-th place, put an edge between C_i^l and x_j^T and an edge between C_i^l and x_j^F. If x_j occurs in the l-th place of C_i, put an edge between C_i^l and x_j^T. If $\overline{x_j}$ occurs in the l-th place of C_i, put an edge between C_i^l and x_j^F. These are all the edges.

---

[8] Generally this is done by mapping 3SAT in polynomial fashion into specific instances of X. This definition has be be considered somewhat "informal".

We claim that the instance of 3SAT has a satisfying assignment iff there is a graph homomorphism f from G to H such that $f(C_i^*) \in \{C_i^1, C_i^2, C_i^3\}$ and $f(x_j^*) \in \{x_j^T, x_j^F\}$. Indeed, if there is such a map, then the assignment of "true" to x_j iff $f(x_j^*) = x_j^T$ gives a satisfying assignment. For the converse, fix a satisfying assignment. Then for each C_i there is a literal or a negation thereof in C_i that must evaluate to being true, say one is in the l_i-th place. Map C_i^* to $C_i^{l_i}$, map x_j^* to x_j^T iff x_j is assigned the value "true" and map x_j^* to x_j^F otherwise. This map is as desired.

Now we have to add a structure to change the conditions $f(C_i^*) \in \{C_i^1, C_i^2, C_i^3\}$ and $f(x_j^*) \in \{x_j^T, x_j^F\}$ into conditions about adjacencies. To do this we obtain graphs $K_P = (V_P, E_P)$ from G and $K_Q = (V_Q, E_Q)$ from H as follows. To construct K_P from G add vertices $C_i^0, x_i^a, x_i^b, i = 1, \ldots, m$ to G. Each C_i^0 is adjacent to C_i^*, while x_i^a and x_i^b are both adjacent to x_i^*. These are all the additions to G. To construct K_Q from H add vertices $D_i^0, x_i^c, x_i^d, i = 1, \ldots, m$ to H. Each D_i^0 is adjacent to C_i^1, C_i^2, C_i^3 while x_i^a and x_i^b are both adjacent to x_i^T and x_i^F. These are all the additions to H.

It is easy to see that there is a homomorphism from G to H as described above iff there is a homomorphism g from K_P to K_Q such that $g(C_i^0) = D_i^0, g(x_i^a) = x_i^c$ and $g(x_i^b) = x_i^d$. Now let P be the dual of the split of K_P and let Q be the dual of the split of K_Q. Then there is a homomorphism g from K_P to K_Q such that $g(C_i^0) = D_i^0, g(x_i^a) = x_i^c$ and $g(x_i^b) = x_i^d$ iff there is an order-preserving map $h : P \to Q$ such that $h(C_i^0) \geq D_i^0, h(x_i^a) \geq x_i^c$ and $h(x_i^b) \geq x_i^d$. (Note that since the elements in question are maximal, the inequality is in fact an equality). $\blacksquare$

The proof of NP-completeness of checking for the existence of a fixed point free order-preserving map[9] is now accomplished by embedding the above problem into a fixed point setting via the (almost) lexicographic construction in Proposition 9.5.2.

Theorem 12.4.5 (Cf. [62], Theorem 1.1.) *The following decision problem is NP-complete.*

Given. *A finite ordered set P.*
Question. *Is there a fixed point free order-preserving map $f : P \to P$?*

Proof. First note that for the sets P and Q in the proof of Lemma 12.4.4 it is easy to find maps $h' : Q \to P$ and $h'' : P \to P$ such that $h'(D_i^0) \geq C_i^0$, $h'(x_i^c) \geq x_i^a, h'(x_i^d) \geq x_i^b$, and $h''(C_i^0) \geq C_i^0, h''(x_i^a) \geq x_i^a, h''(x_i^b) \geq x_i^b$, respectively.

Now in the construction of Proposition 9.5.2 let X and Y be copies of P and let Z be a copy of Q. Call the resulting ordered set W. Let $t = 2n + m$, let $x_1, \ldots, x_t$ and $y_1, \ldots, y_t$ each be $C_1^0, \ldots, C_m^0, x_1^a, \ldots, x_n^a, x_1^b, \ldots, x_n^b$ in this or-

---

[9]This is equivalent to saying that determining if a given finite ordered set has the fixed point property is co-NP-complete.

der and let $z_1, \ldots, z_t$ be $D_1^0, \ldots, D_m^0, x_1^c, \ldots, x_n^c, x_1^d, \ldots, x_n^d$ in this order. Now if the thus constructed ordered set has a fixed point free order-preserving map, then by Proposition 9.5.2 without loss of generality there is an order-preserving map $f : Y \rightarrow Z$ such that $f(y_i) \geq z_i$. This implies that there is an order-preserving map $h : P \rightarrow Q$ such that $h(C_i^0) \geq D_i^0, h(x_i^a) \geq x_i^c$ and $h(x_i^b) \geq x_i^d$. Conversely, if there is such a map it is simple to construct a fixed point free order-preserving map $f : W \rightarrow W$.

Thus, if there was a polynomial algorithm that decides if a finite ordered set has the fixed point property, then there would be a polynomial algorithm that decides the problem in Lemma 12.4.4.   ∎

We could now continue and show that there are other NP problems that are interesting in order theory and which are NP-complete. In particular we mention

Corollary 12.4.6 *The constraint satisfaction problem given in Section 12.3.1 is NP-complete.*

Proof. It was already remarked that the constraint satisfaction problem is NP. Suppose we had a polynomial algorithm that decides if a given CSP has a solution. Then by Example 12.3.6, we would have an algorithm that decides in polynomial time if a given ordered set has a fixed point free order-preserving self map. By Theorem 12.4.5 this would lead to a polynomial algorithm for 3SAT.   ∎

Instead of proceeding further in this direction, some NP problems are to be shown to be NP-complete in Exercise 12 and the complexity status of some problems is discussed in the remarks.

12.5   So It's NP-complete...

... so what? Defiance in the face of near-insurmountable odds is a distinct human trait.[10] There are several avenues to go. While it has not been proved that there is *no* polynomial algorithm to solve NP-complete problems, it appears that the resolution of this question is a problem comparable to Fermat's last conjecture in its difficulty (consider Remark 3). Still fearless souls should not shrink away from this problem.[11]

On the other hand, one also has to realize that NP-completeness and complexity in general is a worst case notion. Certainly the ordered sets we encountered when proving Theorem 12.4.5 do not appear typical to this author. Many problems have terrible worst case behavior and yet the cases one is likely to encounter are com-

---

[10]For example, even though it is acknowledged to be difficult, people continue to raise children, teach mathematics, explore space ...

[11]One should just remain careful and realistic. For example, the author has found several wrong proofs that the fixed point property can be solved in polynomial time. One of these approaches also shows a typical trap of theoretical work with CSPs is outlined in Exercises 27–29.

paratively easy to solve. Thus it can be fruitful to explore special scenarios in which certain problems are solvable in polynomial time. Two such examples are the checking for fixed point free maps in finite ordered sets of height 1 or width 2. In both classes the existence of a fixed point free order-preserving map can be checked in polynomial time. These examples are consequences of groundwork we laid earlier and they are to be treated by the reader in Exercise 8.

Finally, one may also be interested not just in existence or non-existence of a solution, but also in all solutions to a certain problem or the number of solutions. In this case one may have to deal with a potentially exponential problem by default.

In the following we will present the two main search paradigms for CSPs and the basics of local consistency enforcing algorithms. These algorithms are practical day-to-day tools for researchers in constraint satisfaction. Moreover, their formal structure also allows us to prove that certain types of problems can be solved in polynomial time.

We will first describe the search algorithms and then the consistency enforcing algorithms. Generally one would first preprocess the CSP with a consistency enforcing algorithm and then search with a search algorithm. The art is always to balance the gains in the search obtained through higher consistency against the overhead invested in preprocessing. The description given here can only serve as a brief overview of the vast topic of search and reduction algorithms for CSPs. An excellent introduction to search algorithms is [148].

12.5.1 Search Algorithms

The **search space** for solving a CSP in n variables is the set of all possible instantiations of the variables. This space in particular contains the solutions of the CSP (if any exist). The task of every search algorithm is to find these solutions in the search space or to report that no solutions exist. A very primitive (and inefficient) idea is to first linearly order the variables and their domains[12], then recursively compute all possible instantiations for $\{x_1, \ldots, x_k\}$ from the set of all possible instantiations for $\{x_1, \ldots, x_{k-1}\}$ and finally, when all possible instantiations for all variables are available, check each one if it is a solution or not.

The reader immediately checks that the set thus computed has $\Pi_{i=1}^{n}|D_i|$ elements, which is generally an unacceptably large number. Moreover the indiscriminate generation of instantiations will force us to check many instantiations that we could have ruled out with less effort. To wit, if $\{(x_1, u_1), (x_2, u_2)\}$ is inconsistent, then none of its extensions can be consistent and we need not check any of them. Thus the search space generally is an entity that remains in the background unless one wants to do a theoretical analysis of search algorithms as in, say [252] or [276]. We include its formal definition for completeness' sake.

[12]For our purposes we can always assume that the order of the variables $x_1, \ldots, x_n$ is the order of the indices. We will not explicitly specify the order of the domains.

Definition 12.5.1 *The **search tree** or **search space** of a CSP in n variables is the set*

$$\{\{(x_1, u_1), \ldots, (x_k, u_k)\} : u_i \in D_i, k \leq n\}$$

of all instantiations of the first k variables for $0 \leq k \leq n$, ordered by reverse inclusion.

The search tree as defined above is indeed a tree. The proof of this simple fact is left to the reader. Let us also note here that this definition assumes a static ordering of the variables. The subject of variable ordering heuristics will be touched only briefly in this text.

There are now many ideas for searching trees in a more efficient manner than the above indicated exhaustive search (cf. [148]). In this text we will focus on the two main paradigms for search algorithms. These are backtracking and forward checking. Throughout this subsection we will assume that the variables are ordered as indicated by their indices and that there is a fixed value order on each domain.

Backtracking. The backtracking algorithm maintains a consistent instantiation of the first k variables at all times. (At the start this is the empty set.) Given a consistent instantiation CI of the first k variables, backtracking instantiates x_{k+1} to the first value $u_{k+1,1}$ of D_{k+1}. If $CI \cup \{(x_{k+1}, u_{k+1,1})\}$ is consistent, then CI is replaced with $CI \cup \{(x_{k+1}, u_{k+1,1})\}$ and backtracking tries to instantiate x_{k+2}. If not, the next value in D_{k+1} is tried. If backtracking does not find any instantiation of x_{k+1} that allows consistent extension of the current instantiation, then x_k is uninstantiated. That is, CI is replaced with $CI \setminus \{(x_k, u_{k,\text{current}})\}$. The search then resumes as above by instantiating x_k to the next element of the kth domain, $u_{k,\text{current}+1}$. If backtracking encounters a solution, the algorithm stops if only one solution was to be found. It continues with a backtrack if all solutions were to be found. If backtracking terminates without finding any solutions, then there is no solution for the CSP (cf. Exercise 14).

Pseudocode for a recursive backtracking algorithm is given in Figure 12.1. Note that while this is the most efficient way to talk about backtracking, a non-recursive implementation normally has shorter run-times because of lower overheads in the control structures. Backtracking terminates when a solution is found or after the whole tree of instantiations generated this way has been searched. In the latter case, either there is no solution or all solutions have been listed. The tree generated by backtracking is easily characterized.

Definition 12.5.2 (Compare [148], Figure 3.) *For a CSP that has n variables $\{x_1, \ldots, x_n\}$ with domains $D_1, \ldots, D_n$ we define the **backtracking tree** to be the set of all instantiations $\{(x_1, v_1), \ldots, (x_k, v_k)\}$ such that the parent instantiation $\{(x_1, v_1), \ldots, (x_{k-1}, v_{k-1})\}$ is consistent. The backtracking tree is assumed to be ordered by reverse inclusion.*

For an example of a backtracking tree, cf. Figure 12.2. We shall say a search algorithm **visits** an instantiation I iff at some time during the execution the al-

recursive algorithm `backtrack`(CI, depth)

If there is another instantiation u for x_{depth} that was not checked against CI

    If $CI \cup \{(x_{\text{depth}}, u)\}$ is consistent and depth $< n$

        Replace depth with depth+1, CI with $CI \cup \{(x_{\text{depth}}, u)\}$

        Call `backtrack`(CI, depth)

    If $CI \cup \{(x_{\text{depth}}, u)\}$ is consistent and depth $= n$

        Output $CI \cup \{(x_{\text{depth}}, u)\}$ as a solution

        If the algorithm only needs to find the first solution, stop

        If the algorithm is to find all solutions

            Go back to the start of this routine

    If $CI \cup \{(x_{\text{depth}}, u)\}$ is not consistent

        Go back to the start of this routine

If all instantiations for x_{depth} have already been tested

    If depth$= 1$, stop; otherwise

    Remove the instantiation of $x_{\text{depth}-1}$ from CI

    Replace depth with depth-1

    Return to the previous level of execution

Figure 12.1: Pseudocode for a recursive backtracking algorithm. The algorithm is called with $CI = \emptyset$ and depth $= 1$. Instantiations for x_{depth} that have not been checked against CI are normally "detected" by checking all domains in a fixed order. If the recently checked or removed instantiation is the last in this order, we need to backtrack.

gorithm checks if I can become the algorithm's current instantiation CI. The reader will prove in Exercise 14 that the instantiations of the backtracking tree are exactly the instantiations visited by backtracking.

Forward Checking. Backtracking is a paradigm that in its crucial step considers the *past*, not the future. Backtracking only checks if the new instantiation for x_{k+1} is consistent with the already recorded instantiations for $x_1, \ldots, x_k$. The paradigm for algorithms that consider the *future* is forward checking.

Definition 12.5.3 (Compare [148], Definition 2.) *Let A and B be two instantiations of disjoint sets of variables. Then A is called* **consistent with** *B iff* $A \cup B$ *is consistent.*

Definition 12.5.4 *A consistent instantiation* $\{(x_1, u_1), \ldots, (x_k, u_k)\}$ *of the first k variables in a CSP will be called* **forward consistent** *iff for all* $i > k$ *there is a* $u \in D_i$ *such that* (x_i, u) *is consistent with* $\{(x_1, u_1), \ldots, (x_k, u_k)\}$.

Example 12.5.5 To describe the connection between consistent and forward consistent instantiations refer to Figure 1.4 e). For parts 1 to 3 we shall consider the task of enumerating all order-preserving maps of this set. We assume that the variables (points) are ordered alphabetically.

1. The instantiation $\{(a, a), (b, c), (c, b), (d, e)\}$ is consistent.

2. The instantiation $\{(a, d), (b, e), (c, f)\}$ is consistent, but not forward consistent. Indeed, there is no instantiation for the variable e that is consistent with $\{(a, d), (b, e), (c, f)\}$. (There is one for d, though. This is the strength of forward consistency. Problems that are "(far) ahead" can be recognized "early".)

3. Every forward consistent instantiation of $\{a, b, c\}$ actually is part of an order-preserving map.

4. Forward consistent instantiations need not be part of solutions. As an example, consider the problem of finding automorphisms for the same set with the variables (points) ordered in reverse alphabetical order. The instantiation $\{(g, h), (h, k), (k, g)\}$ is forward consistent for this problem, but not part of a solution (that is, an automorphism).

The reader will generate similar examples for the search for fixed point free order-preserving maps in Exercise 15.

The forward checking algorithm maintains at all times a forward consistent instantiation CI of the first k variables and a list of remaining possible extensions $(x_i, c_{i,j})$, $i > k$ that are consistent with CI. (We will assume this list is updated as needed as we describe the algorithm.) Given a forward consistent instantiation CI of the first k variables, forward checking instantiates x_{k+1} to the first value $c_{k+1,1}$ of D_{k+1} that is consistent with CI. If $CI \cup \{(x_{k+1}, c_{k+1,1})\}$ is forward consistent,

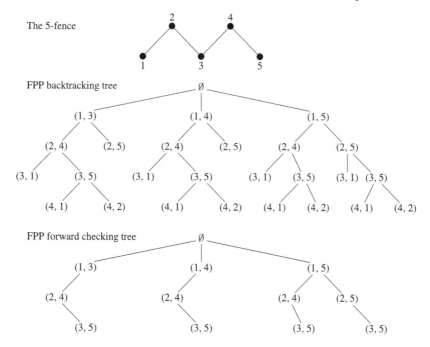

Figure 12.2: The backtracking and forward checking trees for determining if the 5-fence has a fixed point free order-preserving self map. The unary constraint is that no element is to be comparable to its image.

then CI is replaced with $CI \cup \{(x_{k+1}, c_{k+1,1})\}$ and forward checking instantiates x_{k+2}. If not, the next consistent value in D_{k+1} is tried. If forward checking does not find any instantiation of x_{k+1} that allows us to extend the current instantiation to a forward consistent instantiation of $x_1, \ldots, x_k, x_{k+1}$, then x_k is uninstantiated. That is, CI is replaced with $CI \setminus \{(x_k, c_{k,\text{current}})\}$. The search then resumes as above by instantiating x_k to the next consistent element of D_k, $c_{k,\text{current}+1}$.

If forward checking encounters a solution, the algorithm stops if only one solution was to be found. It continues with a backtrack if all solutions were to be found. If forward checking terminates without finding any solutions, then there is no solution for the CSP (cf. Exercise 16).

Definition 12.5.6 (Compare [148], Figure 3.) *For a CSP that has n variables* $\{x_1, \ldots, x_n\}$ *with domains* $D_1, \ldots, D_n$ *we define the* **forward checking tree** *to be the set of all consistent instantiations* $\{(x_1, v_1), \ldots, (x_k, v_k)\}$ *such that the parent* $\{(x_1, v_1), \ldots, (x_{k-1}, v_{k-1})\}$ *is forward consistent, ordered by reverse inclusion.*

The reader will immediately notice that the forward checking tree is a subtree of the backtracking tree. In Figure 12.2 the forward checking tree is the backtracking tree without the leaves. Indeed, the reader will prove in Exercise 17b that all nodes of the forward checking tree are interior nodes of the backtracking tree. The removal of the leaves of the backtracking tree is already a large reduction in the number of nodes visited. However, forward checking usually prunes the search tree even more. In Exercise 17c the reader is asked to order the variables for the 5-fence in such a way that the backtracking tree has an interior node that is not in the forward checking tree.

Thus if one's only concern is the size of the tree searched, then forward checking is the better algorithm. However when traversing the respective search trees, the algorithms encounter overhead of different sizes at every node. Just consider the fact that to check if an extension $CI \cup \{(x_{k+1}, u_{k+1})\}$ of a consistent instantiation CI is still consistent, backtracking has to check k pairs of instantiations $\{(x_i, u_i), (x_{k+1}, u_{k+1})\}$, $i < k$ for consistency. In contrast, to check if a consistent extension $CI \cup \{(x_{k+1}, u_{k+1})\}$ of a forward consistent instantiation CI is still forward consistent, forward checking for each variable x_i with $i > k$ has to check if (x_{k+1}, u_{k+1}) is consistent with at least one of the consistent extensions of CI. In fact, to update the list of remaining possibilities, forward checking has to check consistency between (x_{k+1}, u_{k+1}) and all consistent extensions of CI, erasing those that are not consistent with (x_{k+1}, u_{k+1}). Note however, that the overhead that forward checking encounters *decreases* as we go deeper into the tree, while the overhead that backtracking encounters *increases*.

The large overhead appears to make forward checking slow. Still the reduction in the tree size generally makes up for the larger overhead encountered for instantiations of the first few variables (cf. [204]). There is an implementation of forward checking that reduces the overhead at every extension $CI \cup \{(x_{k+1}, u_{k+1,c})\}$ to $n - k - 1$ bitwise ANDs. For details the reader is referred to [176], Sec. 4 for the representation idea and [105], p. 270–271 for the application to forward checking. We outline this idea in Exercise 20. This bit-parallel forward checking algorithm is very efficient.

As noted, an example of a forward checking tree is given in Figure 12.2. The reader should also check the proof of Proposition 1.4.6 and his/her solution to Exercise 18b in Chapter 1 to see how backtracking and forward checking ideas are present in these proofs.

Unfortunately even the most efficient implementations of either algorithm are still in the worst case exponential. While this problem cannot be resolved by the following preprocessing algorithms, the idea of reducing the CSP before starting a search can greatly reduce the search time.

12.5.2   Algorithms That Enforce Local Consistency

Complementing the idea of differently structured search algorithms is the idea to pre-process the CSP to obtain an equivalent CSP that is easier to search.

Definition 12.5.7 *Two CSPs are called* **equivalent** *iff they have the same solutions.*

The underlying idea is to establish a certain level of local consistency defined as follows.

Definition 12.5.8 (Cf. [91], p. 26.) *A CSP is called k-**consistent** iff for each consistent instantiation of* $(k - 1)$ *variables and any choice of a* k^{th} *variable there is an instantiation of the* k^{th} *variable such that the k instantiations together also are consistent.*

In particular (cf. [173]), 1-consistency is also called **node consistency**, *2-consistency is also called* **arc consistency**, *3-consistency is also called* **path consistency**[13].

A CSP is **strongly** *k-**consistent** iff for all* $1 \le j \le k$ *the CSP is* j*-consistent.*

From the definition we can see that local consistency only guarantees certain subsets of the set of all variables to have consistent instantiations. This type of local consistency is probably all one can hope for if one wants the preprocessing algorithm to run in polynomial time.

The simplest idea to enforce k-consistency is to compute all consistent instantiations CI of $k - 1$ variables $\{x_{i_1}, \dots, x_{i_k}\}$ such that for each variable x_j there is an instantiation $\{(x_j, u_j)\}$ such that $CI \cup \{(x_j, u_j)\}$ is consistent. The equivalent CSP would have to have $(k - 1)$-ary constraints that only admit instantiations as given above. This very simple-minded approach takes polynomial time for fixed k. Refinements to this idea can be found in Chapter 4 in [265]. It is to be noted that the order of the polynomial that bounds the effort for enforcing k-consistency increases with k. It was observed in [53] that enforcing a high order of consistency need not be efficient. Another problem with high levels of k-consistency is that one is forced to work with k-ary constraints, which require a lot of storage space.

Yet for well-behaved instances of a CSP, local consistency can lead to a fast solution. Essentially what is needed is that for the given level of consistency the constraints are not "too dense". The idea of density of constraints is expressed in the width of the constraint graph.

Definition 12.5.9 *A nontrivial constraint between two variables is a constraint that declares at least one instantiation of the two variables as inconsistent. The* **constraint graph** *of a binary CSP has the variables* $\{x_1, \dots, x_n\}$ *as its vertex set* V. *There is an edge between* x_i *and* x_j *iff there is a nontrivial constraint between these two variables, that is,* $E = \{\{x_i, x_j\} : C_{ij} \text{ is not trivial }\}$.

The **width** *of the constraint graph is the smallest integer* k *such that there is a linear ordering* $x_{i_1}, \dots, x_{i_n}$ *of the vertices such that for each d the vertex* x_{i_d} *is adjacent to at most k vertices* x_{i_j} *with* $j < d$.

Example 12.5.10 In Example 12.3.6 the constraint graph is the comparability graph of the ordered set.

Using the ideas of strong k-consistency and width, one can identify some situations in which search does not run into any dead ends.

---

[13]This name will be motivated in Exercise 21.

Theorem 12.5.11 (Cf. [91], Theorem 1, part 2.) *If for a CSP the level of strong k-consistency is larger than the width of the constraint graph, then there exists a backtrack-free search order for the CSP.*

Proof. Assume without loss of generality that the ordering $x_1, \ldots, x_n$ of the variables realizes the width of the constraint graph. Backtracking starts by assigning an allowed value to x_1. Now assume that variables $x_1, \ldots, x_d$ have already been instantiated. Then, since x_{d+1} only shares constraints with $l < k$ variables among $x_1, \ldots, x_d$, say $x_{i_1}, \ldots, x_{i_l}$, by strong k-consistency there is an instantiation for x_{d+1} such that it plus the present instantiation of $x_{i_1}, \ldots, x_{i_l}$ form a consistent instantiation of $x_{i_1}, \ldots, x_{i_l}, x_{d+1}$. Assign x_{d+1} this value and continue. This procedure either does not find any instantiations for x_1 or the above procedure finds a consistent overall instantiation in n steps. ∎

A tempting trap. While Theorem 12.5.11 is strong, it is easy to read too much strength into it. It is tempting to argue that for any CSP whose constraint graph is of width k, we simply form a strongly $(k + 1)$-consistent CSP with the same solutions and then search it without backtracking. For example, we can modify Example 12.3.6 to an equivalent problem by erasing all constraints that involve nonadjacent points in the ordered set. The constraint graph of this problem would be the diagram of the ordered set and its width would be bounded by the width of the ordered set. Strong $(k + 1)$-consistency can be achieved in polynomial time. Thus it is now tempting to conclude that for ordered sets of width w there is a polynomial algorithm (the order of the polynomial would depend on the width) that decides if the ordered set has a fixed point free order-preserving self map.

Unfortunately (as is noted frequently in the literature, cf. [52], p. 280 or [91], p. 32) an algorithm that generates an equivalent strongly k-consistent CSP can (and normally does) change the constraint graph and increase its width. The increased width then makes a higher level of local consistency necessary. Achieving this next level of local consistency may again increase the width, etc. This trap is illustrated in Exercises 27, 28 and 29. Along similar lines, experimental results (cf. [53], top of p. 213) with what is called adaptive consistency suggest that enforcing high levels of consistency may be less efficient than brute force search because of the overhead encountered.

We shall conclude this section with a more detailed discussion of 3-consistency. The author found this notion helpful and efficient in working with the fixed point property. The language we use will be the graph-theoretical language of the expanded constraint network.

Example 12.5.12 (A simple 3-consistency algorithm.)

Given. An expanded constraint network with r variables.
Task. Compute an equivalent 3-consistent expanded constraint network.

The algorithm simply removes superfluous edges until no further such edges can be found.

Repeat

    `changed := FALSE`

    For every edge $\{(x, u), (y, v)\}$

        For every color $z \in C \setminus \{x, y\}$

            Check if there is an instantiation $\{(z, w)\}$ such that $\{(x, u), (y, v), (z, w)\}$ is consistent. If there is no such instantiation, remove $\{(x, u), (y, v)\}$ from the expanded constraint network and set `changed := TRUE`

        end for z

    end for $\{(x, u), (y, v)\}$

until `changed = FALSE`.

The above algorithm runs in polynomial time. Indeed, all for-loops go over polynomially many objects and the repeat-loop is carried out at most as many times as there are edges in the expanded constraint network. Moreover, as the reader will show in Exercise 23, the reduced network the algorithm produces does not depend on any of the orderings in the for-loops. It is also easy to see that the above removals do not affect any solutions that might be present. Indeed if $\{(x, u), (y, v)\}$ is an edge in an r-clique, then for every color $z \in C \setminus \{x, y\}$ there is a w in the domain of z such that (z, w) is in the r-clique. This means that $\{(x, u), (y, v), (z, w)\}$ is consistent and $\{(x, u), (y, v)\}$ is not removed.

For more efficient implementations of 3-consistency algorithms cf. [104, 176, 183] or Section 4.3 of [265].

We leave it to the reader to devise an algorithm that enforces 2-consistency in Exercise 24. The strength of 2- and 3-consistency algorithms is seen when applied to some familiar scenarios in deciding if an ordered set has the fixed point property. Suppose one wants to use only the outcome of a consistency enforcing algorithm to decide whether a solution exists or not. Since enforcing consistency does not remove any solutions, the result is unambiguous if the algorithm returns a graph without edges. In this case there are no solutions. The natural conjecture would be that if there are edges left, there is a solution. This is of course not always the case. Thus to establish a consistency enforcing algorithm as a decision making tool, one has to prove that for all unsolvable CSPs in a given class the algorithm will return a graph without edges.

Proposition 12.5.13 (Cf. [278], proof of Theorem 5.) *Let P be a finite ordered set. Then applying a 2-consistency algorithm to the CSP for determining if P has a fixed point free order-preserving self map yields a graph without edges iff P is $\mathcal{I}$-dismantlable to a singleton.*

Proof. For the direction "$\Leftarrow$" let $G = (V, E, \gamma)$ be the graph as in Theorem 12.3.8 that is associated with the CSP for determining if the $\mathcal{I}$-dismantlable ordered set P has a fixed point free order-preserving map. That is

$$
\begin{aligned}
V &= \{(x, u) : x, u \in P, x \nsucceq u\}, \\
E &= \{\{(x, u), (y, v)\} : x \mapsto u, y \mapsto v \text{ is order} - \text{preserving}\}, \\
\gamma((x, u)) &= x.
\end{aligned}
$$

Assume that applying a 2-consistency enforcing algorithm to this graph produces the 2-consistent subgraph $G' = (V, E', \gamma)$ with $E' \neq \emptyset$. Recall that an isolated vertex in a graph is a vertex that is not contained in any edges. We define $\Phi : P \to \mathcal{P}(P) \setminus \{\emptyset\}$ by

$$
\Phi(x) := \{u : (x, u) \text{ is not isolated in } G'\}.
$$

We claim that Φ is an isotone relation as defined in Section 4.6. Indeed, let $x < y$ and let $u \in \Phi(x)$. We need to find a v such that $u \le v \in \Phi(y)$. Since (x, u) is not isolated in G' and since G' is 2-consistent, we have that there is a $v \in P$ such that $\{(x, u), (y, v)\} \in E'$. However this means that $v \in \Phi(y)$ and $u \le v$, which was to be proved. The second half of the definition of $\Phi(x) \sqsubseteq \Phi(y)$ is verified similarly.

Now since $x \nsucceq u$ for all u such that $(x, u) \in V$, we have that Φ is a fixed point free isotone relation on P. By Theorem 4.6.8 this means that P is not $\mathcal{I}$-dismantlable to a singleton, contradicting our hypothesis. Thus E' must be empty.

To prove the direction "$\Rightarrow$" we prove the contrapositive. So assume that P is not $\mathcal{I}$-dismantlable to a singleton. Then there is a fixed point free isotone relation $\Psi : P \to \mathcal{P}(P) \setminus \{\emptyset\}$. This means that the induced subgraph G'' of the graph G as above that has vertices (x, u) with $u \in \Psi(x)$ is such that for all (x, u) and all $y \in P \setminus \{x\}$ there is a v such that $\{(x, u), (y, v)\} \in E$. However this means that any reduced graph obtained from G via a 2-consistency algorithm will contain G''.   ∎

Proposition 12.5.14 *Let P be a connectedly collapsible ordered set of height ≤ 2. Then applying a 3-consistency algorithm to the CSP for determining if P has a fixed point free order-preserving map yields a graph without edges.*

Proof. This is a proof by induction on $|P|$ with the case $|P| = 1$ being trivial.

For the induction step, let P be connectedly collapsible and assume the result has been proved for all connectedly collapsible ordered sets of height ≤ 2 that have fewer elements than P. Again let $G = (V, E, \gamma)$ be the graph as in Theorem 12.3.8 that is associated with the CSP for determining if the ordered set P has a fixed point free order-preserving map. Let $G' = (V, E', \gamma)$ be the 3-consistent subgraph obtained by applying a 3-consistency enforcing algorithm to this graph G. We will need to show that $E' = \emptyset$. For a contradiction, we assume $E' \neq \emptyset$.

Let $a \in P$ be retractable to $b \in P$. Then, since P was connectedly collapsible of height 2, we have that $(\uparrow a) \setminus \{a\}$ or $(\downarrow a) \setminus \{a\}$ has the fixed point property and it is thus by Theorem 4.4.6, $\mathcal{I}$-dismantlable to a singleton. Assume without loss of generality that $(\uparrow a) \setminus \{a\}$ is $\mathcal{I}$-dismantlable to a singleton.

We will first show that (b, a) (if it is even in G') is an isolated vertex in G'. Suppose (b, a) was not an isolated vertex. Then for every $z > a$ we have $z \geq b$ and thus there would be a $w > a$ such that $\{(b, a), (z, w)\} \in E'$. However then by 3-consistency of G' for each $z' > a$, there is a $w' > a$ such that $\{(b, a), (z', w')\} \in E'$ and $\{(z', w'), (z, w)\} \in E'$. Thus the subgraph with vertices $V'' := \{(z, w) \in V : z, w > a, \{(b, a), (z, w)\} \in E'\}$ and edges $E'' := E' \cap (V'' \times V'')$ is 2-consistent with $E'' \neq \emptyset$. By Proposition 12.5.13 this is a contradiction to the fact that $(\uparrow a) \setminus \{a\}$ is $\mathcal{I}$-dismantlable to a singleton.

Thus (b, a) is isolated in G' (or not in G' at all). Moreover we claim that for all $\{(x, u), (y, a)\} \in E'$ with $u \neq a$, $y \not> a$ and $y \neq b$ we have that $\{(x, u), (y, b)\} \in E'$. Indeed, first we must have that $y \not> b$, since there must be a $v \in P$ with $\{(b, v), (y, a)\} \in E'$. If y was comparable to b, then the only choice for v would be a, which contradicts fact that (b, a) is isolated (or not in G'). This shows that $\{(x, u), (y, b)\} \in E$. To see that $\{(x, u), (y, b)\} \in E'$ simply note that any instantiation (w, z) with $z \neq a$ such that $\{(x, u), (y, a), (w, z)\}$ is consistent is such that $\{(x, u), (y, b), (w, z)\}$ is consistent. In case $z = a$, if $\{(x, u), (y, a), (w, a)\}$ is consistent, then $\{(x, u), (y, b), (w, b)\}$ is consistent.

Now let H be the graph with vertices $V_H := V \setminus [\gamma^{-1}(a) \cup \{(x, a) : x \in P\}]$ and edges $E_H := E' \cap (V_H \times V_H)$. Then H is a 3-consistent graph with $E_H \neq \emptyset$ contained in any graph that can be obtained via a 3-consistency algorithm from the graph as in Theorem 12.3.8 that is associated with the CSP for determining if the ordered set $P \setminus \{a\}$ has a fixed point free order-preserving map. Since $P \setminus \{a\}$ is connectedly collapsible, this contradicts the induction hypothesis. Thus E' must have been empty. ∎

The above shows that in the class of collapsible ordered sets of height ≤ 2 the graph returned by a 3-consistency algorithm can be used to decide if there is a fixed point free order-preserving self map. Similarly, in the class of ordered sets of height 1, a 2-consistency algorithm can be used. A fixed point free order-preserving self map exists iff the algorithm returns a nonempty graph. The same statement holds for ordered sets of width 2. Thus some very general tools can be applied to the fixed point property and they will in some situations produce the correct answer.

However, the reader will also notice that there are specific order-theoretical results that already allow us to make this decision in polynomial time in the above situations. Experimental evidence suggests that much more is the case (cf. Open Question 11 at the end of this chapter). That is, results such as Proposition 12.5.14 should be true in much larger classes of ordered sets. This seems to be a consistent problem in constraint satisfaction. There is a gap between provable performance and observed performance, with the latter being greater than the former. Several of the open questions at the end of this chapter address this observation.

12.6   A Polynomial Algorithm for the Fixed Point Property in Graded Ordered Sets of Bounded Width

We conclude this chapter on an upbeat note for the fixed point property. There is at least one other class for which existence of a fixed point free order-preserving map can be established in polynomial time. Theorem 12.6.2 can be viewed as inspired by [55]. The key to our algorithm is the following simple observation.

Proposition 12.6.1 *Let P, Q be ordered sets and let P be graded by g_P. Then $f : P \to Q$ is order-preserving iff for all $k \in g_P[P]$ the restriction $f|_{g_P^{-1}[\{k, k+1\}]}$ of f to grade levels k and $k+1$ is order-preserving.*

Proof. Exercise 6 in Chapter 11.   ∎

Theorem 12.6.2 *Let P be a graded ordered set with $|P| = n$ and let the width of P be bounded by w. Then it can be determined if P has a fixed point free order-preserving self map in $O(n^{2w+1}w^2)$ time and $O(n^{2w+1})$ space.*

Proof. Let P be an ordered set, graded by g_P. As stipulated in Definition 11.1.1 we have $\min g_P[P] = 1$. Let $h := \max g_P[P]$. For all $k \in \{1, \ldots, h\}$ let $\{p_1^k, \ldots, p_{w_k}^k\} = g_P^{-1}(k)$. Define a clustered CSP for the fixed point property as follows.

1. For $k = 1, \ldots, h$, the variable x_k is a stand-in for the w_k-tuple $(p_1^k, \ldots, p_{w_k}^k)$.

2. For $k = 1, \ldots, h$, let $D_k := \Pi_{i=1}^{w_k}\{y \in P : y \not> p_i^k\}$.

3. Our constraints are:

   - No explicit unary constraints ($p \not> f(p)$ are already built into the way the domains are set up).

   - A binary constraint $C_{i(i+1)}$ for every set $\{x_i, x_{i+1}\}$. The pair of vertices $((y_1^i, \ldots, y_{w_i}^i), (y_1^{i+1}, \ldots, y_{w_{i+1}}^{i+1}))$ is in $C_{i(i+1)}$ iff the map

   $$p_1^i \mapsto y_1^i, \ldots, p_{w_i}^i \mapsto y_{w_i}^i, p_1^{i+1} \mapsto y_1^{i+1}, \ldots, p_{w_{i+1}}^{i+1} \mapsto y_{w_{i+1}}^{i+1}$$

   is order-preserving.

Every overall solution of the CSP represents a function $f : P \to P$. Indeed, just define $f(p_j^i) := y_j^i$ for all appropriate i, j. The function f satisfies the condition in Proposition 12.6.1, so f is order-preserving. Moreover, f is fixed point free as no point could ever be mapped to itself. Conversely, if $f : P \to P$ is a fixed point free order-preserving self map, then $\{(x_i, (f(p_1^i), \ldots, f(p_{w_i}^i))) : i \in \{1, \ldots, h\}\}$ is a solution of the above CSP. Hence the above CSP has an overall solution iff P has a fixed point free order-preserving self map.

The constraint graph of the CSP is a tree (in fact a path). Thus after establishing arc consistency (which can be done without changing the constraint graph) by Theorem 12.5.11 we could obtain a solution (if one exists) in a backtrack free manner. Existence of a solution can be established a little faster, as there will be a solution iff there is a consistent instantiation of x_1 in the arc-consistent CSP.

While this effectively finishes the proof of polynomial solvability, in the following we will establish the bounds of the run time and space given in the result using the weaker (and thus more easily achieved) notion of directional arc consistency. Since directional arc consistency can be established faster than arc consistency, we obtain a faster algorithm. This is another typical situation in algorithmic treatment of problems. Having a fast algorithm is nice, having a faster one is better.

The directional arc consistency algorithm for the fixed point problem is easily seen to be an adaptation of DAC in [54], p. 11.

Algorithm DAC for FPP

For $i = h - 1$ to 1, step -1

        Remove all elements $a_i \in D_i$ for which there is *no* element $a_{i+1} \in D_{i+1}$ such that $\{(x_i, a_i), (x_{i+1}, a_{i+1})\}$ is consistent.

end for i

DAC for FPP erases assignments of grade levels that cannot be extended to the next higher grade level. After DAC for FPP terminates, if there is a consistent instantiation for x_1 (i.e., grade level 1), there is an instantiation for x_2 that extends the original instantiation to a consistent instantiation for x_1 and x_2. Now continue with x_2 in place of x_1. After $h - 1$ steps we will have a fixed point free order-preserving self map of P. If there is no consistent instantiation of x_1, then P has the fixed point property, since DAC for FPP cannot erase any tuples induced by a fixed point free order-preserving self map.

As we can see, the directional arc consistency makes use of the fact that variables are instantiated in a certain order. Therefore consistency may only be needed in a certain direction. In order-theoretical terms, since we build our map starting at grade level 1 and going up, we only need to make sure that we can extend to the next higher level.

Having established the correctness of the algorithm, we need to analyze its time and space requirements. Let w be the width of the graded ordered set P. Then for all i we have $w_i \leq w$. Each domain D_i has at most n^w elements and each constraint can be stored in $\leq n^{2w}$ space. Generating D_i takes $O(n^w)$ time and generating $C_{i(i+1)}$ takes $O(n^{2w} w^2)$ time (since for each of the $n^w n^w$ combinations of assignments we have to check if up to w^2 comparabilities are preserved). Thus the CSP can be generated in $O(n^{2w} w^2 h)$ time and stored in $O(n^{2w} h)$ space. The run time of DAC for FPP is bounded by $O(hn^{2w})$, since at each level we have to check each of the up to $O(n^w)$ instantiations against each of the up to $O(n^w)$ remaining instantiations of the next-higher level.

Finally note that h is always less than the number of elements of the ordered set.   ∎

Exercises

1. Turn the description of a correctness proof in Example 12.1.3 into a solid proof by induction on n.

2. Prove that the algorithm that suggests itself from the definition of the rank (cf. Definition 2.4.1) takes $O(|P|^2)$ time. Then prove that the height of an ordered set can be determined in $O(|P|^2)$ time.

3. Prove that the width of a finite ordered set P need *not* be given by the quantity $\max\{|S| : (\forall s, t \in S)\mathrm{rank}_P(s) = \mathrm{rank}_P(t)\}$, which is the size of the largest set of points with the same rank.

4. Prove that any algorithm that computes the transitive closure of a relation can be extended to compute the Hasse diagram of an order. (Hint: Erase all comparabilities between points whose ranks differ by more than k. Compute the transitive closure of this relation. All edges in this relation that are between points whose ranks differ by more than k are not in the Hasse diagram.)

5. Prove that in an ordered set of width w the transitive closure of the Hasse diagram is computable in $O(w|P|^2)$ steps. (Hint: Construct from the top down. Use the fact that each element has at most w upper/lower covers.)

6. Prove that there is an algorithm that checks in $O(|P|^2)$ steps if a given function $f : P \to Q$ between ordered sets P and Q is order-preserving.

7. Let P be a finite ordered set. Give an $O(|P|^2)$ algorithm that computes $|\uparrow x|$ and $|\downarrow x|$ for all $x \in P$.

8. Let P be a finite ordered set of which we have the Hasse diagram and the order relation.

   (a) Prove that for any $x \in P$ it can be checked in $O(|P|)$ time if x is irreducible.

   (b) Prove that it can be verified in $O(|P|^3)$ time if P is $\mathcal{I}$-dismantlable.

   (c) Conclude that for finite ordered sets of width 2 or height 1 it can be determined in polynomial time if the set has the fixed point property.

   (d) Prove that for fixed k there is a polynomial algorithm that verifies if a given finite ordered set is $\mathcal{R}_k$-dismantlable.

9. Use Exercise 9 in Chapter 8 to give an algorithm that is more efficient than the algorithm given in Proposition 12.3.3 to check if two interval orders are isomorphic.

10. For each of the following problems formulate a CSP for which the solutions are the maps in question.

    (a) Does P have a fixed point free automorphism?

    (b) Given P and $A \subseteq P$, is A a retract of P?

    (c) Given a graph, does this graph have a Hamiltonian Cycle?

    (d) Given sets P and Q and first-order-logic propositions $\sigma(x, u)$ and $\tau(x, u, y, v)$ with two and four unquantified variables respectively, is there a function $f : P \rightarrow Q$ such that $\forall x \in P : \sigma(x, f(x))$ and $\forall x, y \in P : \tau(x, f(x), y, f(y))$?

    (e) Given a finite ordered set P, find all antichains in P. (Hint: Use Proposition 2.6.7.)

11. Consider the two problems:

    Given. Two finite graphs G and H.

    Question. Is there an isomorphism between G and H?

    and

    Given. Two finite ordered sets P and Q.

    Question. Is there an isomorphism between P and Q?

    Prove that these two problems have the same complexity. That is, every algorithm that solves one can be turned into an algorithm that solves the other via a polynomial translation.

12. Prove that the following problems are NP-complete.

    (a) **Given.** A finite ordered set P.
        Question. Is there a fixed point free automorphism of P?

        You may use as a lemma that the problem

        Given. A finite graph G.
        Question. Is there a fixed point free automorphism of G?

        is NP-complete as shown in [171].

(b) (More on the weak fixed point property.)

Given. A finite ordered set P.

Question. Is there a fixed point free order-preserving, rank-preserving map $f : P \to P$?

(Use the construction in Theorem 12.4.5.)

13. Draw the backtracking and forward checking trees for the following problems.

   (a) Decide if the six crown has a fixed point free order-preserving self map.

   (b) Find all fixed point free order-preserving self maps of the six crown.

   (c) Find all order-preserving self maps of the four crown.

14. Nodes visited by backtracking.

   (a) Prove that a backtracking algorithm that is to find all solutions of a CSP visits exactly the instantiations that are vertices of the backtracking tree.

   (b) Prove that backtracking set to find all solutions will visit all solutions of a CSP. (This means the algorithm can be called "correct".)

15. Consider the ordered set in Figure 1.4 e). We shall consider the task of searching for fixed point free order-preserving maps of this set.

   (a) Find a consistent instantiation of $\{a, b, c\}$.

   (b) Find a consistent instantiation of $\{a, b, c, d\}$ that is not forward consistent.

   (c) Is there a consistent instantiation of $\{a, b, c\}$ that is not forward consistent?

   (d) Find a forward consistent instantiation of $\{a, b, c\}$ that is not part of a solution. (Why is this trivial?)

16. Nodes visited by forward checking.

   (a) Prove that a forward checking algorithm that is to find all solutions of a CSP visits exactly the instantiations that are vertices of the forward checking tree.

   (b) Prove that forward checking set to find all solutions will visit all solutions of a CSP. (This means the algorithm can be called "correct".)

17. Comparing forward checking and backtracking.

    (a) Prove that the forward checking tree for a problem is always a subtree of the backtracking tree.

    (b) Prove that the forward checking tree for a problem is contained in the interior nodes of the backtracking tree.

    (c) Find an ordering of the variables in the 5-fence such that the FPP backtracking tree as in Figure 12.2 contains an interior node that is not in the FPP forward checking tree.

18. Give an example that shows that for a given problem the backtracking trees corresponding to two different orderings of the variables need not have the same size. Do the same for forward checking trees.

    (Variable pre-ordering is an important topic in CSPs, since a better variable order can sometimes drastically speed up computations, cf. [199].)

19. Write the pseudocode for a forward checking algorithm.

20. (Bit-parallel implementation of backtracking and forward checking, cf. [105, 176].) Given a graph as in Theorem 12.3.8, for each vertex (x, u) and each color y, let $b(x, u, y)$ be a vector of $|D_y|$ zeroes and 1's such that a 1 in the ith place indicates that $\{(x, u), (y, v_i)\}$ is consistent, where v_i is the ith element of D_y. (A zero indicates that $\{(x, u), (y, v_i)\}$ is inconsistent.)

    (a) Prove that the componentwise AND of $b(x, u, y)$ and $b(x', u', y)$ encodes exactly those instantiations (y, v) that are consistent with (x, u) and with (x', u').

    (b) Use part 20a to write a 3-consistency algorithm that uses componentwise ANDs to check for each edge $\{(x, u), (x', u')\}$ if there is an instantiation of y that is consistent with $\{(x, u), (x', u')\}$.

    (c) (Bit-parallel backtracking.) Use part 20a to write a backtracking algorithm that for each consistent instantiation CI computes the instantiations for the next variable that are consistent with CI using at most $|CI|$ componentwise ANDs.

    (d) (Bit-parallel forward checking.) Use part 20a to write a forward checking algorithm that for each visited instantiation CI computes the instantiations for all future variables that are consistent with CI using at most $n - |CI|$ componentwise ANDs. (Hint: At each level store the consistent instantiations for the future variables in a separate vector.)

    (e) (Bit-parallel Backjumping.) Modify the algorithm in part 20c as follows. Whenever the bitwise ANDs show that there is no consistent instantiation at the next level, backtrack not to the previous level, but to the first level k for which $b(x_1, u_1, x_{\text{current}})$ compAND $\cdots$ compAND $b(x_k, u_k, x_{\text{current}})$ is a vector of zeroes. Prove that this algorithm does not miss any solutions.

(f) (A comparison between bit-parallel forward checking and regular back-tracking.) Prove that for CSPs with n variables and domain sizes at least n, bit-parallel forward checking performs fewer componentwise AND operations than regular backtracking visits nodes when search-ing for all solutions.

21. (Why is path consistency called path consistency?) In a path consistent CSP, let $\{(x, u), (y, v)\}$ be a consistent instantiation of two variables that share a constraint. Let $x = x_0, \ldots, x_k = y$ be a path from x to y in the constraint graph. Prove that there are instantiations $\{(x_i, y_i)\}$ (with $y_0 = u$ and $y_k = v$) such that for all $i \in \{1, \ldots, n\}$ we have that $\{(x_{i-1}, y_{i-1}), (x_i, y_i)\}$ is consistent.

22. Give an example of a CSP that is k-consistent, but not strongly k-consistent. (Examples exist for k as small as 3.)

23. Prove that the simple 3-consistency algorithm in Example 12.5.12 always produces the same reduced CSP independent of the ordering of the edges and the colors in the loops. (Hint: Order all subgraphs that can be stages in a 3-consistency algorithm by inclusion, consider removal of a single edge and use Theorem 3.2.3.)

24. Devise an algorithm that reduces a given expanded constraint network to an equivalent arc-consistent expanded constraint network.

25. Let P be a finite ordered set and let P_k be the set of elements of rank k. The ordered set P_{ps} is obtained as follows.

- Replace every adjacency of an element $p_k \in P_k$ with an element $p_{k+j} \in P_{k+j}$ with $j \geq 2$ with a chain (a "string of pearls") $p_k < p_{k+1} < \cdots < p_{k+j-1} < p_{k+j}$, where the $p_{k+1}, \ldots, p_{k+j-1}$ are new elements to be added to P.

- Add only the indicated comparabilities plus those forced by transitiv-ity.

Call P_{ps} the **pearl-strung version** of P.

We define the **edge-width** e of an ordered set P to be the width of the pearl-strung version P_{ps}.

Prove that for each $e > 0$ there is a polynomial p_e and an algorithm that for ordered sets of edge-width $\leq e$ decides in $\leq p_e(|P|)$ steps if P has a fixed point free order-preserving self map.

26. Let P and Q be graded ordered sets with grade functions g_P and g_Q re-spectively. Let $|P| = |Q| = n$ and let the width of P and Q be bounded by w. Prove that isomorphism of P and Q can be checked in $O(w!^4 n)$ time and $O(w!^2 n)$ space.

27. For any finite ordered set P we define the ordered set P_{2lc} as follows.

   (a) For every $x \in P$ let c_x be the number of lower covers of x and let $l_1^x, \ldots, l_{c_x}^x$ be an enumeration of these lower covers.

   (b) To build the underlying set for P_{2lc}, for every $x \in P$ add points $d_1^x, \ldots, d_{c_x-2}^x$ to P.

   (c) To build the order relation for P_{2lc}, add the following comparabilities to the comparabilities in P:

      i. The unique upper cover of d_1^x is x.

      ii. For $i \in \{2, \ldots, c_x - 2\}$ the unique upper cover of d_i^x is d_{i-1}^x.

      iii. For $i \in \{1, \ldots, c_x - 3\}$, d_i^x has exactly two lower covers, namely l_{i+1}^x and (as forced by part 27(c)ii) d_{i+1}^x.

      iv. $d_{c_x-2}^x$ has exactly two lower covers, namely $l_{c_x-1}^x$ and $l_{c_x}^x$.

      v. Plus all the comparabilities forced by transitivity and reflexivity.

Prove that:

   (a) P is an ordered subset of P_{2lc}.

   (b) P has the fixed point property iff P_{2lc} has the fixed point property.

   (c) Every element of P_{2lc} has at most two lower covers.

   (d) $|P_{2lc}| \leq |P|^2$.

28. (Another CSP for FPP.) For determining if the ordered set P has a fixed point free order-preserving self map, let the variable set be equal to P, that is, $\{x_1, \ldots, x_n\} := P$. For each x_i the domain D_i is $D_i := \{p \in P : p \not\leq x_i \text{ and } p \not\geq x_i\}$.

The constraints C_{ij} are

   • If $x_i \not< x_j$ and $x_i \not> x_j$, there is no constraint between x_i and x_j.

   • If $x_i \prec x_j$, let C_{ij} be $C_{ij} := \{(u_i, u_j) \in D_i \times D_j : u_i \leq u_j\}$.

Prove that

   (a) Any consistent instantiation of all the variables corresponds to a fixed point free order-preserving self map of P.

   (b) When applied to the set P_{2lc} of Exercise 27, the constraint graph of the above CSP has at most width 2.

29. Explain why going from P to P_{2lc} as in Exercise 27, then enforcing 3-consistency for the corresponding CSP from Exercise 28 and then using backtracking does not lead to a polynomial algorithm to determine the fixed point property.

Remarks and Open Problems

As the reader may have anticipated, open questions abound in an area as vast as algorithms in general and constraint satisfaction in particular. In the following we concentrate on remarks and questions that connect with the main themes of the text.

1. For a start on the body of work on transitive closures, the reader should consider [101].

2. There is an $O(n^{2.5})$ algorithm to compute the width of an ordered set. This is done via Dilworth's Theorem through a translation to a matching problem as in [271], p.274 and then using an $O(n^{2.5})$ algorithm for matching, cf. [126]. (This is also mentioned in [25]. p. 259.) [271] provides many connections between graph-theoretical and order-theoretical parameters.

3. The Clay Mathematics Institute has chosen the question if P=NP as one of the seven leading questions in mathematics for the future, cf. [39].

4. For an excellent exposition of search algorithms for CSPs the reader should consider [148].

   Results and references in [9, 148, 204], as well as in some measure Exercise 20f, indicate why forward checking is generally (though not always) preferable to backtracking.

   There is also another search paradigm that is currently gaining in importance. It is called Maintaining Arc Consistency, cf. [95, 176, 238]. In this algorithm the consistent future assignments are generated and stored just as in forward checking. However, to achieve further pruning of the tree an arc-consistency algorithm is run on the set of consistent future assignments every time we go deeper in the tree.

5. For some results on variable ordering heuristics, cf. [105, 199, 265].

6. In Proposition 12.5.14 we have shown that applying a path consistency algorithm to the CSP associated with determining existence of a fixed point free order-preserving map always returns a graph without edges for connectedly collapsible ordered sets of height 2. Is the same true if the height 2 condition is dropped? This would show that in the class of collapsible ordered sets, existence of a fixed point free order-preserving self map can be established in polynomial time.

7. Is it possible to determine in polynomial time if a given finite ordered set is (connectedly) collapsible? The author was only able to show this for fixed height (cf. [243]).

8. Regarding isomorphism it still is not not known if checking isomorphisms for graphs is polynomial, NP-complete or a separate complexity class altogether. For more information on isomorphism problems and their complexity theory the reader is referred to [147]. For the special case of graphs with bounded valence, cf. [172].

9. By [279] (also cf. Chapter 4, Theorem 2.7 in [263]) for any $k \geq 3$ it is NP-complete to decide if a given ordered set has dimension $\leq k$.

   For further examples of NP-complete problems, the reader is referred to [94, 225] and reminded of Remark 3 in Chapter 9.

10. Interval orders have a nice short proof of reconstructibility that uses the fact that isomorphism of interval orders is easily verifiable. Does polynomially verifiable isomorphism imply reconstructibility in general?

11. To date the smallest ordered set with the fixed point property for which applying a 3-consistency algorithm does not return an empty graph is constructed through the NP-completeness proof in Theorem 12.4.5. Take all eight possible clauses on three literals and construct the ordered set as in the proof of Theorem 12.4.5. Applying 3-consistency does not return an empty graph. Yet the ordered set has the fixed point property, since of course the eight clauses cannot all be satisfied simultaneously. This set has over 400 elements. Are there smaller such examples that are more "natural" in the sense that they are not derived from the NP-completeness proof? Insights into this question would shed some further light on how difficult the fixed point property is to decide.

12. Propositions 12.5.13 and 12.5.14 and also results in [59] show that there are results in order theory that required intelligent insights into the structure to allow a proof that certain sets have the fixed point property. Yet they can be rephrased as "under these circumstances a path consistency algorithm answers the question". How many more results of discrete mathematics can be translated into performance guarantees for k-consistency type algorithms? How interesting would such performance guarantees be?

13. In [125] and other papers in the same journal volume the **phase transition** for constraint satisfaction problems is described. For the following it is helpful to assume there is a parameter that is linked to how constrained a problem is, say, the "average instance" of the problem becomes more constrained as the parameter grows. (Such a parameter often exists.)

    In essence, for any type of constraint satisfaction problem, problems with "few" constraints or "many" constraints are easy to solve. A solution is either found almost immediately or the problem is so overconstrained, that the backtracking tree is of very small size. Hard problems can be found in abundance in the region in between, where the problem changes from highly underconstrained to highly overconstrained. This is the region of the phase transition. It often coincides with the region where the probability of being solvable changes from 1 to 0. (Plotting the probability of being solvable against the parameter shows a steep drop.) This region is located near the point where the expected number of solutions is 1. Problems in this region have few enough solutions that a search algorithm is not likely to find a solution fast and their backtracking trees are large.

What is the region of the phase transition for determining if a given ordered set has a fixed point free order-preserving self map? What parameter would we use?

14. It makes sense to conjecture that to decide if a given finite truncated lattice has the fixed point property (or equivalently if a given finite simplicial complex has the fixed simplex property) is NP-complete. To date this is unproved.

15. We have seen in the proof of Proposition 12.5.13 that in fact for any finite ordered set P the following is true. If G is the expanded constraint network for the problem of determining if P has a fixed point free order-preserving map and G' is a 2-consistent subgraph that has a nonempty edge set, then

$$\Phi(x) := \{u : (x, u) \text{ is not isolated in } G'\}.$$

is a fixed point free isotone relation. What additional properties does Φ have if G' is obtained via a 3-consistency algorithm? Can these properties be used to characterize those ordered sets P which have no fixed point free order-preserving map and for which a 3-consistency algorithm applied to G erases all edges?

16. In [130] it is shown that any constraint satisfaction problem corresponds to a pair of relational structures in such a way that the solutions correspond to the homomorphisms between the structures. As we have seen, existence of specific homomorphisms and the number of homomorphisms are studied extensively in order theory. Can advances in this direction be translated into results on general constraint satisfaction problems? (Also cf. Remark 19.)

17. Since the property of being order-preserving can be formulated as a constraint, one can use any algorithm that finds all solutions to a constraint satisfaction problem and turn it into an algorithm that enumerates all endomorphisms of an ordered set. This can then be used to compute for example the spectrum and other data on the ordered set in question. Because the number of endomorphisms grows rapidly, this is only feasible for sets of small size. Still, experiments could point the way towards progress on the automorphism conjecture or towards finding the ordered set with the fewest endomorphisms for a given size.

18. In Exercise 25 in Chapter 9 we give a condition that ensures that convex subsets of ordered sets intersect. The fact that any set of linear intervals such

that any two intersect has a nonempty intersection over all has been used in [267]. It is shown there in Theorem 3.2 that for so-called row-convex constraints 3-consistency implies consistency. (Careful though. Both conditions must be satisfied. There are row-convex CSPs that lose row-convexity when 3-consistency is enforced.)

Is there an adaptation of the above result (together with Exercise 25 in Chapter 9) that can be used to prove polynomial decidability of the fixed point property in sets of small width? This would require a way to address the problem outlined in and after Theorem 5.6 in [267].

19. Algorithms for constraint satisfaction problems can be used to enumerate the order-preserving self maps of an ordered set. Thus for small sets it may be possible to get some bounds on the quotient in Open Question 11.5.1. For larger sets estimation algorithms such as proposed in [37, 105, 146, 205] might at least give an estimate of the number of solutions. The algorithms in [37, 146, 205] perform a randomized partial search. Their primary purpose is to estimate the run time. However, the estimate for the number of consistent nodes at maximum depth could be used as an estimate for the number of solutions. This estimate through randomized algorithms is normally bad in the region of the phase transition (for phase transitions cf. [125]). When counting order-preserving maps, solutions abound and the problem is well away from the phase transition region. Thus there is hope estimates may be reasonable, especially for sets of small height. (Chains and tall sets can be problematic for these algorithms.)

The idea for [105] is probabilistic. We consider an arbitrary constraint satisfaction problem. Let the number of variables be n and let all variable domains be of size m. For each constraint $C_{i,j}$, the number of consistencies divided by the total number of entries in the matrix representation of $C_{i,j}$ gives a probability $p_{i,j}$ for this constraint to be satisfied. The expected number of solutions then is $m^n \prod_{i=1}^{n} \prod_{j=i+1}^{n} p_{ij}$. Unfortunately, this model only applies to randomly seeded problems where every entry of $C_{i,j}$ is equally likely to be a 1. Even refinements so far have only been useful in randomly seeded problems of different types. How promising is the refinement below in terms of making the probabilistic approach more widely applicable?

Refining the model one can add all matrices $C_{i,j}$ with $i < j$ to obtain the sum C^* (recall all $C_{i,j}$ are $m \times m$ matrices). Rescale C^* to obtain the matrix C by dividing C^* by its largest entry. C gives a measure of how much the consistencies in the problem are clustered. (In the uniform probability model all entries of C would be close to 1.) One can then replace the number of entries of $C_{i,j}$ in the computation of $p_{i,j}$ with the sum of the entries of C. The factor m^n in the computation needs to be replaced with an estimate on the number of solutions of a constraint satisfaction problem "where all constraints are equal to C". Since C has rational numbers as entries, this can be done with a probabilistic variation on the methods of [146]. Small entries in C give small probability to there being a consistency. There is little known about this idea of the author's. However, experiments in prediction and analysis of randomly seeded constraint satisfaction problems beyond the simple uniform model are encouraging. Indeed, the probabilistic estimates give a certain likelihood whether the problem is solvable. This likelihood remains a fairly good predictor even in some regions where a randomized search algorithm fails.

The main place for improvement of the probabilistic approach currently seems to be the computation of the analogue of the factor m^n through C. This is a chance for feedback from ordered sets. If an ordered set is labelled in such a way that $i < j$ implies $x_i \not> x_j$, then all $C_{i,j}$ with $i < j$ have the same matrix representation. Thus these problems are among the simplest prototypes for trying to estimate the analogue of the factor m^n through C. Indeed, this factor is nothing but the total number of solutions. Therefore, conversely to the idea at the beginning of this remark, any advance on estimating the number of endomorphisms of an ordered set could help in estimating the number of solutions of a constraint satisfaction problem in the above fashion. The first step consists of problems where all constraints have the same matrix (maps from chains to ordered sets if there are no trivial constraints). Subsequently the method would need to be adapted to allow for rational entries using probabilistic modifications.

20. A multitude of reduction and search algorithms has been implemented by the author in C++ with a WINDOWS front end. The interested reader is encouraged to contact the author and obtain the code.

For more implemented algorithms for constraint satisfaction problems and for more background information, the reader should consider the constraints archive at http://www.cs.unh.edu/ccc/archive/

Appendix A
A Primer on Algebraic Topology

This appendix is a review of the basic notions of algebraic topology that are needed for Sections 6.5 and 6.6. To keep the necessary new vocabulary limited we will not use the language of category theory here, though those versed in it will easily be able to identify the functors etc. behind the results. Most of the (standard) concepts from algebraic topology have been taken from [253], Chapter 4, Sections 1-4. Proofs that were omitted are either short, or a reference to the proof in the literature is given.

Notation A.0.1 We know how to assign to an ordered set P a graph $G_C(P)$ (the comparability graph, cf. Definition 6.3.4), to a graph G a simplicial complex $K(G)$ (cf. Example 6.3.9), and to a simplicial complex K a topological space $|K|$ (cf. Definition 6.4.1). We will see how to assign to a simplicial complex K a chain complex $C(K)$ (cf. Definition A.1.3). All these notions allow constructions that naturally arise in the respective settings, which then can also be executed for the induced structures. (For example, the homology complex for a chain complex induced by a simplicial complex induced by a comparability graph of an ordered set.) To reduce the amount of necessary notation and without spelling out all the necessary definitions (which would be a large task indeed) we will sometimes use language formally "out of turn". (For example, we talk about the homology complex of an ordered set.) We shall also abbreviate symbols accordingly. (For example we will write $H(P)$ rather than $H(C(K(G_C(P))))$.) This should not cause any confusion as we will always assume the construction is performed for the appropriate induced structure.

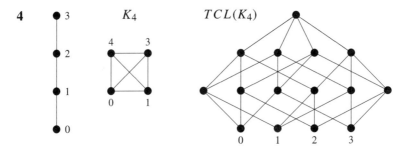

Figure A.1: The four chain **4**, the corresponding comparability graph K_4 and the corresponding clique complex visualized as $TCL(K_4)$.

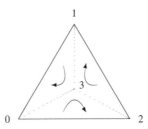

Figure A.2: The topological realization of a four chain, a tetrahedron. The orientations of the "hidden" boundary pieces (triangles) of the tetrahedron are indicated. The front triangle is positively oriented (counterclockwise).

A.1   Chain Complexes

The key idea to using homology in fixed point theory is to connect algebra with simplicial complexes. In this section we will translate/embed simplicial complexes and their morphisms into the theory of chain complexes, which is the first step. Throughout this section let us consider a four chain $P = \{0, 1, 2, 3\}$ with its natural order. The chain P, its comparability graph and the clique complex of the comparability graph (which is also referred to as the "P-chain complex") are pictured in Figure A.1.

Definition A.1.1 *A **chain complex** $C = (\{C_n\}_{n\in\mathbb{Z}}, \{\partial_q\}_{q\in\mathbb{Z}})$ is an ordered pair of a family $\{C_n\}_{n\in\mathbb{Z}}$ of Abelian groups together with a family of functions $\partial_q : C_q \to C_{q-1}$ (also called **boundary maps**) such that $\partial_q \partial_{q+1} = 0$ for all $q \in \mathbb{Z}$.*
   *C is called **finitely generated** iff $C_q = 0$ for all but finitely many q and all nonzero C_q have a finite set of generators.*

To obtain a chain complex from a simplicial complex, we consider the topological realization of the simplicial complex. For the four chain in Figure A.1 this is the tetrahedron given in Figure A.2.

Definition A.1.2 *Let* $S = \{v_0, \ldots, v_q\}$ *be a finite set. Two linear orders* $v_{j_0} < v_{j_1} < \cdots < v_{j_q}$ *and* $v_{k_0} < v_{k_1} < \cdots < v_{k_q}$ *are* **equivalently oriented** *iff the permutation* σ *such that* $\sigma \circ (k_0, \ldots, k_q) = (j_0, \ldots, j_q)$ *is even. (Equivalent orientation is an equivalence relation and it has two equivalence classes for* $q > 0$.) *The equivalence class of* $v_{j_0} < v_{j_1} < \cdots < v_{j_q}$ *will be denoted by* $[v_{j_0}, v_{j_1}, \ldots, v_{j_q}]$ *and will be called an* **orientation** *of* S. *If* $v_0, \ldots, v_q$ *are vertices of a simplex and* $v_i = v_j$ *for some* $i \neq j$, *then we let* $[v_0, \ldots, v_q] := 0$.

While the orientations of the geometric objects one visualizes are certainly subject to some choices, the connections between higher dimensional and lower dimensional objects via the boundary map are determined by the following.

Definition A.1.3 *Let* $K = (V, S)$ *be a simplicial complex and let* $S \in \mathcal{S}$. *For* $q \in \mathbb{Z}$ *let* $C_q(K)$ *be the free Abelian group generated by the orientations of the q-simplexes with different orientations of the same simplex being additive inverses of each other. Let* $\partial_q^K : C_q(K) \to C_{q-1}(K)$ *be the homomorphism defined on the generators by*

$$\partial_q^K([v_0, \ldots, v_q]) = \sum_{i=0}^{q} (-1)^i [v_0, \ldots, \hat{v}_i, \ldots, v_q].$$

(The hat indicates, as is often customary, that the vertex under the hat is to be dropped. We will show that ∂_q^K *is well-defined via this definition.)* $C(K) := (\{C_q(K)\}_{q \in \mathbb{Z}}, \{\partial_q^K\}_{q \in \mathbb{Z}})$ *is a chain complex, called the* **oriented chain complex** **of** K.

The orientations of $\partial_2(\{0, 1, 2, 3\})$ are indicated in Figure A.2. Note that the boundaries of the tetrahedron (the triangles) are oriented in such a way that their boundaries (sides) in turn are traversed once in each direction by anyone who travels the boundaries of the triangles in the direction indicated in the triangle. For the author's (limited) view of this subject, the motivation for this particular definition of the boundary operator seems to lie deeply in considerations of differential geometry somewhere near Stokes' theorem. Algebraically, the following proof shows that the right way to line up alternating signs is what makes the oriented chain complex a chain complex.

Proof that the oriented chain complex really is a chain complex. To see that the ∂_q^K are well-defined note that a transposition of adjacent elements v_j, v_{j+1} and a subsequent transposition of elements v_k, v_{k+1} (in the new indexing after the first transposition) in the preimage does not affect the image. Since even permutations are compositions of an even number of transpositions this shows the ∂_q^K are well-defined.

To prove that $C(K) = (\{C_q(K)\}_{q \in \mathbb{Z}}, \{\partial_q^K\}_{q \in \mathbb{Z}})$ truly is a chain complex, we need to show that the iteration of the boundary maps gives an operator identical to zero. To do this we can limit ourselves to investigating the action of $\partial_{q-1}^K \partial_q^K$ on the generators. We can assume that $q \geq 1$, since for lower q the maps $\partial_{q-1}^K \partial_q^K$

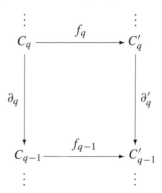

Figure A.3: A chain map.

automatically will be zero.

$$\partial_{q-1}^K \partial_q^K ([v_0, \ldots, v_q])$$

$$= \partial_{q-1}^K \left(\sum_{i=0}^q (-1)^i [v_0, \ldots, \hat{v}_i, \ldots, v_q] \right)$$

$$= \sum_{i=0}^q (-1)^i \partial_{q-1}^K ([v_0, \ldots, \hat{v}_i, \ldots, v_q])$$

$$= \sum_{i=0}^q (-1)^i \left(\sum_{j=0}^{i-1} (-1)^j [v_0, \ldots, \hat{v}_j, \ldots, \hat{v}_i, \ldots, v_q] \right.$$

$$\left. + \sum_{j=i+1}^q (-1)^{j-1} [v_0, \ldots, \hat{v}_i, \ldots, \hat{v}_j, \ldots, v_q] \right)$$

$$= \sum_{i=0}^q \sum_{j=0}^{i-1} (-1)^{i+j} [v_0, \ldots, \hat{v}_j, \ldots, \hat{v}_i, \ldots, v_q]$$

$$+ \sum_{i=0}^q \sum_{j=i+1}^q (-1)^{i+j-1} [v_0, \ldots, \hat{v}_i, \ldots, \hat{v}_j, \ldots, v_q]$$

$$= \sum_{i=1}^q \sum_{j=0}^{i-1} (-1)^{i+j} [v_0, \ldots, \hat{v}_j, \ldots, \hat{v}_i, \ldots, v_q]$$

$$+ \sum_{j=1}^q \sum_{i=0}^{j-1} (-1)^{i+j-1} [v_0, \ldots, \hat{v}_i, \ldots, \hat{v}_j, \ldots, v_q]$$

$$= 0.$$

Definition A.1.4 *A **chain map** (cf. Figure A.3) $f : C \to C'$ from the chain complex $C = (\{C_q\}_{q \in \mathbb{Z}}, \{\partial_q\}_{q \in \mathbb{Z}})$ to the chain complex $C' = (\{C'_n\}_{n \in \mathbb{Z}}, \{\partial'_q\}_{q \in \mathbb{Z}})$ is a family $f = \{f_q\}_{q \in \mathbb{Z}}$ of group homomorphisms $f_q : C_q \to C'_q$ such that for all $q \in \mathbb{Z}$ we have $f_{q-1} \circ \partial_q = \partial'_q \circ f_q$.*

Chain maps can be induced by simplicial maps as shown in the following.

Proposition A.1.5 *Let $K = (V, S)$ and $K' = (V', S')$ be simplicial complexes and let $f : K \to K'$ be a simplicial map. Then f induces a chain map $f^{\mathrm{ch}} : C(K) \to C(K')$ between the corresponding oriented chain complexes, which is defined by $f_q^{\mathrm{ch}}([s_0, \dots, s_q]) := [f(s_0), \dots, f(s_q)]$.*

Proof. First note that f_q^{ch} is well-defined. Indeed, an even permutation of the $s_0, \dots, s_q$ translates into an even permutation of the $f(s_0), \dots, f(s_q)$.

Now we need to show that $\partial_q^{K'} \circ f_q^{\mathrm{ch}} = f_{q-1}^{\mathrm{ch}} \circ \partial_q^K$. Yet this is quite trivial, since for all generators we have

$$
\begin{aligned}
f_{q-1}^{\mathrm{ch}} \circ \partial_q^K ([v_0, \dots, v_q]) &= f_{q-1}^{\mathrm{ch}} \left(\sum_{i=0}^{q} (-1)^i [v_0, \dots, \hat{v}_i, \dots, v_q] \right) \\
&= \sum_{i=0}^{q} (-1)^i [f(v_0), \dots, \widehat{f(v_i)}, \dots, f(v_q)] \\
&= \partial_q^{K'} ([f(v_0), \dots, f(v_q)]) \\
&= \partial_q^{K'} f_q^{\mathrm{ch}}([v_0, \dots, v_q]).
\end{aligned}
$$

∎

At this point we have succeeded in embedding/translating the objects that we care about in this context (simplicial complexes and their morphisms) into the theory of chain complexes and chain maps. The next step is to use algebraic tools to obtain fixed point results.

A.2   The Lefschetz Number

A possible motivation for the Lefschetz number of a map is the following. The entries on the diagonal of a matrix tell "how much" of a vector that is multiplied with the matrix will still point in the original direction. The trace of a matrix is the sum of the entries on the diagonal. Thus if we have a large trace, then lots of vectors have images with large components in the original direction. This simplistic description will be made more precise in the following.

Definition A.2.1 *Let $C = (\{C_q\}_{q \in \mathbb{Z}}, \{\partial_q\}_{q \in \mathbb{Z}})$ be a finitely generated chain complex and let $f : C \to C$ be a chain map. The **Lefschetz number** $\Lambda(f)$ of f is* *defined to be*

$$\Lambda(h) := \sum_{q \in \mathbb{N}} (-1)^q Tr(f_q),$$

where Tr denotes the trace.

As we tried to motivate above, the Lefschetz number of a map can be used as a sufficient criterion for the existence of fixed points or cliques or simplexes.[1] Let $f : V \to V$ be a simplicial map on the finite simplicial complex $K = (V, S)$. If $\Lambda(f) \neq 0$, then there is a $q \in \mathbb{N}$ such that $Tr(f_q) \neq 0$. Thus there is a q-dimensional simplex $S = \{v_0, \ldots, v_q\} \in S$ such that $f_q([v_0, \ldots, v_q]) = k[v_0, \ldots, v_q] + h$ for some $k \in \mathbb{Z} \setminus \{0\}$ and h a combination of the generators other than S. However since

$$f_q([v_0, \ldots, v_q]) = [f(v_0), \ldots, f(v_q)]$$

this means

$$f_q([v_0, \ldots, v_q]) \in \{[v_0, \ldots, v_q], -[v_0, \ldots, v_q]\}.$$

Hence we have that $f[S] = S$, i.e., f has a fixed simplex. Now if K is a clique complex and f is induced by a graph endomorphism, this means that there is a clique C such that $f[C] = C$. If K is the clique complex of the comparability graph of the finite ordered set P and f is an order-preserving map on P, then f maps a chain to itself, which naturally means that f has a fixed point.

A.3   (Integer) Homology

Here (specifically in Lemma A.3.5) is "where the miracle of algebraic topology occurs" (at least for our purposes). Through Lemma A.3.5 many Lefschetz numbers become computable and thus a host of combinatorially surprising fixed point theorems (many of which still have no combinatorial proof) enters the theory.

Lemma A.3.1 *Let K be a simplicial complex and let L be a subcomplex of K. Then $C_q(L)$ is a subgroup of $C_q(K)$ and the quotient group $C_q(K, L) := C_q(K)/C_q(L)$ is well-defined. Moreover*

$$\tilde{\partial}_q : C_q(K, L) \to C_{q-1}(K, L); \qquad (c + C_q(L)) \mapsto \partial_q(c) + C_{q-1}(L)$$

is a chain map and $C(K, L) := (\{C_q(K, L)\}_{q \in \mathbb{Z}}, \{\tilde{\partial}_q\}_{q \in \mathbb{Z}})$ is a chain complex. ∎

Definition A.3.2 *Let $C = (\{C_q\}_{q \in \mathbb{Z}}, \{\partial_q\}_{q \in \mathbb{Z}})$ be a chain complex. We will call the subgroups $Z_q(C) := \ker(\partial_q)$ the* **subgroups of cycles** *of C and the subgroups $B_q(C) := \partial_{q+1}[C_{q+1}]$ the* **subgroups of boundaries** *of C. Two cycles whose*

---

[1] The following short argument was shown to the author by J.D. Farley.

difference is a boundary are called **homologous**. *The* **homology groups** *are the quotient groups*

$$H_q(C) := Z_q(C)/B_q(C).$$

Lemma A.3.3 (Cf. [253], Chapter 4, Section 1, Theorem 1.) *Let a chain complex* $C = (\{C_q\}_{q \in \mathbb{Z}}, \{\partial_q\}_{q \in \mathbb{Z}})$ *be given. Then*

$$
\begin{aligned}
Z(C) &:= (\{Z_q(C)\}_{q \in \mathbb{Z}}, \{\partial_q|_{Z_q(C)}\}_{q \in \mathbb{Z}}), \\
B(C) &:= (\{B_q(C)\}_{q \in \mathbb{Z}}, \{\partial_q|_{B_q(C)}\}_{q \in \mathbb{Z}})
\end{aligned}
$$

are chain complexes. If we define

$$\partial_q^* : H_q(C) \to H_{q-1}(C); \qquad (c + B_q(C)) \mapsto \partial_q(c) + B_{q-1}(C),$$

then ∂_q^* *is a group homomorphism and* $H(C) := (\{H_q(C)\}_{q \in \mathbb{Z}}, \{\partial_q^*\}_{q \in \mathbb{Z}})$ *is a chain complex. We will call these complexes the* **cycle**, **boundary** *and* **homology complex**. ∎

Lemma A.3.4 (Cf. [253], Chapter 4, Section 1, Theorem 1.) *Let chain complexes* $C = (\{C_q\}_{q \in \mathbb{Z}}, \{\partial_q\}_{q \in \mathbb{Z}})$ *and* $C' = (\{C'_q\}_{q \in \mathbb{Z}}, \{\partial'_q\}_{q \in \mathbb{Z}})$ *be given and let* $f : C \to C'$ *be a chain map. Let*

$$f_q^* : H_q(C) \to H_q(C'); \qquad (c + B_q(C)) \mapsto f_q(c) + B_q(C').$$

Then all f_q^* *are well-defined and* f^* *is a chain-map.* ∎

Lemma A.3.5 (Cf. [253], Chapter 4, Section 7, Theorem 6.) *Let* C *be a finitely generated chain complex and let* $f : C \to C$ *be a chain map. Then*

$$\Lambda(f) = \Lambda(f^*).$$

Proof. First note that $f_q[Z_q(C)] \subseteq Z_q(C)$ and $f_q[B_q(C)] \subseteq B_q(C)$, since for $x \in Z_q(C)$ we have

$$\partial_q f_q(x) = f_{q-1} \partial_q(x) = f_{q-1}(0) = 0,$$

while for $y \in B_q(C)$ there is an $x \in C_{q+1}$ with $y = \partial_{q+1}(x)$ and thus

$$f_q(y) = f_q(\partial_{q+1}(x)) = \partial_{q+1} f_{q+1}(x) \in B_q(C).$$

Now since $Z_q(C)$ is isomorphic to $H_q(C) \oplus B_q(C)$ and $C_q/Z_q(C)$ is isomorphic to $B_{q-1}(C)$ (with ∂_q being the isomorphism) we have

$$
\begin{aligned}
Tr(f_q|_{Z_q(C)}) &= Tr(f_q^*) + Tr(f_q|_{B_q(C)}), \\
Tr(f_q) &= Tr(f_q|_{Z_q(C)}) + Tr(f_{q-1}|_{B_{q-1}(C)}).
\end{aligned}
$$

Adding the two equations, multiplying with $(-1)^q$ and summing over q gives the equality of the Lefschetz numbers. ∎

We thus have translated Lefschetz numbers of chain maps in the chain complex associated with a simplicial complex into Lefschetz numbers of their associated maps in the homology complex. The advantage in this translation is that the homology groups tend to be smaller and more manageable than the original groups in the oriented chain complex. The zeroeth homology group can be described geometrically.

Definition A.3.6 *Let* $K = (V, S)$ *be a simplicial complex. A* **path** *is a subset* $\{p_0, \ldots, p_n\}$ *such that for* $k \in \{1, \ldots, n\}$ *the set* $\{p_{k-1}, p_k\}$ *is a simplex of* K. *A* **component** *of* K *is a maximal full subcomplex* C *such that any two vertices of* C *are in a path.* K *is called* **connected** *iff* K *has exactly one component.*

Lemma A.3.7 *Let* $K = (V, S)$ *be a finite simplicial complex and let* k *be the number of components of* K. *Then* $H_0(K)$ *is isomorphic to* $\mathbb{Z}^k$.

Proof. The generators for $B_0(K)$ are the differences $v - w$, where v, w are vertices of K with $\{v, w\} \in S$. Thus if u, z are vertices in the same component of K, then $u - z \in B_0(K)$. On the other hand if $u - z \in B_0(K)$, then $u - z$ must the sum of finitely many generators of $B_0(K)$ and thus u and z must be in the same component of K.

Since $Z_0(K) = C_0(K)$ is generated by the vertex set of K, this implies that any two vertices x, y in the same component of K satisfy $x + B_0(K) = y + B_0(K)$. On the other hand, if x' and y' are not in the same component of K we have $x' + B_0(K) \neq y' + B_0(K)$. Hence $H_0(K)$ has exactly k generators with no further conditions to be satisfied. That is, it is isomorphic to $\mathbb{Z}^k$. ∎

In particular this means that if K is connected, then $H_0(K)$ is isomorphic to $\mathbb{Z}$. If this is the only interesting homology group of K, we will say that K is acyclic. Acyclicity is the key property in using homology to prove fixed point theorems so far (cf. Lemma A.3.9).

Definition A.3.8 *Let* $C = (\{C_q\}_{q \in \mathbb{Z}}, \{\partial_q\}_{q \in \mathbb{Z}})$ *be a chain complex. Then* C *is called* **acyclic** *iff*

$$H_q(C) = \begin{cases} \{0\}; & \text{if } q \neq 0, \\ \mathbb{Z}; & \text{if } q = 0. \end{cases}$$

Lemma A.3.9 *Suppose* $K = (V, S)$ *is a finite acyclic simplicial complex. Then for each simplicial map* $f : V \to V$ *we have that* $f^* = id_{\mathbb{Z}}$ *and thus* $\Lambda(f) = 1$. *Hence for every simplicial map* $f : K \to K$ *there is a simplex* V *such that* $f[V] = V$. ∎

A.4   A Homological Reduction Theorem

What is left to be done after the above excursion into algebraic topology is to re-connect with combinatorial properties that are sufficient to assure acyclicity.

The following is a translation of the results in [177] from their original version for modulo 2 homology to integer homology.

Acyclicity means that homology groups are trivial. Thus we will need lemmas about cancelations.

Lemma A.4.1 (Cf. [177], Lemma 5.) *Let $G = (V, E)$ be a graph, let $v \in V$ and let $[v, v_1^1, \ldots, v_q^1], [v, v_1^2, \ldots, v_q^2], \ldots, [v, v_1^t, \ldots, v_q^t], [v_0^{t+1}, v_1^{t+1}, \ldots, v_q^{t+1}], \ldots, [v_0^p, v_1^p, \ldots, v_q^p]$ be q-dimensional oriented simplexes in the chain complex associated with G such that no v_i^j is equal to v. If v occurs in no oriented simplex of*

$$\partial_q \left(\sum_{j=1}^{t} [v, v_1^j, \ldots, v_q^j] + \sum_{j=t+1}^{p} [v_0^j, v_1^j, \ldots, v_q^j] \right),$$

then we have that $\partial_{q-1} \left(\sum_{j=1}^{t} [v_1^j, \ldots, v_q^j] \right) = 0$.

Proof. In case $q \leq 1$, there is nothing to prove (the boundary of a sum of zero-dimensional simplexes is 0). Thus in the following we assume $q \geq 2$. Let $[w_1, \ldots, w_{q-1}]$ be a $(q - 2)$-dimensional oriented simplex that occurs in one of the $(q - 1)$-dimensional simplexes $[v_1^j, \ldots, v_q^j]$, with $j \in \{1, \ldots, t\}$. Then $[v, w_1, \ldots, w_{q-1}]$ occurs once (positive or negative) in $\partial_q([v, v_1^j, \ldots, v_q^j])$. However, by assumption we have that $[v, w_1, \ldots, w_{q-1}]$ does not occur in the image

$$\partial_q \left(\sum_{j=1}^{t} [v, v_1^j, \ldots, v_q^j] + \sum_{j=t+1}^{p} [v_0^j, v_1^j, \ldots, v_q^j] \right). \text{ Therefore } [v, w_1, \ldots, w_{q-1}]$$

occurs in an even number, say $2k$, of the $[v, v_1^j, \ldots, v_q^j]$ ($j \in \{1, \ldots, t\}$). Moreover it occurs in such a way that in k of the $\partial_q([v, v_1^j, \ldots, v_q^j])$ the summand $[v, w_1, \ldots, w_{q-1}]$ is positive and in the other k boundaries it is negative. If the simplex $[v, w_1, \ldots, w_{q-1}]$ is a positive summand of $\partial_q([v, v_1^j, \ldots, v_q^j])$, then $[w_1, \ldots, w_{q-1}]$ is a negative summand of $\partial_{q-1}([v_1^j, \ldots, v_q^j])$ and vice versa.

Thus $[w_1, \ldots, w_{q-1}]$ occurs in $\partial_{q-1} \left(\sum_{j=1}^{t} [v_1^j, \ldots, v_q^j] \right)$ k times as a positive summand and k times as a negative summand. Since $[w_1, \ldots, w_{q-1}]$ was arbitrary, the conclusion follows.   ∎

We are now in a position to prove the main theorem about homology in this appendix.

Theorem A.4.2 (Cf. [177], Theorem 6.) *Let $G = (V, E)$ be a graph and let $v \in V$ be such that $H_q(G[N(v) \setminus \{v\}]) = \{0\}$. Then:*

1. *If $q = 1$ and $H_0(G[N(v) \setminus \{v\}]) = \mathbb{Z}$, then $H_1(G) = H_1(G[V \setminus \{v\}])$.*

2. If $q \geq 2$ and $H_{q-1}(G[N(v) \setminus \{v\}]) = \{0\}$, then $H_q(G) = H_q(G[V \setminus \{v\}])$.

Proof. Assume $q \geq 1$, $H_q(G[N(v)\setminus\{v\}]) = \{0\}$. If $q = 1$ assume $H_0(G[N(v)\setminus\{v\}]) = \mathbb{Z}$, if $q \geq 2$ assume $H_{q-1}(G[N(v) \setminus \{v\}]) = \{0\}$.

First, we show that every homology class of q-dimensional cycles has a representative that does not contain v. Assume $q \geq 2$. Let

$$C := \sum_{j=1}^{t} [v, v_1^j, \ldots, v_q^j] + \sum_{j=t+1}^{p} [v_0^j, v_1^j, \ldots, v_q^j]$$

be a representative of the homology class $C + B_q(G)$, with $v_i^j \neq v$. If $t = 0$, we are done, so we can assume $t \geq 1$. By Lemma A.4.1,

$$\partial_{q-1} \left(\sum_{j=1}^{t} [v_1^j, \ldots, v_q^j] \right) = 0$$

in G and hence also in $G[N(v) \setminus \{v\}]$. Since $H_{q-1}(G[N(v) \setminus \{v\}]) = \{0\}$ there are q-simplexes $[w_0^1, \ldots, w_q^1], \ldots, [w_0^s, \ldots, w_q^s]$ of $G[N(v) \setminus \{v\}]$ such that

$$\partial_q \left(\sum_{i=1}^{s} [w_0^i, \ldots, w_q^i] \right) = \sum_{j=1}^{t} [v_1^j, \ldots, v_q^j].$$

But then

$$\partial_{q+1} \left(\sum_{i=1}^{s} [v, w_0^i, \ldots, w_q^i] \right) = \sum_{i=1}^{s} [w_0^i, \ldots, w_q^i] - \sum_{j=1}^{t} [v, v_1^j, \ldots, v_q^j].$$

Thus

$$\partial_q \left(\sum_{i=1}^{s} [w_0^i, \ldots, w_q^i] + \sum_{j=t+1}^{p} [v_0^j, v_1^j, \ldots, v_q^j] \right)$$

$$= \partial_q \left(\sum_{i=1}^{s} [w_0^i, \ldots, w_q^i] - \sum_{j=1}^{t} [v, v_1^j, \ldots, v_q^j] \right)$$

$$= \partial_q \partial_{q+1} \left(\sum_{i=1}^{s} [v, w_0^i, \ldots, w_q^i] \right) = 0.$$

Thus $C' := \sum_{i=1}^{s} [w_0^i, \ldots, w_q^i] + \sum_{j=t+1}^{p} [v_0^j, v_1^j, \ldots, v_q^j]$ forms a q-dimensional cycle with vertices in $V \setminus \{v\}$. Finally C and C' are homologous, since

$$C' - C$$

$$= \sum_{i=1}^{s} [w_0^i, \ldots, w_q^i] + \sum_{j=t+1}^{p} [v_0^j, v_1^j, \ldots, v_q^j]$$

$$- \left(\sum_{j=1}^{t} [v, v_1^j, \ldots, v_q^j] + \sum_{j=t+1}^{p} [v_0^j, v_1^j, \ldots, v_q^j] \right)$$

$$= \sum_{i=1}^{s} [w_0^i, \ldots, w_q^i] - \sum_{j=1}^{t} [v, v_1^j, \ldots, v_q^j] = \partial_{q+1} \left(\sum_{i=1}^{s} [v, w_0^i, \ldots, w_q^i] \right).$$

We have thus shown that C' is a representative of the homology class of C that does not contain any occurrences of v.

To show the same is possible in case $q = 1$, let

$$C := \sum_{j=1}^{t} (-1)^{s_j} [v, v_1^j] + \sum_{j=t+1}^{p} [v_0^j, v_1^j]$$

be a representative of the homology class $C + B_q(G)$, with $v_i^j \neq v$. Then t must be even, say $t = 2k$ and k of the s_j must be 0, the rest 1 (otherwise the boundary of C contains a multiple of $[v]$). Since $H_0(G[N(v) \setminus \{v\}]) = \mathbb{Z}$, we have by Lemma A.3.7 that $G[N(v) \setminus \{v\}]$ is connected. However then $\sum_{j=1}^{t} (-1)^{s_j+1} [v_1^j] + B_0(G[N(v) \setminus \{v\}]) = 0 + B_0(G[N(v) \setminus \{v\}])$, since in a connected graph any sum of differences of zero-dimensional simplexes is the boundary of a sum of 1-dimensional simplexes (just use the path that connects the two). This part is now finished just like the part $q \geq 2$.

The remainder of the proof is the same for $q = 1$ and for $q \geq 2$.

We claim that for any q-cycles C, C' that do not contain v we have

$$C + B_q(G) = C' + B_q(G) \text{ iff } C + B_q(G[V \setminus \{v\}]) = C' + B_q(G[V \setminus \{v\}]).$$

Provided the claim holds, we obtain an isomorphism from $H_q(G)$ to $H_q(G[V \setminus \{v\}])$ via the map that maps every homology class $\mathcal{C}$ in $H_q(G)$ to the corresponding homology class in $H_q(G[V \setminus \{v\}])$ that is defined by the representatives of $\mathcal{C}$ that do not contain v. This would finish the proof.

To prove the claim, first note that the direction "$\Leftarrow$" is trivial. For the other direction let $C + B_q(G) = 0 + B_q(G)$ with $C = \sum_{j=1}^{p} [v_0^j, v_1^j, \ldots, v_q^j]$ with $v_i^j \neq v$.

Then

$$C = \partial_{q+1} \left(\sum_{i=1}^{t} [v, w_1^i, \ldots, w_q^i, w_{q+1}^i] + \sum_{i=t+1}^{p} [w_0^i, w_1^i, \ldots, w_q^i, w_{q+1}^i] \right),$$

with $w_j^i \neq v$. By Lemma A.4.1 $\sum_{i=1}^{t}[w_1^i, \ldots, w_q^i, w_{q+1}^i]$ is a q-cycle and since $H_q(G[N(v)\backslash\{v\}]) = \{0\}$ there must be $(q+1)$-dimensional simplexes in $G[N(v)\backslash\{v\}]$ such that

$$\sum_{i=1}^{t}[w_1^i, \ldots, w_q^i, w_{q+1}^i] = \partial_{q+1}\left(\sum_{i=1}^{l}[u_0^i, \ldots, u_q^i, u_{q+1}^i]\right).$$

But then

$$\partial_{q+1}\left(\sum_{i=1}^{l}[u_0^i, \ldots, u_q^i, u_{q+1}^i] + \sum_{i=t+1}^{p}[w_0^i, w_1^i, \ldots, w_q^i, w_{q+1}^i]\right)$$

$$= \partial_{q+1}\left(\sum_{i=1}^{l}[u_0^i, \ldots, u_q^i, u_{q+1}^i]\right) + \partial_{q+1}\left(\sum_{i=t+1}^{p}[w_0^i, w_1^i, \ldots, w_q^i, w_{q+1}^i]\right)$$

$$= \sum_{i=1}^{t}[w_1^i, \ldots, w_q^i, w_{q+1}^i] + \partial_{q+1}\left(\sum_{i=t+1}^{p}[w_0^i, w_1^i, \ldots, w_q^i, w_{q+1}^i]\right)$$

$$+\partial_{q+1}\left(\sum_{i=1}^{t}[v, w_1^i, \ldots, w_q^i, w_{q+1}^i]\right) - \partial_{q+1}\left(\sum_{i=1}^{t}[v, w_1^i, \ldots, w_q^i, w_{q+1}^i]\right)$$

$$= C + \sum_{i=1}^{t}[w_1^i, \ldots, w_q^i, w_{q+1}^i] - \partial_{q+1}\left(\sum_{i=1}^{t}[v, w_1^i, \ldots, w_q^i, w_{q+1}^i]\right) = C,$$

with the cancelation in the last step happening since $\sum_{i=1}^{t}[w_1^i, \ldots, w_q^i, w_{q+1}^i]$ is a cycle in $G[N(v) \setminus \{v\}]$. We have shown that $C \in B_q(G[V \setminus \{v\}])$. For equal nonzero homology classes in G, we can show in the same fashion that their difference is zero in $G[V \setminus \{v\}]$. This finishes the proof. ∎

Theorem A.4.2 allows us to compute the homology group $H_q(G)$ from the homology group $H_q(G[V \setminus \{v\}])$ provided the homology groups $H_q(G[N(v) \setminus \{v\}])$ and $H_{q-1}(G[N(v)\setminus\{v\}])$ are trivial. Note that Theorem A.4.2 does not need any hypotheses on $H_q(G)$ and $H_q(G[V \setminus \{v\}])$. We shall now use Theorem A.4.2 to prove our main tool for work with acyclicity in ordered sets.

Corollary A.4.3 *Let* $G = (V, E)$ *be a finite connected graph and let* $v \in V$ *be such that* $G[N(v) \setminus \{v\}]$ *is acyclic. Then for all* $q \in \mathbb{N}$ *we have* $H_q(G) = H_q(G[V \setminus \{v\}])$. *In particular,* $G[V \setminus \{v\}]$ *is acyclic iff* G *is acyclic.*

Proof. By Theorem A.4.2 for all $q \geq 1$ we have $H_q(G) = H_q(G[V \setminus \{v\}])$. Since G is connected we have $H_0(G) = \mathbb{Z}$. Since $G[N(v) \setminus \{v\}]$ is acyclic we have that $G[N(v) \setminus \{v\}]$ is connected. This means that $G[V \setminus \{v\}]$ is connected also. Hence by Lemma A.3.7 we have $H_0(G[V \setminus \{v\}]) = \mathbb{Z}$.   ∎

The above together with the removal of escamotable points as discussed in Section 6.5 gives a proof of Baclawski and Björner's result on truncated noncomplemented lattices (cf. Theorem 6.6.4) that is entirely algebraic. Contractibility of the neighborhood was needed in [12] to guarantee removability of the point without affecting the homology. The contractibility of escamotable graphs was a nice consequence, which is however stronger than what was needed. We have seen that the notion "vertex is weakly escamotable iff its pointed neighborhood is acyclic" is strong enough for our purposes. We conclude this appendix by connecting the notion of dismantlability to acyclicity.

Lemma A.4.4 *Call a graph $G = (V, E)$ **dismantlable** iff either G has only one vertex or there is a vertex $v \in V$ such that*

- *There is a $w \in V \setminus \{v\}$ such that $N(v) \subseteq N(w)$, and*

- *$G[V \setminus \{v\}]$ is dismantlable.*

Every finite dismantlable graph is acyclic.

Proof. This is a proof by induction on $n = |V|$. For $n = 1$ there is nothing to prove. For the induction step $n \to (n+1)$ let G have $n+1$ vertices and let v, w be as in the definition of dismantlable graphs. Then $N(v) \setminus \{v\}$ contains a point w that is adjacent to all other points in $N(v) \setminus \{v\}$. Thus $G[N(v) \setminus \{v\}]$ is dismantlable and hence acyclic. By induction hypothesis $G[V \setminus \{v\}]$ is dismantlable and hence acyclic. Thus by Corollary A.4.3 G is acyclic.   ∎

Remarks and Open Problems

This appendix can only shed the smallest of lights on the vast subject of algebraic topology. The arguments were held entirely algebraic, which may not be as intuitive as possible. However, it allowed the leanest possible complete presentation of the results that we need. For further guidance into the use of algebraic topology the reader is referred to the papers [12] and [41] and to standard texts on algebraic topology such as [253].

The first arguments by Baclawski and Björner that link algebraic topology to fixed points in ordered sets (cf. [12], Theorems 1.1, 1.2; [41], Théorème 1.1) actually prove Lefschetz type theorems that state that for an order-preserving map $f : P \to P$ the Euler characteristic of the set $\mathrm{Fix}(f)$ of fixed points of f is equal to the Lefschetz number of f. The **Euler(–Poincaré) characteristic** $\chi(C)$ of C is defined to be

$$\chi(C) := \sum_{q \in \mathbb{N}} (-1)^q \rho(C_q),$$

where ρ denotes the rank. It can also be seen as the Lefschetz number of the identity function.

With these results one also gains an insight into the structure of Fix(f) (also cf. Section 5.2.2). For brevety's sake we did not pursue these arguments. We will be satisfied with the existence of a fixed point.

The main open problem in the direction of fixed points certainly is to find entirely combinatorial proofs of the results that so far depend on algebraic topology. A recent paper that announces progress in that direction is [11]. The argument in [11] proves a result similar to Corollary A.4.3 about structures called "pseudo cones" using neither algebra, nor topology. The fixed point algorithm for pseudo cones does not use any previously used method.

Appendix B
Order vs. Analysis

The use of order in analysis starts with the realization that spaces of real valued functions carry a natural "pointwise" order (cf., e.g., Example 1.1.2, part 5). As nonlinear problems gain in importance, order emerges as one of the tools that can help tackle some problems. Order-theoretical tools can, in the right situation, replace the use of functional analytical tools. For an introduction to the analytical background the reader is referred to standard references such as [40]. What is presented here was inspired by Heikkilä's work (cf., e.g., [35, 113, 114, 115, 117]).

B.1   The Spaces $L^p(\Omega, \mathbb{R})$

In the following we will assume that (Ω, Σ, μ) is a measure space with Ω being the underlying set, Σ a σ-algebra of subsets of Ω and $\mu : \Sigma \to \mathbb{R}$ a measure. Integration will be the usual (Lebesgue-) integration on measure spaces. For $p \geq 1$ the spaces $L^p(\Omega, \mathbb{R})$ are the usual spaces

$$\{f : \Omega \to \mathbb{R} : |f|^p \text{ is } \mu - \text{integrable }\}$$

with the usual L^p-norm

$$\|f\|_p = \left(\int_\Omega |f|^p d\mu \right)^{\frac{1}{p}} .$$

Definition B.1.1 *Let* (Ω, Σ, μ) *be a measure space. For two* Σ-*measurable functions* $f, g : \Omega \to \mathbb{R}$ *we define* $f \leq g$ *iff* $f(x) \leq g(x)$ μ-*almost everywhere* (μ-*a.e.*).

Proposition B.1.2 $\leq$ *as in Definition B.1.1 defines an order on $L^p(\Omega, \mathbb{R})$. With this order, $L^p(\Omega, \mathbb{R})$ is a lattice.*

Proof. Easy exercise. ∎

Proposition B.1.3 *Let (Ω, Σ, μ) be a measure space. If $W \subseteq L^p(\Omega, \mathbb{R})$ is well-ordered without a largest element, then W is countable and there is a cofinal subchain $N \subseteq W$ that is isomorphic to $\mathbb{N}$.*

Proof. We shall first prove that W has a cofinal subchain $N \subseteq W$ that is isomorphic to $\mathbb{N}$. Let $w_0 := \bigwedge W$. Since $f \mapsto [f - (w_0 \wedge 0)]$ is an order-automorphism of $L^p(\Omega, \mathbb{R})$, we can assume without loss of generality that $w_0 \geq 0$ μ-a.e..

Now for all $w \in W$ we have $w^+ \geq w \geq 0$ μ-a.e., and $w^+ \neq w$ in L^p. Therefore $w \mapsto \|w\|_p \in \mathbb{R}$ is strictly order-preserving on W. Let the supremum of the norms be denoted $s := \bigvee_{w \in W} \|w\|_p \in \mathbb{R} \cup \{\infty\}$. Then $\|w\|_p < s$ for all $w \in W$. If $s < \infty$, the elements $w_n := \min_W \left\{ w : \|w\|_p > s - \frac{1}{n} \right\}$ are a cofinal subchain as desired. If $s = \infty$, the elements $w_n := \min_W \left\{ w : \|w\|_p > n \right\}$ are a cofinal subchain as desired.

This proves that every well-ordered set W in $L^p(\Omega, \mathbb{R})$ that does not have a largest element has a cofinal subchain $N \subseteq W$ that is isomorphic to $\mathbb{N}$. Now assume W was an uncountable well-ordered subset of $L^p(\Omega, \mathbb{R})$. Then W contains a copy W_1 of the first uncountable ordinal ω_1. By what we just proved, W_1 must have a countable cofinal subchain, which is a contradiction. Thus all well-ordered chains in $L^p(\Omega, \mathbb{R})$ must be countable. ∎

Theorem B.1.4 *Let (Ω, Σ, μ) be a measure space. The space $L^p(\Omega, \mathbb{R})$ is a conditionally complete ordered set.*

Proof. By Proposition B.1.2 we know that $L^p(\Omega, \mathbb{R})$ is a lattice.

Now let $C \subseteq L^p(\Omega, \mathbb{R})$ be a chain that has an upper bound in $L^p(\Omega, \mathbb{R})$. By Proposition 2.2.9, C has a well-ordered cofinal subchain $W \subset C$. By Proposition B.1.3, W is countable and W has a cofinal subchain N that is isomorphic to $\mathbb{N}$. Consider N as an increasing sequence of measurable functions. By the dominated convergence theorem N has a limit. This limit is the pointwise μ-a.e. supremum of the functions in N. Hence it is the supremum of N (and thus of W and C) in the order of $L^P(\Omega, \mathbb{R})$.

Analogous to Proposition 5.1.7 we conclude that $L^P(\Omega, \mathbb{R})$ is conditionally complete. ∎

B.2   Fixed Point Theorems

Much of analysis is concerned with the solution of equations $F(f) = 0$. Simple rewriting to $F(f) + f = f$ turns such a problem into a fixed point problem. This type of problem then may be accessible to us. In this section we present analytical versions of order-theoretical fixed point theorems.

Theorem B.2.1 (An Abian–Brown theorem.) *Let (Ω, Σ, μ) be a measure space and let $F : L^p(\Omega, \mathbb{R}) \to L^p(\Omega, \mathbb{R})$ be a mapping that is order-preserving on $\uparrow f \subseteq L^p(\Omega, \mathbb{R})$. If $F(f) \geq f$, then F has a smallest fixed point above f or F has no fixed points at all in $\uparrow f$.*

Proof. Let $L_*^p(\Omega, \mathbb{R})$ be the space $L^p(\Omega, \mathbb{R})$ with a largest element **T** and a smallest element **B** attached to it. Define $F_* : L_*^p(\Omega, \mathbb{R}) \to L_*^p(\Omega, \mathbb{R})$ as the extension of F that maps **B** to **B** and **T** to **T**. By the Abian–Brown theorem F_* has a smallest fixed point g above f in $L_*^p(\Omega, \mathbb{R})$. If $g \in L^p(\Omega, \mathbb{R})$ we are done. If not, then $g = \mathbf{T}$ and F has no fixed points above f that are in $L^p(\Omega, \mathbb{R})$.   ∎

While existence of fixed points is nice, one might also want to have an algorithm that finds a fixed point. With this in mind we can record the following.

Scholium B.2.2 *Let (Ω, Σ, μ) be a measure space and let $F : L^p(\Omega, \mathbb{R}) \to L^p(\Omega, \mathbb{R})$ be a mapping that is order-preserving on $\uparrow f \subseteq L^p(\Omega, \mathbb{R})$. If $F(f) \geq f$ and F has a smallest fixed point above f, then this fixed point can be found via a transfinite iteration scheme. Moreover, if F is continuous, then the fixed point can be found with an iteration that stops (at the latest) at the first infinite ordinal number.*

Proof. The transfinite iteration scheme is of course nothing but the construction of the maximal F-chain (cf. Definition 3.4.4) that starts at f. By Proposition B.1.3, the chain must be at most countable.

To prove the "moreover"-part, suppose the maximal F-chain C that starts at f is not finite and not isomorphic to $\mathbb{N} \cup \{\infty\}$. Let ω be the smallest limit ordinal in C. Let $\{f_n\}_{n \in \mathbb{N}}$ be the sequence of elements of C that are strictly smaller than ω, listed such that $n < m$ implies $f_n < f_m$. Then $\omega = \bigvee_{L^p(\Omega, \mathbb{R})} \{f_n : n \in \mathbb{N}\}$. Let $\gamma := \lim_{n \to \infty} f_n$ in $L^p(\Omega, \mathbb{R})$, which exists by the dominated convergence theorem ($|f_n| \leq |f_0| \vee |\omega|$). Since γ, being the pointwise μ-a.e. limit of the increasing sequence $\{f_n\}_{n \in \mathbb{N}}$, is an upper bound for all f_n, we have that $\omega \leq \gamma$. But then

$$\|\omega - f_n\|_p \leq \|\gamma - f_n\|_p \to 0 \qquad (n \to \infty),$$

which implies that $\omega = \lim_{n \to \infty} f_n$ in $L^p(\Omega, \mathbb{R})$. Now since C is not isomorphic to $\mathbb{N} \cup \{\infty\}$, we have $F(\omega) > \omega$ and

$$\lim_{n \to \infty} F(f_n) = \lim_{n \to \infty} f_{n+1} = \omega < F(\omega),$$

which is a contradiction to the continuity of F.   ∎

The above shows that fixed points (or their absence) can be detected once a point that is comparable to its image is found. This leaves us with the task of finding these starting points. As a simple observation we record the following.

Proposition B.2.3 *Let* (Ω, Σ, μ) *be a measure space and let* $F : L^p(\Omega, \mathbb{R}) \to L^p(\Omega, \mathbb{R})$ *be an order-preserving mapping. If there is an* $f \in L^p(\Omega, \mathbb{R})$ *and an* $n \in \mathbb{N}$ *such that* $F^n(f) \geq f$, *then for* $g := \bigvee_{i=0}^{n} F^i(f) \in L^p(\Omega, \mathbb{R})$ *we have* $F(g) \geq g$. *Moreover* F *has a fixed point above* f *iff* F *has a fixed point above* g.

Proof. First note that

$$F(g) = F\left(\bigvee_{i=0}^{n} F^i(f)\right) \geq \bigvee_{i=1}^{n+1} F^i(f) \geq \bigvee_{i=0}^{n} F^i(f) = g.$$

For the "moreover" part note that "$\Leftarrow$" is trivial. For "$\Rightarrow$" we record that if $h \geq f$ is a fixed point, then $h \geq \bigvee_{i=0}^{n} F^i(f) = g$.   ∎

What is left to investigate (as often in fixed point theory in infinite ordered sets) is the situation in which there is no n such that $F^n(f)$ is comparable to any $F^m(f)$ with $m \neq n$. If this situation we are at least able to rule out the existence of fixed points in certain parts of the set.

Proposition B.2.4 *Let* (Ω, Σ, μ) *be a measure space and let* $F : L^p(\Omega, \mathbb{R}) \to L^p(\Omega, \mathbb{R})$ *be an order-preserving mapping. If* $\{F^n(f) : n \in \mathbb{N}\}$ *is an infinite antichain with no upper bounds, then there is no fixed point of* F *above* f.

Proof. If $g \geq f$ was a fixed point of F, then $g \geq F^n(f)$ for all $n \in \mathbb{N}$, which contradicts our assumption.   ∎

It is tempting to hope that if $\{F^n(f) : n \in \mathbb{N}\}$ has an upper bound, then the supremum of $\{F^n(f) : n \in \mathbb{N}\}$ could play the same role as the finite supremum in Proposition B.2.3. Unfortunately, this is not the case, since the inequality

$$F\left(\bigvee_{n=1}^{\infty} F^n(f)\right) \geq \bigvee_{n=2}^{\infty} F^n(f) \text{ does not guarantee that } F\left(\bigvee_{n=1}^{\infty} F^n(f)\right) \text{ is above}$$

$$\bigvee_{n=1}^{\infty} F^n(f).$$

B.3   An Application

To give an idea how order-theoretical results can be applied in analysis, we shall consider a certain type of integral equation.

Proposition B.3.1 (For consideration in a more general context, cf. [113], Section 3.) *Consider the Hammerstein integral equation*

$$u(t) = v(t) + r \int_{\Omega} k(t, s) f(s, u(s)) \, ds, \qquad t \in \Omega.$$

Ω *is a closed and bounded subset of* $\mathbb{R}^m$ *and all functions assume values in* $\mathbb{R}$. *Assume that:*

1. $k : \Omega \times \Omega \to \mathbb{R}_+$ *only assumes nonnegative values and is continuous.*

2. $f : \Omega \times \mathbb{R}_+ \to \mathbb{R}_+$ *only assumes nonnegative values and is such that*

   (a) $f(\cdot, u(\cdot))$ *is measurable for each* $u \in C(\Omega, \mathbb{R}_+)$,

   (b) $f(t, \cdot)$ *is increasing for almost every* $t \in \Omega$,

   (c) *There are* $h, h_0 \in L(\Omega, \mathbb{R}_+)$ *such that for all* $x \in \mathbb{R}_+$ *and almost every* $t \in \Omega$ *we have*

$$f(t, x) \le h_0(t) + h(t)x,$$

3. $r > 0$ *is such that* $r \int_\Omega k(t, s)h(s)\, ds =: b < 1$ *for each* $t \in \Omega$.

Then for each $v \in C(\Omega, \mathbb{R}_+)$ *the above integral equation has a solution.*

Proof. We shall work with the operator

$$Gu(t) := v(t) + r \int_\Omega k(t, s) f(s, u(s))\, ds, \qquad t \in \Omega$$

on $C(\Omega, \mathbb{R}_+)$ equipped with the L^1-norm. The fact that k and f map into $\mathbb{R}_+$ shows that $Gv \ge v$ and condition 2b assures that G is order-preserving. Now we need to prove that the G-chain starting at v has an upper bound. To do this, we consider the operator

$$Hz(t) = v(t) + r \int_\Omega k(t, s)h_0(s)ds + r \int_\Omega k(t, s)h(s)z(s)ds.$$

By condition 3 we have that $\|Hz - Hz'\|_\infty \le b\|z - z'\|_\infty$, where the norm is the uniform norm $\|f\|_\infty = \sup\{f(t) : t \in \Omega\}$. Moreover H maps $\uparrow v$ to itself. By Banach's fixed point theorem (cf. [119], 111.11 "Banachscher Fixpunktsatz" or [56], Theorem 10.1.2; the result in Dieudonné would need to be adjusted slightly to fit our purposes) this means there is a unique continuous function $w \in\uparrow v$ such that $Hw = w$. By condition 2c we have $G \le H$, which means $Gw \le w$.

Thus G maps $[v, w] \subseteq C(\Omega, \mathbb{R}_+)$ to itself. Unfortunately this interval is not chain-complete in the L^1-norm, so we are not trivially done.

Let C be the maximal G-chain in $C(\Omega, \mathbb{R}_+)$ that starts at v. Let u_* be the pointwise a.e. supremum of C, which exists in $L^1(\Omega, \mathbb{R}_+)$. We shall now show that u_* is continuous. So assume u_* is not continuous. Then in particular $u_* \notin C$ and C has no supremum in $C(\Omega, \mathbb{R}_+)$. Since $C(\Omega, \mathbb{R}_+) \subseteq L^1(\Omega, \mathbb{R}_+)$, there is an increasing sequence $\{u_n\}_{n \in \mathbb{N}}$ that is cofinal in C. Then $\{G(u_n)\}_{n \in \mathbb{N}}$ is cofinal in C also. Let $c_n := \int_\Omega f(s, u_n(s))ds$ and let $c := \bigvee_{n \in \mathbb{N}} c_n$. Let $k_0 := \max\{rk(t, s) : t, s \in \Omega\}$. Then for all $m > n$ we have

$$0 \le G(u_m) - G(u_n) \le k_0(c_m - c_n) \le k_0(c - c_n).$$

This means that $G(u_n)$ is an increasing uniform Cauchy sequence, which converges uniformly to its pointwise supremum u_*. However this means that u_* is continuous.

Since u_* is the supremum of the maximal G-chain C we must have $G(u_*) = u_*$. Therefore u_* is a solution of the equation as desired. ∎

It is notable that conditions 2c and 3 are used exclusively to assure that the G-chain starting at v has an upper bound. Without them, the conclusion of Proposition B.3.1 could be re-phrased as "the integral equation has a solution iff the G-chain starting at v has a supremum". This modification widens the scope of the result, while the verification if a solution exists becomes dependent on the outcome of a potentially transfinite (though countable) iteration scheme.

Remarks and Open Problems

1. For more on the connection between order and analysis the reader is referred to [116, 168].

2. Is it possible to use some of the more sophisticated techniques to find fixed points that were presented in this text in analysis?

   Consider the following example. Define $\mathbb{Q}_m := \left\{ \dfrac{k}{2^m} : k \in \mathbb{Z} \right\}$ and define
   $\mathbb{Q}_d := \bigcup_{m \in \mathbb{N}} \mathbb{Q}_m$. $\mathbb{Q}_d$ is called the set of **dyadic rational numbers**. If Ω is $[0, 1)$, then one can approximate any real valued function on Ω with a sequence of step functions $s_m = \sum_{i=1}^{j_m} q_i^m \mathbf{1}_{[l_i^m, r_i^m)}$ such that $q_i^m, l_i^m, r_i^m \in \mathbb{Q}_m$.
   For each $m \in \mathbb{N}$, these sets of step functions are forming an (albeit infinite) truncated noncomplemented lattice. Moreover any two elements of such a truncated noncomplemented lattice have a distance (in L^1) of at least $\left(\dfrac{1}{2^m} \right)^2$ from each other. This means these truncated noncomplemented lattices have a rather discrete structure.

   Is there a way to prove infinitary fixed point results for truncated noncomplemented lattices that allow application to analysis (possibly through the above construction)?

3. Brouwer's fixed point theorem says that every continuous self map of $[0, 1]^d$ must have a fixed point.

   Let $n \in \mathbb{N}$. Subdivide $[0, 1]^d$ into n^d subcubes $\Pi_{k=1}^{d} \left[\dfrac{x_k - 1}{n}, \dfrac{x_k}{n} \right]$, where $x_k \in \{1, \dots, n\}$ and consider the cubic complex

$$C_n^d := \left\{ \Pi_{k=1}^d \left[\frac{x_k}{n}, \frac{y_k}{n} \right] : y_k \in \{1, \ldots, n\}, \right.$$

$$\left. x_k \in \{y_k - 1, y_k\} \text{ or } x_k = y_k = 0 \right\}.$$

as an ordered set ordered by inclusion. Since

$$C_n^1 = \left\{ \left[\frac{x}{n}, \frac{y}{n} \right] : y \in \{1, \ldots, n\}, x \in \{y - 1, y\} \text{ or } x = y = 0 \right\}$$

it follows that $C_n^d = \Pi_{k=1}^d C_n^1$, where the product is a product of ordered sets. The set C_n^1 is a fence with $n + 1$ minimal and n maximal elements, so C_n^d is a product of fences, which is a very nice ordered set.

If $f : [0, 1]^d \to [0, 1]^d$ is a continuous map, we can define $F_n^f : C_n^d \to \mathcal{P}(C_n^d)$ by

$$F_n^f(c) := \left\{ k \in C_n^d : f[c] \cap k \neq \varnothing \right\}.$$

This is a multivalued function on C_n^d. Moreover $x \subseteq y$ implies that for each $a \in F_n^f(x)$ we have $a \in F_n^f(y)$ also. Thus F_n^f is almost an isotone relation.

Suppose we could prove that any such function F_n^f (or some similarly defined function) has a fixed point in C_n^d. Then for each $n \in \mathbb{N}$ there is an $x_n \in [0, 1]^d$ with $|f(x_n) - x_n| \leq 2\frac{\sqrt{d}}{n}$. By compactness of $[0, 1]^n$ there is a subsequence $\{x_{n_k}\}_{k \in \mathbb{N}}$ that converges to a point x. x would have to be a fixed point of f and the argument would prove Brouwer's theorem.

For $d = 1$ this is indeed possible, but the proof relies more on the linear structure of $[0, 1]$ than on order theory. For higher dimensions any multivalued functions the author could come up with were lacking some feature. Hence, even though C_n^d has very nice order-theoretical structure, a proof of Brouwer's theorem was not possible this way (yet?).

What kinds of fixed point results for multivalued maps can be proved for products of fences? Could any of them be used to prove Brouwer's theorem?

A combinatorial proof of Brouwer's fixed point theorem that relies on graph theory and Sperner's lemma can be found in [272].

References

[1] A. Abian (1971), Fixed point theorems of the mappings of partially ordered sets, *Rendiconti del circolo mathematico di Palermo* 20, 139–142

[2] S. Abian and A. B. Brown (1961), A theorem on partially ordered sets with applications to fixed point theorems, *Canad. J. Math.* 13, 78–82

[3] J. Adámek, H. Herrlich and G. Strecker (1990), *Abstract and concrete categories*, John Wiley & Sons, New York

[4] R. Aeschlimann and J. Schmidt (1992), Drawing orders using less ink, *Order* 9, 5–13

[5] A. V. Aho, J. E. Hopcroft and J. D. Ullman (1983), *Data structures and algorithms*, Addison-Wesley, Reading, MA

[6] L. Alvarez (1965), Undirected graphs realizable as graphs of modular lattices, *Canad. J. Math.* 17, 923–932

[7] J. C. Arditti (1976), Graphes de comparabilité et dimension des ordres, *Notes de Recherches* CRM 607, Centre de Rech. Math. Univ. Montréal

[8] J. C. Arditti and H. A. Jung (1980), The dimension of finite and infinite comparability graphs, *J. London Math. Soc.* (2) 21, 31–38

[9] F. Bacchus and A. Grove (1995), On the forward checking algorithm, in: Lecture Notes in Computer Science 976, *Principles and Practices in Constraint Programming (CP-95)*, Springer Verlag, p. 292–309, available at `http://www.cs.toronto.edu/~fbacchus/on-line.html`

[10] K. Baclawski (1977), Galois connections and the Leray spectral sequence, *Adv. Math.* 25, 191–215

[11] K. Baclawski (1996), A fixed point algorithm for ordered sets, preprint

[12] K. Baclawski and A. Björner (1979), Fixed points in partially ordered sets, *Adv. Math.* 31, 263–287

[13] H. J. Bandelt and M. van de Vel (1987), A fixed cube theorem for median graphs, *Discrete Math.* 67, 129–137

[14] M. Basso-Gerbelli and P. Ille (1993), La reconstruction des relations définis par interdits, *C. R. Acad. Sci. Paris* t. 316, sér. I, 1229–1234

[15] M. F. Bélanger, J. Constantin, G. Fournier (1994), Graphes et ordonnés démontables, propriété de la clique fixe, *Discrete Math.* 130, 9–17

[16] C. Bergman, R. McKenzie and Sz. Nagy (1982), How to cancel a linearly ordered exponent, *Colloq. Math. Soc. Janos Bolyai* 21, 87–94

[17] G. Birkhoff (1967), *Lattice theory (third edition)*, AMS Colloquium Publications XXV, Providence, Rhode Island

[18] A. Björner (1981), Homotopy type of posets and lattice complementation, *J. Comb. Theory Ser. A* 30, 90–100

[19] K. P. Bogart, R. Freese, J. P. S. Kung, editors (1990), *The Dilworth theorems: selected papers of Robert P. Dilworth*, Birkhäuser, Boston

[20] K. P. Bogart (1994), Intervals and orders: What comes after interval orders?, in: V. Bouchitté and M. Morvan (eds.), *Orders, Algorithms and Applications (Proceedings of the ORDAL '94 in Lyon)*, Lecture Notes in Computer Science, 831, Springer, Berlin, p. 13–32

[21] B. Bollobás (1978), *Extremal graph theory*, Academic Press, London-New York-San Francisco

[22] B. Bollobás (1979), *Graph theory*, Graduate Texts in Mathematics nr. 63, Springer Verlag, New York

[23] J. A. Bondy and R. L. Hemminger (1976), Graph reconstruction – a survey, *J. Graph Theory* 1, 227–268

[24] J. P. Bordat (1992), Sur l'algorithmique combinatoire d'ordres finis, Thèse de Docteur d'état, Université Montpellier II

[25] V. Bouchitté and M. Habib (1989), The calculation of invariants for ordered sets, in: I. Rival (ed.), *Algorithms and Order*, Kluwer Academic Publishers, p. 231–279

[26] V. Bouchitté and M. Morvan (1994), *Orders, Algorithms and Applications (Proceedings of the international workshop ORDAL '94 in Lyon)*, Lecture Notes in Computer Science 831, Springer Verlag, Berlin

[27] G. Brightwell (1988), Linear extensions of infinite posets, *Discrete Mathematics* 70, 113–136

[28] G. Brightwell (1989), Semiorders and the $\frac{1}{3}$-$\frac{2}{3}$ conjecture, *Order* 5, 369–380

[29] G. Brightwell (1993), On the complexity of diagram testing, *Order* 10, 297–303

[30] G. R. Brightwell, S. Felsner, W. T. Trotter (1995), Balancing pairs and the cross product conjecture, *Order* 12, 327–349

[31] Graham Brightwell and Sarah Goodall (1990), The number of partial orders of fixed width, *Order* 15, 315–337

[32] G. Brinkmann and B.McKay (2002), Posets on up to 16 points, *Order* 19, 147–179

[33] R. Brown (1982), The fixed point property and cartesian products, *Amer. Math. Monthly* November issue, 654–678

[34] R. A. Brualdi, H. C. Jung and W. T. Trotter (1994), On the poset of all posets on n elements, *Discrete Applied Math.* 50, 111–123

[35] S. Carl and S. Heikkilä (1992), An existence results for elliptic differential inclusions with discontinuous nonlinearity, *Nonlinear Anal.* 18, 471–479

[36] C. Chang, B. Jonsson, A. Tarski (1964), Refinement properties for relational structures, *Fund. Math.* 55, 249–281

[37] P. C. Chen (1992), Heuristic sampling: A method for predicting the performance of tree searching programs, *SIAM J. Comput.* 21, 295–315

[38] C. Chaunier and N. Lygerōs (1992), The number of orders with thirteen elements, *Order* 9 203–204,

[39] Clay Mathematics Institute web site, `http://www.claymath.org/prize_problems/index.htm`

[40] D. L. Cohn (1980), *Measure Theory*, Birkhäuser, Boston

[41] J. Constantin and G. Fournier (1985), Ordonnés escamotables et points fixes, *Discr. Math.* 53, 21–33

[42] S. A. Cook (1971), The complexity of theorem-proving procedures, in: Association for Computing Machinery, *Proc.* 3rd *Ann. ACM Symp. on Theory of Computing*, New York, p. 151–158

[43] E. Corominas (1990), Sur les ensembles ordonnés projectifs et la propriété du point fixe, *C. R. Acad. Sci. Paris* 311 Série 1, 199–204

[44] P. Cousot and R. Cousot (1979), Constructive Versions of Tarski's Fixed Point Theorems, *Pacific J. Math.* 82, 43–57

[45] P. Cousot and R. Cousot (1979), A Constructive Characterization of the Lattices of all Retractions, Preclosure, Quasi-closure and Closure Operators on a Complete Lattice, *Portugaliae Mathematica* 38, 185–198

[46] H. H. Crapo (1982), Ordered Sets: Retracts and Connections, *Journal of Pure and Applied Algebra* 23, 13–28

[47] P. Crawley and R. P. Dilworth (1973), *Algebraic theory of lattices*, Prentice Hall, Englewood Cliffs, NJ

[48] S. K. Das (1977), A machine representation of finite t_0 topologies, *J. ACM* 24, 676–692

[49] B. Davey and H. Priestley (1990), *Introduction to lattices and order*, Cambridge University Press, Cambridge

[50] J. L. Davidson (1986), Asymptotic Enumeration of Partial Orders, *Congressus Numerantium* 53, 277–286

[51] Anne C. Davis (1955), A Characterization of Complete Lattices, *Pacific J. Math.* 5, 311–319

[52] R. Dechter (1992), Constraint networks, in: S. Shapiro (ed.), *Encyclopedia of Artificial Intelligence*, Wiley, New York, p. 276–284

[53] R. Dechter and I. Meiri (1994), Experimental evaluation of preprocessing algorithms for constraint satisfaction problems, *Artificial Intelligence* 68, 211–241

[54] R. Dechter and J. Pearl (1988), Network-based heuristics for constraint-satisfaction problems, *Artificial Intelligence* 34, 1–38

[55] R. Dechter and J. Pearl (1989), Tree clustering for constraint networks, *Artificial Intelligence* 38, 353–366

[56] J. Dieudonné (1960), *Foundations of modern analysis*, Academic Press, New York, London

[57] R. P. Dilworth (1950), A decomposition theorem for partially ordered sets, *Ann. of Math.* 51, 161–166

[58] R. P. Dilworth (1990), Chain partitions in ordered sets, in: K. P. Bogart, R. Freese, J. P. S. Kung (eds.), *The Dilworth theorems: selected papers of Robert P. Dilworth*, Birkhäuser, Boston, p. 1–6

[59] M. Donalies and B. Schröder (2000), Performance guarantees and applications for Xia's algorithm, *Discrete Mathematics* 213, 67–86 (Proceedings of the Banach Center Minisemester on Discrete Mathematics, week on Ordered Sets)

[60] B. Dreesen, W. Poguntke and P. Winkler (1985), Comparability invariance of the fixed point property, *Order* 2, 269–274

[61] D. Duffus (1984), Automorphisms and products of ordered sets, *Algebra Universalis* 19, 366–369

[62] D. Duffus and T. Goddard (1996), The complexity of the fixed point property, *Order* 13, 209–218

[63] D. Duffus, H. A. Kierstead and W. T. Trotter (1991), Fibres and Ordered Set Coloring, *J. Comb. Theory, Series A* 58, 158–164

[64] D. Duffus, T. Łuczak, V. Rödl and A. Ruciński (1998), Endomorphisms of partially ordered sets, *Combin. Probab. Comput.* 7, 33–46

[65] D. Duffus, W. Poguntke and I. Rival (1980), Retracts and the fixed point problem for finite partially ordered sets, *Canad. Math. Bull.* 23, 231–236

[66] D. Duffus and I. Rival (1978), A logarithmic property for exponents of partially ordered sets, *Canad. J. Math.* 30, 797–807

[67] D. Duffus and I. Rival (1981), A structure theory for ordered sets, *Discr. Math.* 35, 53–118

[68] D. Duffus, I. Rival and M. Simonovits (1980), Spanning retracts of a Partially Ordered Set, *Discr. Math.* 32, 1–7

[69] D. Duffus, V. Rödl, B. Sands, R. Woodrow (1992), Enumeration of order-preserving maps, *Order* 9, 15–29

[70] D. Duffus, B. Sands, N. Sauer and R. E. Woodrow (1991), Two-coloring all Two-Element Maximal Antichains, *J. Comb. Theory, Series A* 57, 109–116

[71] D. Duffus and N. Sauer (1987), Fixed points of products and the strong fixed point property, *Order* 4, 221–231

[72] B. Dushnik and E. W. Miller (1941), Partially ordered sets, *Amer. J. of Math.* 63, 600–610

[73] P. Edelman (1979), On a fixed point theorem for partially ordered sets, *Discr. Math.* 15, 117–119

[74] M. El-Zahar (1989), Enumeration of ordered sets, in: I. Rival (ed.), *Algorithms and Order*, NATO Adv. Sci. Inst. Ser. C: Math. Phys. Sci., Kluwer Acad. Publ., p. 327–352

[75] M. Erné (1981), open question on p.843 of [218]

[76] M. Erné and K. Stege (1991), Counting finite posets and topologies, *Order* 8, 247–265

[77] K. Ewacha, W. Li and I. Rival (1991), Order, genus and diagram invariance, *Order* 8, 107–113

[78] R. Fagin (1976), Probabilities on finite models, *J. Symbolic Logic* 41, 50–58

[79] J. D. Farley (1993), The uniqueness of the core, *Order* 10, 129–131

[80] J. D. Farley (1995), The number of order-preserving maps between fences and crowns, *Order* 12, 5–44

[81] J. D. Farley (1997), Perfect sequences of chain-complete posets, *Discrete Mathematics* 167/168, 271–296

[82] J. D. Farley (1997), The fixed point property for posets of small width, *Order* 14, 125–143

[83] J. D. Farley and B. Schröder (2001), Strictly order-preserving maps into $\mathbb{Z}$, II: A 1979 problem of Erné, *Order* 18, 381–385

[84] S. Felsner and W. T. Trotter (2000), Dimension, Graph and Hypergraph Coloring, *Order* 17, 167–177

[85] P. C. Fishburn (1985), *Interval orders and interval graphs: a study of partially ordered sets*, Wiley, New York

[86] P. C. Fishburn and W. T. Trotter (1999), Geometric Containment Orders: A Survey, *Order* 15, 168–181

[87] T. Fofanova and A. Rutkowski (1987), The fixed point property in ordered sets of width two, *Order* 4, 101–106

[88] T. Fofanova, I. Rival, A. Rutkowski (1994), Dimension 2, fixed points and dismantlable ordered sets, *Order* 13, 245–253

[89] W. Fouché (1996), Chain partitions of ordered sets, *Order* 13, 255–266

[90] R. Freese, J. Ježek, J. Nation (1995), *Free lattices*, American Mathematical Society, Mathematical Surveys and Monographs 42, Providence, Rhode Island

[91] E. C. Freuder (1982), A sufficient condition for backtrack-free search, *Journal of the ACM* 29, 24–32

[92] E. C. Freuder (1985), A sufficient condition for backtrack-bounded search, *Journal of the ACM* 32, 755–761

[93] T. Gallai (1967), Transitiv orientierbare Graphen, *Acta Math. Hung.* 18, 25–66

[94] M. R. Garey and D. S. Johnson (1979), *Computers and intractability: A guide to the theory of NP-completeness*, Freeman, San Francisco

[95] J. Gaschnig (1978), Experimental Case Studies of Backtrack vs. (Waltz)-Type vs. New Algorithms for Satisfying Assignment Problems, *Proceedings of the Second Canadian Conference on Artificial Intelligence*, Toronto, Ont., 268–277

[96] M. Gikas (1986), Fixed points and structural problems in ordered sets, Ph. D. dissertation, Emory University

[97] M. Golumbic (1980), *Algorithmic graph theory and perfect graphs*, Academic Press, New York

[98] K. Grant, R. Nowakowski, I. Rival (1995), The endomorphism spectrum of an ordered set, *Order* 12, 45–55

[99] R. Gysin (1977), Dimension transitiv orientierbarer Graphen, *Acta Math. Acad. Sci. Hungar.* 29, 313–316

[100] G. Grätzer (1978), *General lattice theory*, Birkhäuser, Basel

[101] M. Habib, M. Morvan, and J.-X. Rampon (1993), On the calculation of transitive reduction-closure of orders, *Discrete Math.* 111, 289–303

[102] M. Habib, M. Morvan, M. Pouzet, and J.-X. Rampon (1993), Interval dimension and MacNeille completion, *Order* 10, 147–151

[103] P. R. Halmos (1974), *Naive set theory*, Undergraduate Texts in Mathematics, Springer Verlag, New York

[104] C.-C. Han and C.-H. Lee (1988), Comments on Mohr and Henderson's path consistency algorithm, *Artificial Intelligence* 36, 125–130

[105] R.M. Haralick and G.L. Elliott (1980), Increasing tree search efficiency for constraint satisfaction problems, *Artificial Intelligence* 14, 263–313

[106] J. Hashimoto (1948), On the product decomposition of partially ordered sets, *Math. Japon.* 1, 120–123

[107] J. Hashimoto (1951), On direct product decomposition of partially ordered sets, *Ann. Math. (2)* 54, 315–318

[108] J. Hashimoto and T. Nakayama (1950), On a problem of G. Birkhoff, *Proc. Amer. Math. Soc.* 1, 141–142

[109] S. Hazan (1992), The projection property for Orders and Triangle-free Graphs, Ph. D. Dissertation, Vanderbilt University

[110] S. Hazan (1996), On triangle-free projective graphs, *Algebra Universalis* 35, no. 2, 185–196

[111] S. Hazan, V. Neumann-Lara (1995), Fixed points of posets and clique graphs, *Order* 13, 219–225

[112] S. Hazan, V. Neumann-Lara (1998), Two order invariants related to the fixed point property, *Order* 15, 97–111

[113] S. Heikkilä (1990), On fixed points through a generalized iteration method with applications to differential and integral equations involving discontinuities, *Nonlinear Anal.* 14, 413–426

[114] S. Heikkilä (1990), On differential equations in ordered Banach spaces with applications to differential systems and random equations, *Differential Integral Equations* 3, 589–600

[115] S. Heikkilä, V Lakshmikantham, Y. Sun (1992), Fixed point results in ordered normed spaces with applications to abstract and differential equations, *J. Math. Anal. Appl.* 163, 422–437

[116] S. Heikkilä and V. Lakhshmikantham (1994), *Monotone iterative techniques for discontinuous nonlinear differential equations*, Marcel Dekker Inc., New York

[117] S. Heikkilä and V. Lakhshmikantham (1995), On mild solutions of first order discontinuous semilinear differential equations in Banach spaces, *Applicable Analysis* 56, 131–146

[118] J. Heitzig and J. Reinhold (2000), The number of unlabeled orders on fourteen elements, *Order* 17, 333–341

[119] H. Heuser (1983), *Lehrbuch der Analysis, Teil 2*, B. G. Teubner Verlag, Stuttgart

[120] E. Hewitt and K. R. Stromberg (1965), *Real and abstract analysis*, Springer Graduate Texts in Mathematics 25, Springer Verlag, Berlin, Heidelberg

[121] T. Hiraguchi (1955), On the dimension of orders, *Sci. Rep. Kanazawa Univ.* 4, 1–20

[122] H. Höft and M. Höft (1976), Some fixed point theorems for partially ordered sets, *Can. J. Math.* 28, 992–997

[123] H. Höft and M. Höft (1988), Fixed point invariant reductions and a characterization theorem for lexicographic sums, *Houston Journal of Mathematics* 14 (no. 3), 411–422

[124] H. Höft and M. Höft (1991), Fixed point free components in lexicographic sums with the fixed point property, *Demonstratio Mathematica* XXIV, 294–304

[125] T. Hogg, B. Huberman and C. Williams (1996), Phase transitions and the search problem, *Artificial Intelligence* 81, 1–15

[126] J. Hopcroft and R. Karp (1973), A $n^{\frac{5}{2}}$ algorithm for maximum matching in bipartite graphs, *SIAM J. Comput.* 2, 225–231

[127] P. Ille (1993), Recognition problem in reconstruction for decomposable relations, in: N. W. Sauer et. al. (eds.), *Finite and Infinite Combinatorics in Sets and Logic*, Kluwer Academic Publishers, p. 189–198

[128] P. Ille and J.-X. Rampon (1997), Reconstruction of posets with the same comparability graph, *J. Combinatorial Theory (B)* 74, 368–377

[129] E. Jawhari, D. Misane and M. Pouzet (1986), Retracts: Graphs and ordered sets from the metric point of view, in: I. Rival (ed.), *Combinatorics and ordered sets*, Contemp. Math. 57, p. 175–226

[130] P. Jeavons, D. Cohen and J. Pearson (1998), Constraints and universal algebra, *Annals of Mathematics and Artificial Intelligence* 24, 51–67

[131] B. Jónsson (1982), Arithmetic of ordered sets, in: I. Rival (ed.), *Ordered Sets*, D. Reidel, Dordrecht, p. 3–41

[132] B. Jónsson and R. McKenzie (1982), Powers of partially ordered sets: cancellation and refinement properties, *Math. Scand.* 51, 87–120

[133] D. Kelly (1984), Unsolved problems: Removable pairs in dimension theory, *Order* 1, 217–218

[134] D. Kelly (1985), Comparability graphs, in: I. Rival (ed.), *Graphs and Order*, D. Reidel, Dordrecht, p. 3–40

[135] D. Kelly and W. T. Trotter (1982), Dimension theory for ordered sets, in: I. Rival (ed.), *Ordered Sets*, D. Reidel, p. 171–211

[136] P. J. Kelly (1957), A congruence theorem for trees, *Pacific J. Math.* 7, 961–968

[137] H. Kierstead and W. T.Trotter (1991), A note on removable pairs, in: Y. Alavi et. al. (eds.), *Graph Theory, Combinatorics and Applications, vol. 2*, Wiley, New York, p. 739–742

[138] R. Kimble (1973), Extremal problems in dimension theory for partially ordered sets, Ph. D. Dissertation, MIT

[139] S. S. Kislitsin (1968), Finite partially ordered sets and their associated sets of permutations, *Matematicheskiye Zametki* 4, 511–518

[140] D. Klarner (1969), The number of graded partially ordered sets, *J. Comb. Theory* 6, 12–19

[141] D. Klarner (1970), The number of classes of isomorphic graded partially ordered sets, *J. Comb. Theory* 9, 412–419

[142] D.J. Kleitman and B.L. Rothschild (1970), The number of finite topologies, *Proc. Amer. Math. Soc.* 25, 276–282

[143] D.J. Kleitman and B.L. Rothschild (1975), Asymptotic Enumeration of Partial Orders on a Finite Set, *Trans. Amer. Math. Soc.* 205, 205–220

[144] D. J. Kleitman and G. Markowsky (1975), On Dedekind's problem: The number of isotone boolean functions II, *Trans. Amer. Math. Soc.* 213, 373–390

[145] B. Knaster (1928), Un théorème sur les fonctions d'ensembles, *Ann. Soc. Polon. Math.* 6, 133–134

[146] D. E. Knuth (1975), Estimating the efficiency of backtrack programs, *Math. Comp.* 29, 121–136

[147] J. Köbler, U. Schöning and J. Torán (1993), *The graph isomorphism problem: its structural complexity*, Progress in Theoretical Computer Science, Birkhäuser, Boston

[148] G. Kondrak and P. van Beek (1997), A theoretical evaluation of selected backtracking algorithms, *Artificial Intelligence* 89, 365–387

[149] A. Korshunov (1977), On the number of monotone boolean functions (in Russian), *Probl. Kibern.* 38, 5–108

[150] D. Kratsch and J.-X. Rampon (1994), A counterexample about poset reconstruction, *Order* 11, 95–96

[151] D. Kratsch and J.-X. Rampon (1994), Towards the reconstruction of posets, *Order* 11, 317–341

[152] D. Kratsch and J.-X. Rampon (1996), Width two posets are reconstructible, *Discrete Math.* 162, 305–310

[153] V. Kumar (1992), Algorithms for constraint satisfaction problems – a survey, *AI magazine* 13, 32–44

[154] J. P. S. Kung (1999), Möbius Inversion, *Encyclopedia of Mathematics*

[155] J. P. S. Kung (1999), private communication

[156] L. J. Langley (1995), A recognition algorithm for orders of interval dimension 2, ARIDAM VI and VII (New Brunswick, NJ, 1991/1992), *Discrete Appl. Math.* 60, 257–266

[157] B. Larose (1995), Minimal automorphic posets and the projection property, *International Journal of Algebra and Computation* 5, 65–80

[158] B. Li (1990), All retraction operators on a lattice need not form a lattice, *Journal of Pure and Applied Algebra* 67, 201–208

[159] B. Li (1993), The core of a chain complete poset with no one-way infinite fence and no tower, *Order* 10, 349–361

[160] B. Li (1996), The ANTI-order for caccc posets – Part I, *Discrete Mathematics* 158, 173–184

[161] B. Li (1996), The ANTI-order for caccc posets – Part II, *Discrete Mathematics* 158, 185–199

[162] B. Li and E. C. Milner (1992), The PT order and the fixed point property, *Order* 9, 321–331

[163] B. Li and E. C. Milner (1993), A chain complete poset with no infinite antichain has a finite core, *Order* 10, 55–63

[164] B. Li and E. C. Milner (1995), From finite posets to chain complete posets having no infinite antichain, *Order* 12, 159–171

[165] B. Li and E. C. Milner (1997), The ANTI-order and the fixed point property for caccc posets, *Discrete Mathematics* 175, 197–209

[166] B. Li and E. C. Milner (1998), Isomorphic ANTI-cores of caccc posets – an improvement, *Discrete Mathematics* 183, 213–221

[167] B. Li and E. C. Milner (1997), Isomorphic ANTI-cores of caccc posets, *Discrete Mathematics* 176, 185–195

[168] J. Lindenstrauss and L. Tzafriri (1973), *Classical Banach spaces*, Springer Lecture Notes in Mathematics 338, Springer Verlag, New York

[169] W.-P. Liu and H. Wan (1993), Automorphisms and Isotone Self-Maps of Ordered Sets with Top and Bottom, *Order* 10, 105–110

[170] Z. Lonc and I. Rival (1987), Chains, antichains and fibres, *J. Comb. Theory Ser. A* 44, 207–228

[171] A. Lubiw (1981), Some NP-complete problems similar to graph isomorphisms, *SIAM Journal of Computing* 10, 11–21

[172] E. M. Luks (1982), Isomorphism of graphs of bounded valence can be tested in polynomial time, *Journal of Computer and System Science* 25, 42–65

[173] A.K. Mackworth (1977), Consistency in networks of relations, *Artificial Intelligence* 8, 197799–118

[174] A.K. Mackworth (1992), Constraint Satisfaction, in: S. Shapiro (ed.), *Encyclopedia of Artificial Intelligence*, Wiley, New York, p. 284–293

[175] R. Maltby (1992), A smallest-fibre-size to poset-size ratio approaching $\frac{8}{15}$, *J. Combinatorial Theory (A)* 61, 328–330

[176] J.J. McGregor (1979), Relational consistency algorithms and their application in finding subgraph and graph isomorphisms, *Information Sciences* 19, 229–250

[177] T. McKee and E. Prisner (1996), An approach to graph-theoretic homology, in: Y. Alavi et. al. (eds.), *Combinatorics, Graph Theory, and Algorithms*, Proceedings of the Eighth Quadrennial International Conference in Graph Theory, Combinatorics, Algorithms and Applications, Vol. II, p. 631–640

[178] R. McKenzie (1999), Arithmetic of finite ordered sets: cancellation of exponents I, *Order* 16, 313–333

[179] R. McKenzie (2000), Arithmetic of finite ordered sets: cancellation of exponents II, *Order* 17, 309–332

[180] E. C. Milner (1990), Dilworth's Decomposition Theorem in the infinite case, in: K. P. Bogart, R. Freese, J. P. S. Kung (eds.), *The Dilworth theorems: selected papers of Robert P. Dilworth*, Birkhäuser, Boston, p. 30–35

[181] J. Mitas (1992), The structure of interval orders, Doctoral dissertation, TH Darmstadt

[182] J. Mitas (1995), Interval orders based on arbitrary ordered sets, *Discrete Math.* 144, 75–95

[183] R. Mohr and T. C. Henderson (1986), Arc and path consistency revisited, *Artificial Intelligence* 28, 225–233

[184] H. Müller and J.-X. Rampon (1997), Partial orders and their convex subsets, *Discrete Mathematics* 165/166, 507–517

[185] B. Nadel (1989), Constraint satisfaction algorithms, *Computational Intelligence* 5, 188–224

[186] J. Neggers and H. S. Kim (1998), *Basic Posets*, World Scientific Publ. Comp., River Edge, NJ

[187] J. Nešetřil and V. Rödl (1977), Partitions of finite relational and set systems, *J. Comb. Theory (A)* 17, 289–312

[188] J. Nešetřil and V. Rödl (1984), Combinatorial partitions of finite posets and lattices, *Algebra Universalis* 19, 106–119

[189] J. Nešetřil and V. Rödl (1987), Complexity of diagrams, *Order* 3, 321–330

[190] R. Nowakowski (1981), open question on p.842 of [218]

[191] R. Nowakowski and I. Rival (1979), A fixed edge theorem for graphs with loops, *J. Graph Theory* 3, 339–350

[192] A. Pelczar (1961), On the invariant points of a transformation, *Annales Polonici Mathematici* XI, 199–202

[193] M. A. Peles (1963), On Dilworth's theorem in the infinite case, *Israel J. Math.* 1, 108–109

[194] D. Pickering, M. Roddy (1992), On the strong fixed point property, *Order* 9, 305–310

[195] N. Polat (1995), Retract-collapsible graphs and invariant subgraph properties, *J. Graph Theory* 19, 25–44

[196] T. Poston (1971), Fuzzy Geometry, Ph.D. Thesis, University of Warwick

[197] M. Pouzet (1979), Relations non reconstructible par leurs restrictions, *J. Combinatorial Theory (B)* 26, 22–34

[198] L. Pretorius and C. Swanepoel (2000), Partitions of coutable posets, Papers in honour of Ernest J. Cockayne, *J. Combin. Math. Combin. Comput.* 33, 289–297

[199] H. A. Priestley and M. P. Ward (1994), A multipurpose backtracking algorithm, *Journal of Symbolic Computation* 18, 1–40

[200] E. Prisner (1992), Convergence of iterated clique graphs, *Discrete Math.* 103, 199–207

[201] H. J. Prömel (1987), Counting unlabeled structures, *J. Comb. Theory Ser. A* 44, 83–93

[202] J.S. Provan and M.O. Ball (1983), The complexity of counting cuts and of computing the probability that a graph is connected, *SIAM J.Comput.* 12, 777–788

[203] R. Quackenbush (1986), Unsolved problems: Dedekind's problem, *Order* 2, 415–417

[204] P. Prosser (1993), Hybrid algorithms for the constraint satisfaction problem, *Comput. Intell.* 9, 268–299

[205] P. W. Purdom (1978), Tree size by partial backtracking, *SIAM J. Comput.* 7, 481–491

[206] A. Quilliot (1983), Homomorphismes, points fixes, rétractions et jeux de poursuite dans les graphes, les ensembles ordonnés et les espaces métriques, Thèse de doctorat d'état, Univ. Paris VI

[207] A. Quilliot (1983), An application of the Helly property to the partially ordered sets, *J. Comb. Theory (A)* 35, 185–198

[208] A. Quilliot (1985), On the Helly property working as a compactness criterion for graphs, *J. Comb. Theory (A)* 40, 186–193

[209] I. Rabinovitch (1978), The dimension of semiorders, *J. Combinatorial Theory (Ser. A)* 25, 50–61

[210] I. Rabinovitch (1978), An upper bound on the dimension of interval orders, *J. Combinatorial Theory (Ser. A)* 25, 68–71

[211] I. Rabinovitch and I. Rival (1979), The rank of a distributive lattice, *Discrete Math.* 25, 275–279

[212] F. P. Ramsey (1930), On a problem of formal logic, *Proceedings of the London Mathematical Society* 30, 2, 264–286

[213] J.-X. Rampon (2001), What is reconstruction for ordered sets?, in: *Proceedings of the Fraïssé 2000 conference*

[214] V. Reiner, V. Welker (1999), A homological lower bound for order dimension of lattices, *Order* 16, 165–170

[215] K. Reuter (1989), Removing Critical Pairs, *Order* 6, 107–118

[216] I. Rival (1976), A fixed point theorem for finite partially ordered sets, *J. Comb. Theory (A)* 21, 309–318

[217] I. Rival ((1982)), The retract construction, in: I. Rival (ed.), *Ordered Sets*, Dordrecht-Reidel, p. 97–122

[218] I. Rival, ed. (1982), *Ordered sets*, Dordrecht-Reidel, Boston

[219] I. Rival (1984), Unsolved problems, *Order* 1, 103–105

[220] I. Rival, ed. (1984), *Graphs and order*, Dordrecht-Reidel, Boston

[221] I. Rival (1984), The diagram, in: I. Rival (ed.), *Graphs and Order*, Dordrecht-Reidel, Dordrecht, p. 103–136

[222] I. Rival (1985), Unsolved problems: The diagram, *Order* 2, 101–104

[223] I. Rival (1985), Unsolved problems: The fixed point property, *Order* 2, 219–221

[224] I. Rival, ed. (1986), *Combinatorics and Ordered Sets*, Proceedings of the AMS-IMS-SIAM summer research conference at Humboldt State University, Contemporary Mathematics 57, American Mathematical Society, Providence, Rhode Island

[225] I. Rival, ed. (1989), *Algorithms and order*, Kluwer, Dordrecht-Boston

[226] I. Rival and A. Rutkowski (1991), Does almost every isotone self-map have a fixed point?, in: Bolyai Math. Soc., *Extremal Problems for Finite Sets*, Bolyai Soc. Math. Studies 3, Viségrad, Hungary, p. 413–422

[227] I. Rival and N. Zaguia, eds. (1999), *ORDAL '96, Papers from the conference on Orders Algorithms and Applications, Ottawa, 1996*, Theoretical Computer Science 217, Elsevier Science Publishers, Amsterdam

[228] M. Roddy (1994), Fixed points and products, *Order* 11, 11–14

[229] M. Roddy (2001), Fixed points and products: width 3, to appear in *Order*

[230] M. Roddy (2001), On an example of Rutkowski and Schröder, to appear in *Order*

[231] V. Rödl and L. Thoma (1995), The complexity of cover graph recognition for some varieties of finite lattices, *Order* 12, 351–374

[232] G.-C. Rota (1964), On the foundations of combinatorial theory I: Theory of Möbius functions, *Z. Wahrscheinlichkeitstheorie verw. Gebiete* 2, 340–368

[233] A. Rutkowski (1986), Cores, cutsets and the fixed point property, *Order* 3, 257–267

[234] A. Rutkowski (1986), The fixed point property for sums of posets, *Demonstratio Math.* 4, 1077–1088

[235] A. Rutkowski (1989), The fixed point property for small sets, *Order* 6, 1–14

[236] A. Rutkowski, B. Schröder (1994), Retractability and the fixed point property for products, *Order* 11, 353–359

[237] A. Rutkowski, private communication

[238] D. Sabin and E. C. Freuder (1994), Contradicting Conventional Wisdom in Constraint Satisfaction, *Proceedings of the 11th European Conference on Artificial Intelligence*, Amsterdam, 125–129

[239] M. Saks (1985), Unsolved problems: Balancing linear extensions of ordered sets, *Order* 2, 327–330

[240] B. Sands (1985), Unsolved problems, *Order* 1, 311–313

[241] B. Schröder (1993), Fixed point property for 11-element sets, *Order* 10, 329–347

[242] B. Schröder (1995), On retractable sets and the fixed point property, *Algebra Universalis* 33, 149–158

[243] B. Schröder (1996), Fixed cliques and generalizations of dismantlability, in: Y. Alavi et. al., *Combinatorics, Graph Theory, and Algorithms*, Proceedings of the Eighth Quadrennial International Conference in Graph Theory, Combinatorics, Algorithms and Applications, Vol. II, p. 747–756

[244] B. Schröder (1996), Algorithms for the Fixed Point Property, *Theoretical Computer Science* 217, 301–358

[245] B. Schröder (1998), On cc-comparability invariance of the fixed point property, *Discrete Mathematics* 179, 167–183

[246] B. Schröder (2000), The uniqueness of cores for chain-complete ordered sets, *Order* 17, 207–214

[247] B. Schröder (2000), Reconstruction of the neighborhood deck of ordered sets, *Order* 17, 255–269

[248] B. Schröder (2001), Reconstruction of N-free Ordered Sets, *Order* 18, 61–68

[249] B. Schröder (2002), Examples on Ordered Set Reconstruction, to appear in *Order*

[250] B. Schröder (2002), More Examples on Ordered Set Reconstruction, submitted to *Discrete Mathematics*

[251] D. Scott and P. Suppes (1958), Foundational aspects of the theory of measurement, *J. Symbolic Logic* 23, 113–128

[252] B. Smith and M. Dyer (1996), Locating the phase transition in binary constraint satisfaction problems, *Artificial Intelligence* 81, 155–181

[253] E. H. Spanier (1966), *Algebraic Topology*, Springer Verlag, New York

[254] J. Spinrad (1988), Subdivision and lattices, *Order* 5, 143–147

[255] R. P. Stanley (1979), Balanced Cohen-Macauley Complexes, *Trans. Amer. Math. Soc.* 249, 139–157

[256] P.K. Stockmeyer (1977), The falsity of the reconstruction conjecture for tournaments, *J. Graph Theory* 1, 19–25

[257] R. E. Stong (1966), Finite topological spaces, *Trans. Amer. Math. Soc.* 123, 325–340

[258] M. Sysło (1984), A graph theoretic approach to the jump-number problem, in: I. Rival (ed.), *Graphs and order*, Dordrecht-Reidel, Boston, p. 185–215

[259] E. Szpilrajn (1930), Sur l'extension de l'ordre partiel, *Fund. Math.* 16, 386–389

[260] A. Tarski (1955), A lattice-theoretical fixpoint theorem and its applications, *Pacific J. Math.* 5, 285–309

[261] W. T. Trotter (1975), Inequalities in dimension theory for posets, *Proc. Amer. Math. Soc.* 47, 311–316

[262] W. T. Trotter (1976), A forbidden subposet characterization of an order dimension inequality, *Math. Systems Theory* 10, 91–96

[263] W. T. Trotter (1992), *Combinatorics and partially ordered sets: dimension theory*, Johns Hopkins University Press, Baltimore

[264] W. T. Trotter, J. I. Moore, and D. P. Sumner (1976), The dimension of a comparability graph, *Proc. Amer. Math. Soc.* 60, 35–38

[265] E. Tsang (1993), *Foundations of Constraint Satisfaction*, Academic Press, New York

[266] H. Tverberg (1967), On Dilworth's theorem for partially ordered sets, *J. Combinatorial Theory* 3, 305–306

[267] P. van Beek and R. Dechter (1995), On the minimality of row-convex constraint networks, *Journal of the ACM* 42, 543–561

[268] M. von Rimscha (1983), Reconstructibility and perfect graphs, *Discrete Math.* 47, 79–90

[269] S. Wagon (1993), *The Banach-Tarski Paradox*, Cambridge University Press, Cambridge

[270] J. W. Walker (1984), Isotone relations and the fixed point property for posets, *Discrete Math.* 48, 275–288

[271] D. West (1985), Parameters of partial orders and graphs: packing, covering and representation, in: I. Rival (ed.), *Graphs and Orders*, Dordrecht-Reidel, p. 267–350

[272] D. West (1996), *Introduction to Graph Theory*, Prentice Hall, Upper Saddle River, NJ

[273] D. Wiedemann (1991), A computation of the eighth Dedekind number, *Order* 8, 5–6

[274] H. S. Wilf (1994), *Generating Functionology*, Academic Press, New York

[275] S. Willard (1970), *General Topology*, Addison-Wesley, Reading, MA

[276] C. Williams and T. Hogg (1994), Exploiting the deep structure of constraint problems, *Artificial Intelligence* 70, 73–117

[277] S. Williamson (1992), Fixed point properties in ordered sets, Ph. D. dissertation, Emory University

[278] W. Xia (1992), Fixed point property and formal concept analysis, *Order* 9, 255–264

[279] M. Yannakakis (1982), On the complexity of the partial order dimension problem, *SIAM J. Alg. Discr. Math.* 3, 351–358

[280] L. Zádori (1992), Order varieties generated by finite posets, *Order* 8, 341–348

[281] L. Zádori (1998), Characterizing finite irreducible relational sets, *Acta Sci. Math. (Szeged)* 64, 455–462

Index